THE FRONTIERS COLL

THE FRONTIERS COLLECTION

Series Editors:
A.C. Elitzur   L. Mersini-Houghton   M. Schlosshauer   M.P. Silverman   J. Tuszynski   R. Vaas
H.D. Zeh

The books in this collection are devoted to challenging and open problems at the forefront of modern science, including related philosophical debates. In contrast to typical research monographs, however, they strive to present their topics in a manner accessible also to scientifically literate non-specialists wishing to gain insight into the deeper implications and fascinating questions involved. Taken as a whole, the series reflects the need for a fundamental and interdisciplinary approach to modern science. Furthermore, it is intended to encourage active scientists in all areas to ponder over important and perhaps controversial issues beyond their own speciality. Extending from quantum physics and relativity to entropy, consciousness and complex systems – the Frontiers Collection will inspire readers to push back the frontiers of their own knowledge.

Other Recent Titles

Weak Links
Stabilizers of Complex Systems from Proteins to Social Networks
By P. Csermely

The Biological Evolution of Religious Mind and Behaviour
Edited by E. Voland and W. Schiefenhövel
Particle Metaphysics

A Critical Account of Subatomic Reality
By B. Falkenburg

The Physical Basis of the Direction of Time
By H.D. Zeh

Mindful Universe
Quantum Mechanics and the Participating Observer
By H. Stapp

Decoherence and the Quantum-To-Classical Transition
By M. Schlosshauer

The Nonlinear Universe
Chaos, Emergence, Life
By A. Scott

Symmetry Rules
How Science and Nature are Founded on Symmetry
By J. Rosen

Quantum Superposition
Counterintuitive Consequences of Coherence, Entanglement, and Interference
By M.P. Silverman

Series home page – springer.com

Filipe Duarte Santos

Humans on Earth

From Origins to Possible Futures

 Springer

Filipe Duarte Santos
Universidade de Lisboa
Fac. Ciêncas
Depto. Física
Campo Grande
1749-016 Lisboa
Edificio C1, sala 1.4.39
Portugal
fdsantos@siam.fis.fc.ul.pt

ISSN 1612-3018
ISBN 978-3-642-27126-7   ISBN 98-3-642-05360-3 (eBook)
DOI 10.1007/978-3-642-05360-3
Springer Heidelberg Dordrecht London New York

To my grandchildren Constança, Matilde,
Vicente, Frederica, and Sebastião

Foreword

Earth Systems Are in Trouble

"We are at a turning point," writes Filipe Duarte Santos, "as regards the essential resources of food, water, and energy. Demand already exceeds what can be sustained at current levels of consumption. [And] competition between states will be further increased by population growth and climate change impacts."

Professor Filipe Duarte Santos, a physicist and scholar of environmental sciences at the University of Lisbon, provides a sweeping, thoughtful view of the role of humans in shaping our modern world. Beginning with a short history of the Universe, he crafts an easily accessible narrative, weaving together an exploration of the laws of physics, an examination of human evolution, and an illuminating discussion of the roles played by art, religion, science, and technology.

Professor Santos' broad scope brings to mind the work of another physicist, Murray Gell-Mann, who opens his book, *The Quark and the Jaguar*, with the challenge: someone must dare to take a look at the whole. This author's discussion of Bayes' theorem (which enables us to improve our analysis of current situations by incorporating new evidence!) is particularly helpful as we look ahead to the major choices we face, especially regarding the means of generating energy. (The author serves on the United Nations' Intergovernmental Panel on Climate Change.)

As Professor Santos leads us through the era of human dominance, he concisely captures the collapse of civilizations and the social movements that challenged — and continue to challenge — established orders.

The discussion then brings us to a critical juncture: the conflict among *Homo sapiens* in World War II that also marked an acceleration of our conflicts with nature. In a poignant set of passages most relevant to today's world, Professor Santos discusses the Bretton Woods Agreements that were made in 1944 at the end of the war. These agreements marked a rare moment of clear-sightedness, with the implementation of new rules and institutions to steer civilization. (The changes in rules of trade were hammered out under the visible hands of John Maynard Keynes (UK) and Harry Dexter White (US), whose biographies are delightfully recounted in these pages.) The rules — 1. free trade in goods, 2. fixed exchange rates, and 3. constraints on the movement of capital — heralded a period of post-war prosperity. Their unraveling in 1971 gave birth to the modern period of unregulated finance.

In the closing chapters of Professor Santos' panoramic examination of our collective home, Western society's potential decline, paralleled by China's ascendency, emerges as a challenge to all. Can today's leaders reach new agreements that will protect humans and the planet?

Writing this foreword in August 2011, as world markets tumble and financial empires teeter, I can only hope we will choose a route similar to Bretton Woods, and consciously and collectively craft a new world order.

As with the Marshall Plan in 1946, today's world will need new funds to complement the governance structures that guide sustainable development. And, as the call rises from a growing number of Professor Santos' European neighbors — though not yet from the U.S. — the necessary funds would best be raised through levies on over-bloated financial transactions born of deregulation, rather than bled from financially-strapped nations.

With the health of forests, marine life, food systems, and humans threatened by mounting economic and environmental instability, the present book sets a well-lit stage on which to examine today's challenges for our powerful, though not always far-sighted, species.

Boston, Mass.   *Paul R. Epstein, M.D., M.P.H.*
August 2011

Preface

It is our privilege to live in an extraordinary age and to belong to an admirable civilization. Our skill, ingenuity, and determination over many centuries, supported by science and technology, have allowed part of humanity, mainly those who live in the industrialized countries, to enjoy an excellent quality of life compared with previous generations. We have easy access to modern medicine and health care. We have comfortable homes. We have good, drinkable running water, and a ready supply of energy to meet our everyday needs. We have a marvellous freedom of movement on land, sea, and air, and indeed prodigious mobility that allows us to move swiftly from one side of the Earth to the other. There are excellent conditions for reaching the various levels of education and professional training. Information and communication technologies have given us a remarkable facility of access to data, knowledge, and opinion. We are now able to communicate from practically every point on the globe with our families, friends, and colleagues. Generally speaking, we enjoy a good level of security, in our homes and urban areas. Most are convinced that all this has been permanently acquired and that in time it will spread everywhere and reach everyone in the world. We implicitly assume that continuous growth will assure social and economic development and an increasing quality of life for all. As regards the scarcity of natural resources and possible negative impacts on the environment, we are generally convinced that it will always be possible to substitute for exhausted resources and repair environmental degradation.

It looks like a dream. But will it last? Is it really sustainable? Those who do not yet enjoy such well-being are fighting desperately to obtain it. But will it be possible to extend the dream to the whole population of the world?

On the dark side, there are indications that the future is likely to become increasingly uncertain. Poverty, hunger, and serious health problems, such as tuberculosis, malaria, and HIV infections, persist in some parts of the world, and affect an unacceptably large number of people. The staggering social and economic inequality of the world's richest and poorest continues to grow. Over the last two centuries, human activities have begun to have a noticeable impact on global terrestrial systems, in particular on the climate, and also on the biosphere by reduction of biodiversity. The accelerating pace of human activity has created multiple and intertwined global challenges. There is a non-negligible probability that energy, food, and water crises may happen in the current century. The impact of unabated climate change, including rising sea levels and acidification of the oceans, environmental degradation associated with a fast-increasing world population that aims at rapid economic convergence, scarcity of natural resources resulting from increasing per capita use, rising resistance to antibiotics, nuclear proliferation, and international terrorism are examples of the serious challenges facing all of us today and in the near future.

On the bright side, since World War II, we have witnessed a remarkable acceleration in social and economic development worldwide, which has lifted hundreds of millions of people out of poverty and improved the quality of life of many more. This successful spike of global development has mainly been based on the increasing dissemination and application of science and technology, better political and economic structures and institutions, higher levels of production and consumption, increasing globalization through the integration of national economies into the international economy, increased mobility and migration, development of more robust energy systems, and increasingly widespread access to energy services. But what we must ask now is whether it will be possible to maintain this great acceleration in a sustainable way, especially as regards the environment and natural resources. How might we be able to do that?

There are various answers to these questions arising from different economic and environmental outlooks. Some emphasize that there are limits to our present growth paradigm and that we will inevitably encounter crisis if we disregard them. Others are convinced that we can overcome all the projected obstacles, and that our inventiveness and skills, supported by the development of science and technology, will always be able to solve collateral problems that may appear along our path to growth.

Beyond these different approaches, there is a broad consensus that humanity will face crucial challenges in the 21st century regarding the sustainability of its development paradigm. What is the nature and origin of this situation, and why are we experiencing it today? What is the role of science and technology in the relatively long process that has led up to it? What does science have to say about limits and about the future of our environment in the short, medium, and long term?

To answer these questions we must first look into the past. There is a profound and unbreakable link between our past, present, and future. Our current challenges and our ability to deal with them are largely determined by the essential characteristics of human nature, which were forged in the biological evolution that led to the emergence of *Homo sapiens*, and later in the ensuing cultural evolution. The presence of humans on Earth is of course an integral part of the history of the universe and there is no way we can divorce ourselves from it. There is no escape now or in the future from its fundamental physical laws. These are essential and pose insurmountable limits to our dreams.

In order to address our collective future and scrutinize our options, it is thus advisable to begin by reflecting upon and understanding our history. Science has given us the wonderful faculty to reconstruct, often with amazing detail, the past of our species, of life on Earth, and of the universe as a whole. The same scientific methodology has given us the possibility to project the long-term future of the Earth, the solar system and the universe. In this long process, the presence of humans on Earth is just a remarkable epiphenomenon. How sustainable is this episode and how long will it last? What are the driving forces that will determine its duration?

The main aim of the book is to focus on our present and future challenges as humans on Earth in the broad context of our cultural evolution and the unfolding long-term evolution of our earthly and cosmic environments. We are currently at a sensitive time as regards the impact of human activities on the Earth's systems. We may reduce the uncertainty and the risks projected into the future by creating a path of sustainable development. Science gives us the possibility to construct plausible and coherent socio-economic scenarios for the future. It is therefore possible in principle to choose between various options for the development of humanity. We may or may not respect the ethics of intra- and intergenerational solidarity. We may or may not create a future human development path that avoids irreversible environmental degradation. We may or may not increase the risks of future energy, food, and water crises. The following pages are an invitation to reflect upon these questions. They are an invitation to analyse and think about the past, the present, and the seeds that we are sowing for our common future.

The book presupposes that the methodologies of the modern sciences will allow us to observe, interpret, and understand the natural and social phenomena, and constitute a reliable tool to build a sustainable development paradigm. It is also founded on the conviction that the capacity of science and technology to solve our current problems and challenges is limited, whatever the realm of their application.

The present book is a modified and updated version of a previous one published in Portuguese by Gradiva in 2007.

Acknowledgements. Writing this book was only possible through the contributions of many people with whom I have interacted and worked in various institutions. I would like to thank them all for the many discussions, and also for the analysis and creativity of their criticisms and suggestions. Since it is impossible to thank them all individually, I will only mention a few names. I would particularly like to thank Mathilde Bensaúde, who first aroused my interest and curiosity in nature, José Pinto Peixoto, José Moreira Araújo, Eduardo Filipe Duarte Ferreira, Ronald C. Johnson, Peter Hodgson, Willy Haeberli, Hugh T. Richards, Donald Kerst, Gerhardt Graw, Edward J. Ludwig, Stephen Shafroth, Fernando Plácido Real, Mario Ruivo, Mário Baptista Coelho, Luísa Schmidt, and Eugénio Sequeira.

Part of the book was written during a sabbatical at Stanford and Harvard universities. I would like to thank Paul Ehrlich, Paul Epstein, and James McCarthy for their hospitality, the discussions, and the exchange of ideas. I would especially like to thank Viriato Soromenho Marques, Tiago Capela Lourenço and Joana Borges Coutinho for their integral reading of the manuscript, their criticisms, and their suggestions. I would also like to thank Marta Sequeira for her help and critical reading. I thank Paula Teixeira and António Telo for reading part of the book, and Sofia Braga and Duarte Braga for their answers to frequent questions. I thank my university colleagues and friends António Amorim, André Moitinho de Almeida, Ricardo

Aguiar, Elsa Casimiro, Anastasia Svirejeva-Hopkins, Daniel Borrego, and Martin König, Maria João Cruz, Pedro Garrett Lopes, and David Avelar for support and understanding. I would also like to thank Ângela Antunes for her secretarial support. Finally, I would like to thank my family, especially Amparo, for their support during this project.

Lisbon, Portugal   *Filipe Duarte Santos*
June 2010

Contents

Chapter 1
Science and Technology. From the Origins up to the Twenty-First Century

1.1 The Birth of Modern Science

The beginnings of modern science are rooted in a period of great conceptual development, which took place in Western Europe from 1500 to 1750. There were earlier origins, going back to the Greeks; Aristotle developed detailed theories of physics and biology, while Galen contributed to medicine and physiology, and Ptolemy to astronomy. They are coherent theories, based on a robust rationality that only began to slowly crumble when confronted with new facts arising from observation and experimentation. The initial development of modern science was a relatively peaceful process, if we ignore some extreme episodes of opposition offered by the Church, such as the burning of Giordano Bruno at the stake, or the persecution of Galileo. The process was led by a small intellectual minority, and passed largely unnoticed by the vast majority of the population. It was a highly discreet movement in the context of contemporary religious uprisings and the wars of reformation. Despite this contrast, science ended up deeply affecting the mentalities of future generations; even today, certain ways of thinking pioneered by science in the 16th to the 18th centuries still constitute a template for current scientific practice.

Perhaps the greatest legacy we have inherited, and which is now deeply rooted, is an eagerness to seek out relationships between bare and irrefutable facts and general abstract principles. We stress the importance of observation, experimentation, and a detailed analysis of natural phenomena, allied with the search for abstract generalisations that connect them.

In the origins of modern science, there is a new interest in observing phenomena and objects in themselves, stripped of their meaning and utility. It was necessary to develop a capacity to form the idea of an object just as it is, purely physical and devoid of all sensorial qualities. It was necessary to be able, through abstraction, to uncloak objects of their distinct qualities, rough or smooth, coloured or uncoloured, useful or useless, friend or foe. It was necessary to rethink nature in a geometrical form and reduce it to its physical properties. The physics of Aristotle is based on a doctrine characterised by ends, purposes, and meanings. This way of thinking together with the anthropomorphic concepts of nature fail to make sense in modern science. The behaviour of nature, ruled by universal laws and expressed by means of mathematical formulae, is inexorable and completely indifferent to man, to how he feels, and to his worries and anxieties.

This new vision had deep cultural consequences. The human experience became separated from the workings of nature, and this helped to open a conceptual path for the conquest and domination of nature. Only in the 19th and 20th centuries did we start to grasp the meaning of this domination, and to realise that in the end we are conditioned by deep ties of dependence on nature which we need to understand and preserve, rather than break.

The other essential characteristic of modern science is an instinctive conviction about the existence of an intelligible order in nature spanning all levels, from the smallest particle to the immensity of the cosmos. The initial belief in that order was not an attitude that resulted exclusively from an exercise in logic. It sprang from an intuition and a belief that proceeded to feed upon the successes of scientific endeavour, and particularly upon its predictive capacity. It could not be justified by inductive reasoning alone. It had its origin in the perception and understanding of nature's behaviour, revealed by observation and experimentation. The leading actors of the modern science movement from the 16th to the 18th centuries dared to believe in an intelligible nature capable of description through laws expressed in the abstract language of mathematics. For Alfred North Whitehead (Whitehead, 1953) this conviction resulted, in a subconscious way, from the insistence on the rationality of God, which was present in medieval European theology over many centuries. In other regions of the world, God was too impersonal, arbitrary, or despotic to imbue the way of thinking with a belief in logic.

1.2 Philosophy and Science

While for the Greeks science was a branch of philosophy, modern science moved slowly away from it, and became clearly separated in the 19th century. The relationship between them during this process was nevertheless very creative. In the 17th century, philosophical thought found, exposed, and explored various difficulties in the foundations and methodology of science. The philosopher David Hume (1711–1776) considers in his *An Inquiry Concerning Human Understanding* (1784) that each effect is an event distinct from its cause, and as such cannot be found in the cause. In other words, he denies the possibility of knowing causal relationships and considers that there are only connections between events, with no logical guarantee that they are not entirely arbitrary. In this form of empiricism, any rational foundation for prediction and determinism becomes impossible. According to Hume, the use of inductive inference, which plays a central role in the methodology of science, cannot be logically justified. In induction, we use propositions about objects or events that we observe and examine in order to reach conclusions about objects or events that we do not see and examine. Hume considers that this inference is only possible by assuming the existence of what he referred to as the 'uniformity of nature'. But then how could we prove that this theory is true? According to Hume we cannot, because we can imagine non-uniform universes subject to random changes in time. It would therefore be impossible to justify the validity of the inductive method in any rational way. Generations of philosophers have searched systematically for answers to the questions posed by Hume, and this continues to be an active area of research in the philosophy of science. However, and this is what is of most interest to us, these purely philosophical obstacles confronting the inception of modern science did nothing to hinder its evolution. In fact, they helped it to establish its own identity, independently of philosophy. Modern science did not seek to justify its intuitive faith in the intelligibility of nature. In the beginning, it was an anti-rationalist movement based on observation and experimentation that flourished at the end of the Renaissance.

But philosophy also contributed to clarifying and promoting scientific thought. Descartes (1596–1650) established a clear distinction between thought and the material objects of thought that constitute the various forms of matter. The thinking spirit is immaterial, has no extent, and is near to God. With this dualism Descartes created a place for God, a being that he considered absolutely perfect, defending himself against accusations of atheism, and freeing the sciences for the analysis of material objects stripped of their qualities, uses, meanings and symbolisms, that so strongly impressed upon the human spirit. In his *Discours de la Méthode* (1637), he proposes analysis as the privileged method for the spirit to search for truth in the sciences. Faced with a problem, we should break it down into its individual parts, reflecting upon and solving each one separately (which should be easier than treating the problem globally), and finally reconstructing the whole. This abstract methodology of analysis, later called reductionism, has nowadays become a basic habit of thought and constitutes the main instrument of scientific research. Its success is unquestionable.

The physical universe is described as a large set of dynamic galaxies formed mainly by stars made up of various particles including atomic nuclei and electrons, which may form atoms and molecules. The atomic nuclei are themselves made up of neutrons and protons, which in turn are made up of quarks. The structure and dynamics of the various entities on the different spatial scales are determined by the forces they exert among themselves. In one of these galaxies, there is a star with a planetary system in which one of the planets is Earth, built with the same physical entities, molecules and atoms, which in their turn break down into neutrons, protons, electrons, and so on. Reductionism does not apply only to physics, but to all areas of science. It is an essential strategy of science used to systematically explain the behaviour of complex systems that would otherwise be unintelligible. The physicist Steven Weinberg, a great defender and practitioner of reductionism, considers that the search for ever more fundamental physical entities and laws will eventually culminate in a final theory of everything in the Universe, when things cannot be explained in other more basic terms (Weinberg, 1993).

In this context the social sciences have a more balanced approach in practice, because they frequently use the methods of both synthesis and analysis. This difference has inspired some contemporary attacks on the methodology of science, some of which reach the point of considering that its dependence on reductionism constitutes a disorder of obsessive character, possibly indicative of a terminal phase. Nevertheless, we should recognize that the systematic application of abstract analysis that was clearly identified by Descartes in the years 1637 to 1649 and is now contested by some scholars has led to the enormous success of science and technology for more than three centuries.

The Cartesian philosophy has also had a profound cultural influence by contributing to the development of rationalism. This supports the conjecture that nature is governed by intelligible principles that can be reached through the use of reason to interpret the raw facts provided by observation and experimentation. In opposition to this, various lines of recent philosophical enquiry seek to devalue rationalism. They emphasize the absurd and meaningless character of life and claim an essential role for intuition, feelings, and faith, which cannot be reduced to reason. H. Bergson, S. Kierkegaard, M. Heidegger, and M. Marcuse are some of the influential modern philosophers that have contributed most to the analysis and development of such concepts. Where science is concerned, this opposition to rationalism stresses its limits as an expression of Western thought and resists the idea that it is the only path to knowledge.

1.3 The Universality of the Laws of Physics

The first big step toward the modern scientific vision of the cosmos was taken by Nicholas Copernicus (1473–1543) with his heliocentric model of the Solar System. There are records that lead us to believe that Copernicus became convinced of heliocentricism around 1510, but the defining publication — *De Revolutionibus Orbium Coelestium* — only appeared in 1543, the year of his death, and with an anonymous preface where it is stated that "these hypotheses need not be true nor even probable", which certainly does not portray the opinion of Copernicus. For several generations many astronomers reading *De Revolutionibus* thought that Copernicus did not consider the theory that had consumed the greater part of his life as true from the standpoint of physics. The doubtful authenticity of the preface, which was apparently written by the Lutheran theologian Andreas Osiander, was known by a few astronomers but it was Johannes Kepler (1571–1630) who divulged it openly in 1609. These facts clearly reveal the difficulty at that time in affirming that the Earth was in motion: it was an affront to the church, the universities in which Aristotle's model of the world was taught, and also to astronomers who used that model to interpret their observations.

Copernicus's heliocentric model opened the way to the development of modern physics through the works of Johannes Kepler and Galileo Galilei (1564–1642). Kepler discovered that the orbits of the planets around the sun are elliptical rather than circular as Copernicus had thought, and he pronounced two further laws that determine the velocities of the planets in those orbits. Galileo, who was a great defender of the heliocentric model, used a telescope for the first time, and discovered many new facts, such as the existence of mountains on the moon, sunspots, satellites in orbit around Jupiter, and many more stars, all of which conflicted with the tenets of Aristotelian cosmology. However, his most important contribution was to use experimentation as a way to gain knowledge, namely in the study of mechanics, thereby establishing the empirical methodology as a route to understanding nature.

After Galileo, scientific development accelerated, culminating in the works of Isaac Newton (1643–1727), and in particular his main work *Philosophiae Naturalis Principia Mathematica*, published in 1687. The three laws of dynamics and the law of universal gravitation constitute a powerful conceptual base for describing the Universe as a group of particles and bodies in movement. Newton's theory was built with great mathematical rigour and demonstrated that Kepler' laws can be deduced from those of motion and gravitation. This was one of the first demonstrations of the universal and unifying character of the laws of physics that apply to Earth as well as to the entire Solar System. All these advances were made possible by the simultaneous evolution of mathematics, especially the analytical geometry of Descartes and the infinitesimal calculus invented independently by Newton and Gottfried Wilhelm von Leibniz (1646–1716). The rapid development of abstract mathematical thought served as the indispensable support to men of science in their quest to interpret the result of observations of natural phenomena and experimentations. With some surprise they concluded that there were laws underlying their observations and that these laws could be expressed mathematically. The mechanical vision of the Universe inherited from Newton constituted a conceptual framework for science for the next 200 years.

Scientists began to think that it might be possible to give an explanation based on mechanics for all physical processes in nature. This conviction was enhanced by the great successes of the mathematical physicists, which reached its zenith in the *Mécanique Analytique* of Joseph Louis Lagrange (1736–1813) published in 1787, one hundred years after Newton's *Principia*. With the publication by James Clerk Maxwell (1831–1879) of *Electricity and Magnetism* in 1873, almost one hundred years later again, the first cycle in the development of modern physics came to an end.

1.4 Determinism, Uncertainty, and Probability

By constructing a vision of the Universe in which everything can be described by laws of nature of a mechanical character, and always proceeding with mathematical rigour, the pathway to determinism was opened. Its initial and most eloquent expression is due to Pierre Simon Laplace (1749–1827). As he states in the introduction to his *Essai Philosophique sur les Probabilités*, published in 1814:

> An intellect which at a certain moment would know all forces that set nature in motion, and all positions of all items of which nature is composed, if this intellect were also vast enough to submit these data to analysis, it would embrace in a single formula the movements of the greatest bodies of the Universe and those of the tiniest atom, for such an intellect nothing would be uncertain and the future just like the past would be present before its eyes.

The existence of such intelligence, sometimes called Laplace's demon, depends on two conditions: (i) a precise knowledge of the state of all systems that compose nature at a given moment, and (ii) a precise knowledge of the causal relationship between the different states of each of those systems. Twentieth century quantum mechanics invalidated the determinist concept, by proving that condition (i) does not apply in the realm of microphysics. The most frequent situation is a total impossibility to predict the result of a measurement of an observable in an atomic or subatomic physical system. The act of measuring interferes with the system, which responds by 'jumping', in an unpredictable way, to one of the possible final states. Apart from this lack of knowledge of a fundamental nature, the condition (ii) is, in most cases, impossible to put into practice on either the micro- or the macro physical scale.

This impossibility is clearly exemplified in chaotic dynamical systems, in which extremely small differences between various initial states produce very large differences in the corresponding final states. This hypersensitivity renders unpredictable the state of the system at a given moment in the future, since it is impossible to have accurate knowledge of its state at the initial moment. Note, however, that the assumption of indeterminism does not invalidate causality. Science discovers and uses causal relationships, but instead of being univocal, the same cause may produce various effects, with different probabilities.

This raises the question of the concept of probability involved in the relation between an event and preceding events. Sometimes, in a given situation, a conclusion about what will happen is almost certain, while in other situations, all conclusions are equally likely and therefore equivalent to no conclusion at all. The statement suggests that it may be possible to express the differences in the reliability of conclusions in a numerical form, by means of probability calculus. Thomas Bayes (1702–1761) was the first to believe in such a calculation, which according to him was necessary to justify inductive reasoning. The Bayesian theory of probabilities has evolved considerably since its origins and now has important applications in various areas of science.

One of the most interesting propositions of this theory is Bayes' theorem of inverse probability. Let us suppose that we have a theory X that predicts a result Y. The probability that the theory X is true when the prediction of Y is verified (called posterior probability) is equal to the ratio between the probability that the theory X is true (called prior probability) and the probability that Y occurs without presupposing X. The theorem leads to the conclusion that if the result Y in the absence of X is more unexpected then the theory X that predicts Y is more likely to be true. The usefulness of the theorem depends on our assigning prior probabilities with numerical values, and this may prove difficult. Bayesianism allows us to estimate the probabilities of predictions based on a given theory, and has given rise to what is called the Bayesian or subjective interpretation of probabilities.

The concept of probability is so important in contemporary science and technology that it is worth exploring further. Curiously, the first substantial work on the theory of probability was written by Laplace, who was a strong believer in determinism. This somewhat surprising connection is understandable if we bear in mind that, until the end of the 19th century, it was thought that the need to use probabilities was a result of our ignorance and not of fundamental and unavoidable uncertainties, as for instance the uncertainty that manifests itself in quantum phenomena. According to Laplace, probability is the measurement of a property of a sequence, namely, the relative frequency associated with the occurrence of a given event in that sequence. For example, the number of times one gets 'heads' when tossing a coin. However, to be a useful concept, the probability should have a value independent of the number of tosses. Therefore, we reformulate the definition and consider that the value of the probability is the limit to which the relative frequency tends when the sequence becomes infinite. This frequency or objective interpretation of probability presupposes an infinite sequence, which implies that we can repeat the same experiment under identical conditions with the same system, or that we have available an infinite group of identical systems, in each of which we can perform the same experiment independently and under identical conditions. In either case, it is impossible to run infinite sequences of experiences. We avoid this difficulty if we can find convincing arguments to attribute an a priori value to the probability of a determined event. Even so, the conceptual problem is not completely solved because the only way to validate the arguments used is to test the value attributed to the probability through measurements of the relative frequency. In conclusion, probability is not reducible to a measurement or a set of measurements. Ultimately, it expresses the relationship between two propositions — the probability of Y, given X. The nature of this relationship is related to the expectation that we have of finding Y, given X. This argument led various authors to consider a merely subjective interpretation in which the probability of an event Y is only a measure of the degree of confidence that we have in its occurrence, given X. According to Karl Popper (1902–1994), the probability, although subjective, has the capacity to reveal something objective about the outside world, which is the propensity of objects to behave in a certain way.

1.5 L'Esprit Géométrique et l'Esprit de Finesse, by Blaise Pascal

The development of modern science was one of the main forces at the origin of the Enlightenment, the intellectual movement that characterized a period of European history roughly coinciding with the 18th century. This was a cultural and philosophical movement of a rationalistic nature that proposed to 'illuminate' the minds of men with knowledge, obtained through science, and to free them from the darkness of ignorance, superstition, and obscurantism. Founded on a profound confidence in the power of reason, taken as the primary source and basis of authority, it exercised a systematic and critical analysis of laws, customs, institutions, politics, and religion, considered the most powerful and ever-present illusion.

The great success of science in the explanation of nature's phenomena gave origin to the idea that the scientific method could also be applied to other dominions — history, sociology, psychology, archaeology, linguistics — and finish up constituting the 'science of man'. This doctrine, often called scientism, defends the idea that the methods of science can be applied to all the various realms of human experience and that, eventually, after the necessary investigations, it would resolve all uncertainties and controversies. The new crusade of science fed on the expansion of human curiosity to different areas of knowledge, the desire to achieve the unity and objectivity of an all-embracing concept of knowledge, and also the desire to benefit from the high social and intellectual status attributed to scientists in certain circles at that time. An example can be found in the publications of Karl Marx (1818–1883), when he defends the science of history as the only way to understand it. Every historical event should be analysed and interpreted in terms of the economic and social infrastructure that made it possible. According to this perspective, there is a 'mechanics' of history that allows one to predict its future scientifically.

Scientism is based on the belief that the scientific method has a practically unlimited power to interpret phenomena, whatever their nature. There is something fallacious in this argument, because science is based on abstract analysis leading to laws of a quantitative nature whose logical consequences are susceptible to being tested against observation and experiment, whereby they are either rejected or confirmed. However, there are vast dominions of human experience that are impenetrable to abstract analysis.

It was Blaise Pascal (1623–1662) who, for the first time, laid the foundation for that discussion by identifying two fundamental types of temperament and modes of thinking: the geometric spirit — *l'esprit géométrique* — which comes into play when we are dealing with exact definitions and abstract concepts in mathematics or science, and the intuitive spirit — *l'esprit de finesse* — when we are dealing with ideas, perceptions, and feelings that cannot be expressed by means of exact definitions. For Descartes, intuition was the foundation of rational knowledge, but for Pascal it constituted a faculty irreducible to reason, and which, in the final analysis, makes it subordinate in all dominions where it is used. Mathematical reasoning is incapable of learning or understanding the contradictory character of human experience, which lies outside the scope of scientific knowledge. This kind of knowledge is expressed by propositions that will be accepted or rejected in accordance with a specific and consensual methodology. It is understandable that we should try to generalize as far as possible the application of a methodology that leads us to consensus and even unanimity. However, there are other kinds of knowledge, besides scientific knowledge, that come from the 'intuitive spirit', and whose contents are not susceptible to being accepted or rejected via a consensual methodology. Recognizing this limitation is an important step towards developing the necessary tolerance and understanding to face the most diverse conflicts in human society.

Pascal's genius and life experience provided him with unique conditions to illuminate the distinct qualities of intuition, sentiment, and reason. He was a noted mathematician: at the age of 15, he published an essay on conics read by the leading mathematicians of the day, including Leibniz, and later wrote studies on combinatorics, calculus, and probability. He was also a physicist: from the age of 23, he experimented with vacuum and observed the variation of atmospheric pressure with altitude. He was also the inventor of, among other things, the hydraulic press, the

syringe, and a mechanical calculator, called the Pascaline, which was the first prototype of the computer.

With an active intellectual, social, and emotional life, he experienced a deep mystic call and converted to Jansenism in 1655 at the age of 32, living among the 'So-
litaries' of the Port Royal Abbey. He intervened in the theological quarrel between
Jansenists and Jesuits, by writing *Les Provinciales*, which enjoyed enormous success at the time and went on to become the literary model for modern French prose.
Later, he abandoned the fray, and conceived of a project of apology for the catho-
lic religion directed at libertines and atheists. However, he only managed to write
a few notes, published posthumously under the title *Pensées*, which constitutes his
best known and most influential work. There, the distinct faculties of the 'spirit' are
clearly characterized, providing a way to disprove the tenets of scientism and to cir-
cumscribe the limited achievements of science in the explanation and guidance of
human life.

1.6 The Evolution of the Species

In spite of these limitations, later on in the 19th century, science succeeded in identifying the mechanisms of the evolution of life since its origins. This discovery resulted from the patient and systematic observation and analysis of natural phenomena
in conjunction with a quest for the abstract laws that relate them, following the most
exemplary tradition of scientific methodology. The observations were made by the
young biologist Charles Robert Darwin (1809–1882), over fifty seven months on
board the H.M.S. Beagle, particularly in the Madeira and Galapagos Islands and
along the western coast of South America. The great diversity and mutability of the
flora and fauna on the islands led him to postulate that the species evolve as a result of a natural form of selection that favours mutations better adapted for survival.
His book *On the Origin of Species*, in which he expounds this theory, published
on 24 November 1859, sold out in one day, and raised the most violent reactions.
The Church led the opposition, as the concept of natural selection would force one
to abandon the creational thesis of the Bible, and implies the absence of a divine
finality or intention in nature and all its diversity. Nevertheless, Darwin's ideas were
successful, and exercised a strong influence not only in the realm of science, but
also in philosophy, religion, psychology, anthropology, sociology, and politics.

Since we are programmed to deal with processes that can take anything between a
few seconds and a few decades, we have difficulty in understanding the mechanisms
of the evolution of species, because they result from cumulative processes that are
extremely slow, typically taking tens of millions of years. The mutations in genetic
material can generally be put down to chance, but only a few achieve relevance
in evolution, due to the implacable process of natural selection, whereby in each
generation, the individuals least adapted to prevailing conditions are removed from
the population. The combination of these two mechanisms generates an evolution
driven by adaptation to an external environment in permanent change. Apart from

this property, it is completely without purpose or finality. According to Stephen Jay Gould there is no evidence of any privileged direction for the evolution of species, and none of its products, such as *Homo sapiens*, could be considered as inevitable. If it were possible to repeat the 'programme of life' a million times since its beginning on Earth, that peculiar simian with a relatively large brain would never reappear (Gould, 1989).

Although Darwin formulated and developed the theory of evolution strictly within the field of biology, several currents of thought tried to apply it to ethics and the social institutions in the latter half of the 19th century. This movement, known as social Darwinism, arrived at various conclusions, some very controversial, such as the idea of using natural selection to defend a social concept that legitimises the inevitability of social inequalities. This was another example of scientism. Here it was necessary to return to the scientific method and make systematic observations of the behaviour of living organisms to find out to what extent their social behaviour had a biological foundation. This work was undertaken by Edward O. Wilson, who published it in *Sociobiology: The New Synthesis* (Wilson, 1975), based notably on a description of the social behaviour of various animals from ants to antelopes and baboons. Only in the last chapter does he mention sociology, and consider the possibility of turning it into a scientific discipline, asking whether it is possible to subsume it under the Darwinian paradigm. This is an important question, still open, lying at the boundary of science's applicability.

1.7 Symbiosis Between Science and Technology

Since the very beginning, science has had a partner which had already been around for a long time before science came on the scene — technology, or more precisely *techno*, the Greek word for skill, craft, or practical art, without the 'logy', which implies study and the use of reason. The history of technology, in the widest sense, starts with the fabrication of rudimentary artefacts by Hominids, more than two million years ago. The appearance of the first tools in chipped stone, around two and a half million years ago, occurred during a period of large increase in brain size (Leakey, 1994) that led up to *Homo sapiens*. Since then technology has developed prodigiously through a multiplicity of ever more complex inventions, which have in turn led to further innovations that have profoundly transformed human society. Nevertheless, according to Freud (Freud, 1929), technology continues to represent, in its essence, an extension of our limbs and organs that allows us to improve our capacity to act, that is, a means to an end.

In its origins, objectives, and methods, science is clearly distinct from technology, but has benefited from it at the outset. On the other hand, technological innovation has frequently used, in an empirical way, properties and mechanisms that only later became fully explained by science. A revealing example is the water pump, invented and used long before Evangelista Torricelli had measured atmospheric pressure in 1643. Once we were able to get water to rise in a tube with a piston, it was said that nature abhorred the vacuum, but this did not explain the fact that it was quite impossible to get the column of water higher than 10.3 metres, which corresponds to the atmospheric pressure of one atmosphere (1.103×10^3 Pa).

Another example is the steam engine, invented by James Watt and Matthew Boulton, and used in the English textile industry with great success around 1785. The conception, construction, utilization, and commercialization of these machines happened long before the fundamental laws of thermodynamics were discovered. On the other hand, steam driven machines played a very important role in science, because many of the ideas that led to thermodynamics had their origin in the study of the way these machines worked. James Prescott Joule established and measured the equivalence between mechanical and thermal energy only in the 1840s. Around 1851 William Thompson, better known as Lord Kelvin, formulated the second law of thermodynamics, which established the efficiency limits of heat engines. In this case, as in many others, the construction of instruments and machines of great practical value preceded the complete scientific explanation of the processes used. The precedence of practice over theory can be seen not only in science but also in the arts, and it reveals an important characteristic of the creative processes of the human spirit.

The case of James Watt reveals yet another facet of the symbiosis between science and technology. Very early on, he showed great capacity for mathematics, a keen interest in the works of Newton, and an uncommon ability for mechanics. The University of Glasgow employed him as an instrumental technician with the title 'instrument maker to the university'. There he benefited from contact with the science of the day, and especially from acquaintance with Joseph Black, a professor at the university, and one of the pioneers of the study of thermal processes. Called upon to repair an exemplar of a primitive steam driven machine invented by Thomas Newcomen, which belonged to the university, he became very interested in the project and ended up constructing a much more efficient steam engine, thereby becoming one of the most famous mechanical engineers of Scotland. For James Watt, passage through the university was a source of scientific knowledge and inspiration. The construction of the new steam engine was only made possible through a focus on innovative technology, practical skill, and an ability and experience in instrumentation.

Science benefits from and depends on technology because of its fundamental contribution to developing new means of observation and experimentation. This dependence of science on instrumentation has evolved greatly since its origins. Notable examples of such evolution are the telescope and the particle accelerator. We consider the former to have been invented (as far as we know, accidentally) in 1608 by Hans Lippershey, a Dutch optician. In the same year, the instrument, consisting simply of two lenses at the ends of a long tube, was presented to the government in The Hague, and the Captain General of the United Provinces and other high dignitaries were puzzled to see the clock in Delft and the windows of a church in Leiden. It was an immediate success that spread rapidly throughout Europe. In the spring of 1609, Dutch merchants took it to the north of Italy, where in May Galileo saw it and decided to make an improved version. The observations made by Galileo with his telescope opened a new way to explore and understand the Universe, leading to the concept of the universality of the laws of science. After only two more years, in 1611, Kepler constructed a new version of the telescope, replacing the convex lens by an ocular, and obtained an instrument that played an essential role in discovering the laws that govern planetary motions.

Nowadays telescopes are looked upon as collectors of photons, which are the quanta of the electromagnetic field, coming from a chosen direction in space with the most diverse energies, covering practically the whole range of the electromagnetic spectrum: from the radio waves with a long wavelength to the gamma rays with a very short wavelength, passing through the infrared, visible, ultraviolet, and X rays. The concept of the telescope extends as far as detectors of cosmic rays and neutrinos coming from outer space.

To know more about the Universe it is necessary to improve the resolution of the images obtained by telescopes, or, in simpler terms, to see in greater detail and further away, and this across all the wavelengths of the electromagnetic spectrum. This can be achieved with new telescopes capable of capturing photons in a specific band of wavelengths coming from very weak sources. The construction of such telescopes is only possible using the most sophisticated and modern technologies in the fields of radiation detection, materials science, and electronic and computer instrumentation. This type of involvement with technology is absolutely essential to current scientific endeavour in all areas.

Technology had and has its own evolution. An especially creative period was the second half of the 19th century, characterised by the invention of new equipment and products that profoundly changed our everyday lives and the comfort and mobility of human societies from then on. A significant example is the telephone, invented by Alexander Graham Bell in 1875. It seems likely that the telephone was actually invented earlier, in 1860, by the Italian Antonio Meucci, living in New York, but he was unable to raise sufficient funds for the patent application. Other examples are the four-stroke internal combustion engine invented by Nikolaus August Otto in 1876, the long-lasting electric light bulb invented by Joseph W. Swan and Thomas A. Edison around 1880, the electric motor by Nikola A. Tesla and George Westinghouse in 1884, and the pneumatic tyre by John B. Dunlop in 1888.

Practically all of these inventions resulted from the work of engineers and inventors who innovated mainly on the basis of equipment or products which already existed. The capacity and potential for science to stimulate the advance of technology could be felt, but the willingness to assume the development of this relationship was missing. It was only in 1890 that a chemist, Alfred Mond, was able to convince a group of businessmen of the advantages to industry of employing scientists to discover new processes and materials which the engineers could then apply to invent new machines and products — what is known today as research and development.

During the 20th century scientific research became a common tool in economic and social development, used systematically by governments and businesses, especially in the more developed countries, where more resources were available. The main difference between the two funding sources is that, in general, the private sector almost exclusively supports applied research, whereas the public sector also finances fundamental research. The distinction between the two types of research is subtle, residing essentially in the choice of research themes and the predisposition and attitude of the researcher. In fundamental research, the objective is to do science, to discover, to understand, and finally to know nature, its laws, and its complexity. In principle, there is no preoccupation of a utilitarian or technological nature. In applied research the objective is to search for processes, products, or devices that will be useful in a specific functional, economic, or social context. Of course the same researcher may be predisposed to do fundamental or applied research and sometimes in the same laboratory, essentially with the same type of equipment.

Fundamental research frequently requires specific means of experimentation, observation, and computing which lie beyond the limits of available technology. It is then necessary to develop new technologies that may have applications, known as spin-offs, in the most diverse socio-economic sectors. The exploration and utilization of outer space gives us various examples of this transfer of technology. The fuel cells which convert the chemical energy of a fuel such as hydrogen without combustion directly into electricity, were built for the first time in the 1960s to be used on the Gemini and Apollo spacecraft of the US space programme, although the principle of the way they work had been known since 1838. Currently, they are the subject of an intense and expanding research and commercialisation programme, aimed at making them more competitive in a future economy, less dependent on fossil fuels and thus more sustainable.

1.8 Science and Technology in the Military Sphere. The Case of the USA in World War I and II

Beyond the profound relationship and impact of the synergetic development of science and technology in the social, economic, political, and cultural realms, there is another sphere of influence of great importance — military applications. Clear archaeological evidence has been found to suggest that, since the beginning of the Neolithic, rudimentary weapons have been used in collective activities of war. Some authors (Leakey, 1994) associate the start of this warlike behaviour with the need for territorial control, when populations became partially sedentary and dependent on agriculture. Note, however, that the practice of aggression by groups of four or more individuals towards others, usually males, is common in various animal species, including chimpanzees and dolphins, and very probably was also practised by our ancestors before the Neolithic period. All these species are characterized by a large brain development, suggesting that war, in the primordial sense of aggression by an organized group, requires sophisticated brain processes.

The history of civilizations reveals a systematic concern for developing the most varied military technologies in order to improve warfare capacity and subjugate enemies. For about 1000 years, from the 5th century BC, the small states of Greece and Italy were able to dominate their neighbours by military means. The military supremacy of the West was heightened from about 1300 onwards through the use of gunpowder (originally imported from China) in a wide variety of firearms. It was also strengthened by the discoveries and conquests along the coasts of Africa, the Americas, and the East, which began in the 15th century. Finally, it was further enhanced by the industrial revolution in the middle of the 18th century. The success came, above all, from an above average capacity of the European states for developing quality and innovation in the large scale production of explosives and firearms. And this in turn was mainly a result of the predominance of rationalism, the application of scientific methodology, the relatively free dissemination of knowledge, and the progressive secularisation of the states that characterised European culture at the time. The use of firearms on the battlefield tends to destroy hierarchy and promotes egalitarian power-sharing. It is therefore not surprising that feudal Japan considered them to be dangerously revolutionary. Europe had the capacity to invent or import arms and military technology from other countries, and to develop them in a free and pragmatic way, regardless of any concern for social, political, or cultural change that the new technologies frequently gave rise to.

The influence and interdependence of science and technology intensified during the 20th century, after World War I. It became clear that the power to win and dominate depended on the permanent supremacy of the military arsenal on land, at sea, and in the air, and only science and technology could provide this. The history of the United States of America during the 20th century is the most remarkable example of the deep and growing connection between science and war during this period. It is enlightening to review some of the more noted events that reveal the nature and evolution of that link.

At the end of May 1915, when the Lusitania was sunk by the Germans, Thomas Edison wrote in the *New York Times* that Americans are as "clever at mechanics [...] as any people in the world" and would overcome any "engine of destruction", including the U-boats, the military submarines operated by Germany in World War I (Marshall, 1915; Kevles, 1995). At that time the distinguished astronomer George Ellery Hale, inventor of the spectroheliograph, editor of the *Astrophysical Journal*, and director of the Mount Wilson Observatory, was the Secretary of the National Academy of Sciences. On 19 April 1916 he proposed a resolution whereby the Academy would place itself at the disposal of the government in the event of a break in diplomatic relations between the USA and Germany. His proposal was unanimously accepted. One week later President Woodrow Wilson fully approved the engagement in a meeting with a delegation from the Academy. George Hale was ecstatic and said: "I really believe this is the greatest chance we ever had to advance research in America" (Wright, 1966; Kevles, 1995). In June the Academy responded to the formal request of support from President Wilson by forming the *National Research Council*. Its statutory objective was to develop pure and applied research, with the final goal of guaranteeing "national security and welfare". Some contemporary scientists criticised the objectives and military ambitions of the new council, arguing that it would lead to a dangerous centralization of research in the USA. However, the greater part of the best American scientists had very close personal and professional ties with their European peers, who kept them well informed about what was going on in Europe. They were greatly troubled by the extent and intensification of war, and considered it necessary to react against the perverse use of science for the invention and development of powerful weapons of mass destruction, such as toxic gases and submarines.

It is important to note Hale's concerns in this period of transition. Seeking to preserve the unsullied image of science in the face of the logic of war, he wrote to President Wilson saying that: "We must not prepare poisonous gases or debase science through similar misuse; but we should give our soldiers and sailors every legitimate aid and every means of protection" (Kevles, 1995). These two purposes showed themselves to be incompatible during the 20th century; the increasing use of science in defence and war activities removed its fragile candour.

Times were changing fast. On 3 February 1917, the USA severed diplomatic relations with Germany, and the National Research Council was entrusted by the military to discover a means of detecting the German submarines. A commission to achieve that goal was formed, led by the physicist Robert Millikan, famous for the first measurement of the elementary electric charge of the electron. George Hale was happy with such a direct involvement of the National Research Council in the defence effort. His ideas were becoming better defined when he wrote to Cary T. Hutchison, executive secretary of the council, that: "War should mean research [...] and unless we get it started, some other agency will do so" (Kevles, 1995).

1.9 The Manhattan Project

Later, the National Research Council found itself lacking the funds and the autho-rity with regard to military activities, in part due to its own insistence on maintai-ning independence from the hierarchy of the federal institutions. At the beginning of World War II the problem was severe. Vannevar Bush, dean of engineering and vice-president of the Massachusetts Institute of Technology, when appointed Pre-sident of the Carnegie Institute in Washington, expressed concern over the war in Europe and aspired to give a new impulse to the mobilization of science for national defence. He and his colleagues and friends, James B. Conant, Karl T. Compton, and Frank B. Jewett, all strongly aligned with the Allies, proposed to President Frank-lin D. Roosevelt the formation of a new federal agency with that objective. They were all prestigious and influential as regards engineering or fundamental and ap-plied science. James B. Conant was an ambitious chemist, President of Harvard University since 1933, who had helped in the production of chemical weapons for the army. Karl T. Compton was a prominent physicist who maintained an active po-litical profile in the defence of physics, science in general, and the importance of its contribution to national defence. He was President of the Physics Department at Princeton University and later, in 1930, President of the Massachusetts Institute of Technology. His younger brother Arthur H. Compton was awarded a Nobel Prize in Physics for discovering the Compton Effect, the first unequivocal proof of the cor-puscular nature of light and, unlike his brother, was dedicated above all to research. Frank B. Jewett had a PhD degree in physics from the University of Chicago, supervised by Albert A. Michelson, well known for the Michelson–Morley experiment, which invalidated the ether concept and paved the way for the theory of relativity. He believed that research could be profitable and sought to develop applied research and its economic potential, especially in long range telephone communications. In 1925 Jewett founded the Bell Telephone Laboratories and later became President of the National Academy of Sciences, from 1939 to 1947. Unlike Hale, the group was convinced of the need to mobilize science for defence and the war effort under a government-run agency. On the 27th of June 1940, President Roosevelt approved the new federal agency that had been proposed, calling it the National Defence Research Committee, and nominated Vannevar Bush as its president.

One year earlier, in the summer of 1939, Eugene Wigner and Leo Szilard, two Hungarian physicists exiled in the USA, convinced Albert Einstein to write to President Roosevelt, alerting him to the tremendous military potential of nuclear fission of uranium induced by slow neutrons. The first experiments were done by Otto Hahn and Fritz Strassman in the Kaiser Wilhelm Institute in Berlin in December 1938. Probably because they were chemists, they only went as far as identifying some of the elements resulting from the fission, namely, barium and molybdenum. The detailed interpretation of the phenomenon of nuclear fission was established outside Germany, in Sweden, by Lise Meitner and her nephew Otto Frisch, son-in-law of Niels Bohr, and published in an article in Nature (Meitner, 1939). Note that Lise Meitner had collaborated in the experiments at the Wilhelm Institute in Berlin but, against Otto Hahn's will, had been forced to leave Germany because she was Jewish.

In January 1939, before the publication in Nature, Niels Bohr, while visiting New York, mentioned the results of Hahn and Strassman to the Italian physicist Enrico Fermi, who immediately understood the possibility of making a bomb with ^{235}U. Fermi, who lived in Rome and whose wife was Jewish, decided to emigrate to the USA, fleeing Italy, which had recently allied itself with Nazi Germany. He had an invitation for one term as visiting professor at the University of Columbia in New York, starting in the summer of 1939, but was accepted as an immigrant by the American Embassy in Rome in 1938. The proceedings demanded by the embassy included a mental test in which Fermi was asked to add up 15 plus 27, and divide 29 by 2. In the same year of 1938, he received the Nobel Prize for physics and took his family to Stockholm. Then from Sweden, the family moved directly to New York. Here Fermi met Bohr. The linking and overlapping incidents that connected the discovery of nuclear fission with the start of World War II show clearly how chance can be a decisive factor as regards the course of history and the unfolding of civilizations. It is easy to imagine relative delays between the two events which could have had dramatic consequences for the Allies and the free world. As events developed, the important question for the Allies at the beginning of 1939 was what would happen in Germany if they reached the same conclusions as Fermi did on the military implications of nuclear fission. The risk was enormous and it was the scientists who insisted on the need to intensify research on the viability of a nuclear bomb.

On the 28th of June 1941, President Roosevelt created the Office of Scientific Research and Development to coordinate scientific research for military purposes. This superseded the National Defence Research Committee and was also run by Vannevar Bush. Around that time the physicists Ernest Lawrence, inventor of the cyclotron, and Arthur H. Compton convinced Vannevar Bush that the new office should start working immediately on whether it was possible to develop an atomic bomb in time to be used in the war, and also to investigate other applications for nuclear fission. Meanwhile, encouraging scientific results were being obtained. Vannevar Bush was able to convince President Roosevelt who, in June 1942, gave the order to make the bomb. In September, the team responsible for the Manhattan Project was formed, led by Brigadier General Leslie R. Groves, an army engineer who oversaw the construction of the Pentagon. He was a determined and clever man, with great leadership qualities, confident in his capacities to the point of arrogance. About one month later the physicist Robert Oppenheimer was appointed Scientific Director of the project.

The obstacles the project had to overcome were enormous, above all due to the difficulty in separating the isotope ^{235}U from natural uranium. The percentage of ^{235}U is only 0.7204%, whereas ^{238}U comprises 99.2742% and the rest is ^{234}U. The first great success of the Manhattan Project was the construction and operation under the leadership of Enrico Fermi of the first artificial nuclear reactor, the Chicago Pile-1, in the Metallurgical Laboratory of the University of Chicago. The reactor consisted of layers of natural uranium and graphite that served as a moderator to decelerate the neutrons emitted by the ^{235}U. The first self-sustained nuclear chain reaction was obtained on 2 December 1942. At 15:53 the reactor went critical and Fermi shut it down 33 minutes later: it had been proved that nuclear energy could be converted in a controlled way into thermal energy, whereby it became potentially useful. A new path for history had been opened in the 20th century.

Meanwhile, Groves and Oppenheimer recognised that the development of atomic weapons required a specialised laboratory. The installation was built in March 1943 in the Los Alamos Ranch School, a mountainous and remote site, at an altitude of over 2000 metres and 72 km from the nearest town, Santa Fe in the State of New Mexico. This laboratory brought together the largest number of eminent scientists ever achieved, all focused on a very clear and precise objective: to build an atomic bomb as quickly as possible. The high scientific and technological level of the challenge, the proximity of so many excellent and intellectually curious scientists, the availability of research means, the availability of time associated with a fully dedicated activity, the level of mental concentration that an isolated environment affords, and the beauty of the surrounding countryside, all went into making the Manhattan Project a stimulating and gratifying experience, despite the very rigorous discipline and safety rules imposed by the military command. It is said that sometimes as many as eight Nobel Prize winners could be found at the dinner table in Fuller Lodge. I knew several physicists that had been at Los Alamos and they all referred to this experience as one of the most important, gratifying, and unforgettable periods in their lives.

Awareness of the terrible goal of the laboratory did not affect the energy and enthusiasm of its members to overcome the countless scientific and technical problems they encountered. In fact, there was an almost supernatural delight in taking part in an attempt to dominate the most powerful force of nature. Some would think that, in the end, if the prototype bomb exploded as was predicted, it would not be for the scientists to decide upon its usage in the war. On 16 July 1945, the first plutonium atomic bomb rested on top of a metal shack 300 feet above the desert at Alamogordo, about 200 km from Los Alamos. About 9 km away the Manhattan team waited nervously for the explosion. At 05:30 an immense light in the desert illuminated the surrounding mountains and, after a few moments, a shock wave corresponding to 20 000 tons of TNT struck those watching. After the initial enthusiasm, there came a heavy silence: it was the beginning of a much more uncertain future. Kenneth T. Bainbridge, the test's head of planning, turned to Oppenheimer and said: "Now we are all sons of bitches" (Lamont, 1965).

Just twenty-one days later, at 07.30 on 6 August, Captain William S. Parsons finished assembling the triggering device for a uranium atomic bomb aboard the aeroplane Enola Gay, piloted by Paul Tibbets. At 08.50 the bomb, called Little Boy, was dropped on Hiroshima, Japan. Moments later the city was covered by a gigantic rising cloud of smoke that reached a height of 9000 metres in less than three minutes and could still be seen even when the Enola Gay was 600 km away. In the city, 7000 people died immediately and many others would die later. It is estimated that more than 140 000 people died from the explosion. On 9 August, a new bomb, this time made of plutonium, called Fat Man, was dropped on Nagasaki. On 14 August 1945 Japan surrendered unconditionally to the Allies.

President Harry Truman stated afterwards that the atomic bomb had been the "greatest achievement of science in history" (Kevles, 1995). All previous concerns with the misuse of science to produce weapons of mass destruction disappeared from mainstream discourse. In political, military, and civilian circles it was concluded that the new atomic age required a national policy of scientific research with particular effort in the nuclear field. Science became a recognized national instrument of power and therefore inseparable from its politics.

Some of the scientists of the Los Alamos generation became actively engaged in a clarification of the social and moral responsibilities of their research activities in relation with defence and war, and more specifically with the use of the atomic bomb. The Franck Report, led by the Nobel laureate physicist James Franck and drawn up in June 1945, just before the explosion of the first atomic bomb, argued that scientists could not alienate responsibility for the usage that humanity finds for discoveries made in the systematic search for scientific knowledge. The report recommended against using the atomic bomb over Japan, and proposed a demonstration to representatives of the United Nations of the weapon at a desert site. However, other Manhattan Project scientists were not of the same opinion. Robert Oppenheimer, Arthur Compton, Enrico Fermi, and Ernest Lawrence reacted by saying that: "We can propose no technical demonstration likely to bring an end to the war; we can see no acceptable alternative to direct military use" (Smith, 1965).

1.10 Science, Technology, and Defence: From the End of World War II up to the Beginning of the Twenty-First Century

Immediately after the explosion of the atomic bombs in Japan, nuclear scientists began a vigorous public campaign to seek control over the international use of atomic energy. The initial idea was to negotiate a plan through the United Nations under which other countries would not make atomic bombs and the USA would destroy theirs. However, the negotiations failed due to mutual distrust between the USA and the Soviet Union, and the latter exploded its first atomic bomb in 1949. The immediate reaction from the USA was to create an intensive new project, similar to the Manhattan Project, to rapidly produce the first thermonuclear bomb — where the energy comes from fusion of atomic nuclei — which was expected to be much more powerful than the previous nuclear weapons.

As before, there was a small group of scientists who forcibly argued the urgency of the project. Among these, Edward Teller stood out. Despite the opposition of other scientists, such as Enrico Fermi and Isidor I. Rabi, and the contrary opinion of the Atomic Energy Commission, at the beginning of 1950, President Truman signed his authorization to create the first hydrogen thermonuclear bomb. The design was invented by Edward Teller, a Hungarian physicist, and Stanislaw Ulam, a Polish mathematician. The first hydrogen bomb exploded on 1 March 1954, at the Bikini Atoll in the Marshall Islands, and turned out to be about three times more powerful than expected, with a yield of 15 megatons of TNT, which is 750 times more powerful than the first fission bomb from the Manhattan Project. It caused a very serious radioactive contamination in the region and a considerable number of human victims. The radioactive fallout reached Australia, India, Japan, the USA, and parts of Europe. Meanwhile, about one year earlier, on 12 August 1953, the Soviet Union exploded a thermonuclear device which was still not quite a hydrogen bomb. Two years later the first Soviet hydrogen bomb, with a design invented independently by Andrei Sakharov but similar to the Teller–Ulam design, exploded in Kazakhstan. The nuclear arms race intensified and only slowed down at the end of the Cold War, after the collapse of the Soviet Union in December 1991.

During the cold war scientists, and especially physicists, were frequently called upon to participate in the effort to maintain military technological superiority of their countries as superpowers or leading world powers. In return, they thereby benefited from generous financial grants for both applied and fundamental research. Research in various scientific areas was favoured, especially in nuclear and elementary particle physics, sometimes with the justification that there was a chance, however faint or remote, that they would make a breakthrough with important military applications. In spite of this tight bond, many scientists vehemently defended the deceleration and the end of the nuclear arms race and actively participated in movements that led to the nuclear test ban and strategic arms reduction treaties.

At the beginning of the 1990s, with the end of the Cold War, the political priorities were changed and the average budget for research per scientist diminished in the USA. New fields of research and development were opened in areas of increasing interest, such as the exploration and utilization of outer space, biotechnology, nanotechnology, and the study and monitoring of global change.

Not much time went by, however, before the permanent desire to assure military superiority and the rapid evolution of various forms of violent conflict developed a craving for new weapons. The START-II Treaty of 1993, covering the reduction of strategic arms, was ratified but never activated and finally denounced unilaterally by the Russian Federation one day after the USA withdrew from the ABM Treaty. This setback led the USA and the Russian Federation to agree on a new strategic arms reduction treaty which was signed in Moscow in May 2002. This allowed each country to retain a limited number of atomic bombs. Following the terrorist attack on New York on 11 September 2001, and the wars in Afghanistan and Iraq, President George W. Bush decided to start a process of nuclear arms renewal. In 2003, the Department of Defence asked Congress to lift the ten-year-old ban on developing new nuclear weapons of less than 5000 tons of TNT. One of the main objectives was to develop an atomic bomb capable of destroying chemical and biological weapons hidden in fortified, underground bunkers which are inaccessible to conventional non-nuclear weapons.

The justification for such a request for funding presented to Congress stated that the embargo had a "chilling effect" on nuclear weapons research "by impeding the ability of our scientists and engineers to explore the full range of technical options" (Guinnessy, 2003). Furthermore, it stated that: "It is prudent national security policy not to foreclose exploration of technical options that could strengthen our ability to deter, or respond to, new or emerging threats." Thus it was proposed to reactivate the nuclear arms programme to counteract the new perceived risks, and to "train the next generation of nuclear weapons scientists and engineers" (Guinnessy, 2003). The creative capacity of scientists and engineers to produce weapons of mass destruction was a matter of concern and should be guaranteed by the contemporary political process. Presently, we are confronted with the possibility of initiatives, such as the production of mini-nukes and nuclear bunker busters that would blur the distinction between conventional and nuclear warfare and would probably lead to renewed nuclear testing.

As at other times, some scientists and engineers have openly manifested their disagreement with regard to these initiatives. They point out that radioactive waste will be thrown out by the nuclear explosion and also that the chemical and biological agents from the weapons that were meant to be destroyed may just be dispersed into the atmosphere, especially if the target is not hit precisely. Since they are weapons of lesser potential and devised for specific targets, the probability of using them is higher when compared to the use of more powerful atomic bombs, whose main role is dissuasion. In fact, such usage against countries that do not have nuclear arms would violate the Non-Proliferation Treaty of 1970. One should also consider the political aspect that such a programme would weaken the fight against the proliferation of nuclear arms. Finally, there is a real danger of resuscitating the arms race at a time when there are signs that Russia is already responding with plans to renew its nuclear arsenal.

The modernization of nuclear warheads and the construction of new production facilities were put on hold, pending an updated nuclear policy demanded by Congress (Kramer, 2009). Finally, in April 2010, President Obama released the new nuclear strategy of the USA, which has the broader objective of edging the world toward making nuclear weapons obsolete. To set a good example, he renounced the development of any new nuclear weapons. This is good news, but it remains to be seen whether such a strategy will be accepted by all countries. Without international agreements regarding the renewal and proliferation of nuclear weapons worldwide, the tendency to design and produce new nuclear warheads will persist.

Science, in its essence, is a free and creative activity that normally leads to the dissemination of its findings among the scientific community, in publications of peer reviewed journals. However, when these results have potential military applications, conflicts may arise with the community responsible for security. Since the beginning of the Cold War, there have been several instances of conflict in universities and state-run laboratories involved in scientific activities with military implications, particularly in the USA, due to regulations that impose strict procedures regarding security and secrecy. There is frequent tension in the areas of microbiology and biomedicine, where, according to those responsible for defence, the publication of certain specific results would assist terrorists or enemy states to produce chemical and biological weapons of mass destruction. So, today scientific and technological knowledge is not always a worldly commodity, allowed to circulate freely in the world.

1.11 Science, Technology, and Development: The First Reactions

These examples reveal how science and technology have been irreversibly integrated into politics, not only in the social and economic sphere, but also in the military domain. Financial support and development have become dependent on, and inseparable from, political evolution, just as happens with other interest groups. The case of the USA is particularly clear, and indeed exemplary, especially in the military area, due to its role as a superpower. The other developed countries present similar scenarios and a clear tendency to emulate the American model. In developing countries, the situation is quite different. Scientific activities are generally very rudimentary, due to the lack of human and material resources. Science and technology have an increasing impact on the socio-economic life of these countries, but are mainly imported from the more developed nations. The development of endogenous science and technology in developing countries is increasingly accepted as a national priority in spite of limited resources.

Despite these differences, it is important to acknowledge that, nowadays, science and technology constitute an indispensable basis for the exercise of power, the production of goods, the food security, the quality of life, and, generally speaking, the cultural, social, political, and economic transformations that are taking place at national, regional, and global levels. Science and technology are just two of the factors that help explain those transformations. Furthermore, their influence does not guarantee that development will necessarily progress along a path of peace, rationality, and sustainability. With this warning, it is reasonable to believe that science and technology will continue to play a central role in the processes associated with globalization, in the sense that they contribute progressively to the free exchange of ideas, goods, and people. Some technologies have an especially important role in globalization phenomena. A particularly good example is provided by the information and communication technologies (ICT) which began their development in the 1940s. These involve the study, design, development, and support or management of computer-based information and communication systems. Practically all human activities in the vast majority of populated areas of the world now require the application of IC technologies. They thus constitute a powerful force for global transformation. On the other hand, the cultural, social, political, and economic context in which a certain technology is used will influence its local development. Each culture appropriates and integrates the technologies it uses in its own way, and strives to develop them in directions that are more beneficial to the pursuit of its general and specific interests.

The first reactions against the domination of nature and the mechanical vision of progress and development based on the systematic use of science and technology arose with the Romantic Movement, in the transition between the 18th and 19th centuries. From the historical point of view, Romanticism is a complex artistic, intellectual, and philosophical movement against the Enlightenment and the scientific rationalization of nature. The world view purveyed by science is questioned. Preference is given to a more personal, more emotional, and more intuitive view, believed to be perfectly capable of reaching the plenitude of knowledge and truth. It is characterized by a revalorisation of feelings, sentiment, and faith, by a new importance given to history, particularly the Middle Ages, and by the promotion and exaltation of the mystical and magical aspects of life. Romanticism is clearly a precursor of the irrationalistic movements of the beginning of the 20th century.

The danger of an uncontrolled and overwhelming technological development was addressed for the first time in a most cutting way by the romantic writer Mary Shelley in her books *Frankenstein or the Modern Prometheus* (Shelley, 1818) and *The Last Man* (Shelley, 1826). She warned of the danger of an obsessive search for dominion over nature and, more generally, for the 'hubris' of science and technology. Her book *The Last Man* was the first to dramatize the risks of catastrophic global changes. Curiously, Shelley was a contemporary of the Luddites, the name given to the textile artisans in the centre and north of England who, between 1811 and 1816, revolted against their terrible working conditions and poor pay, and more generally against the technological changes brought about by the industrial revolution, which they felt were leaving them without work. They reacted by systematically destroying their tools and machines, regarded as symbols of technological change and a tyrannical and oppressive industrial labour system.

However, it was only in the 20th century that various movements questioned science, forecasting its end, and accusing it of profoundly disturbing the social equilibrium. The beginnings of the century saw the emergence of currents of thought of an irrational tendency that sought to discredit intelligence and the importance of verbal expression, while glorifying action, energy, and strength. They developed as World War I approached and were flourishing as never before by the time it came to an end.

It is during this period that *The Decline of the West* was published (Spengler, 1918). This crucial work served as the intellectual foundation for the disenchantment with, and even hostility to, science. Its author, Oswald Spengler, was a high school teacher in Germany, a little known scholar with a doctoral degree on Heraclitus, and a man of sparse financial means. The book is sombre but fascinating, with some 1200 pages that took ten years to write, in part during the war, and in very difficult circumstances. It presents a comprehensive theory of the past and future of history, in which civilizations rise, then decline and fall, running through similar cycles that end in decadence and extinction. He describes Western civilization as Faustian, since it follows a similar path to Faust. According to him, the cycle of our Faustian civilization is approaching its end, and the next will come from the East. In one of his chronological tables, he indicates the year 2000 as the decisive date for the collapse of Western civilization. Moreover, it is modern science itself that has sown the larger part of the seeds of decline. Spengler attributed a central role to modern science, but considered that Western scientific thought would shortly reach the limits of its evolution. According to his analysis, physics, for instance, was showing signs of decline and disintegration that resulted from the invasion of statistical and probabilistic methodologies, and the dominance of the new theories of relativity and quantum mechanics, steeped in relativism and uncertainty. According to Spengler, the systematic emphasis on scientific thought and the pre-eminence of reason must fail in the fundamental realms of the human condition, where one must instead develop habits of intuition and presentiment.

The book, published in 1918 when the war was dragging tragically to its end, was an immediate success, selling hundreds of copies and rapidly achieving international recognition. Among many others, Ludwig Wittgenstein praised it (Monk, 1990), and Thomas Mann considered it to be "the most important book" of its time (Mann, 1983). The shock of defeat and the realization that Prussian militarism was incapable of assuring victory left many Germans cynical and much more sceptical of the value of science, technology, and politics. The book resonated with Germany's post-war feelings of distress, and did much to promote the ensuing conservative nationalism.

In his lectures on *Science as a Vocation*, the sociologist Max Weber (Weber, 1918) called attention to the extent and depth of the anti-science movement in post-war Germany. Probably the socio-political conjuncture led him to stress that the commitment to science is only compatible with an emotionally uninvolved and detached vision of the world, indifferent to the social and political values that were competing for hegemony at the time. Moreover, Weber considered that the knowledge resulting from the application of scientific methodology would lead to a systematic global disenchantment, and ultimately to the loss of any other values or meanings than those which are strictly practical and technical.

1.12 Sociology of Science and Post-Modernism

Although the pursuit of scientific knowledge is indifferent to human projects and values, according to Weber, the fact is that the practice of science carries the influence of the specific sociological context in which it takes place. Thomas Kuhn first highlighted this exposure in the book *The Structure of Scientific Revolutions* (Kuhn, 1962). He observes that the emergence of a new paradigm, which replaces a previous one — for instance, the replacement of Newtonian mechanics by the theory of relativity — also depends on sociological forces, capable of determining when and how it may be fully accepted. This represents a deep break from the absolute concept of science in which the discourse of scientists bases itself on unquestionable evidence of an experimental or observational nature. As a matter of fact the dialogue and controversy between scientists in the social context in which they find themselves at the very least influences the time required for a new theory to be accepted. Kuhn opened the way for the new field of sociology of science.

Note that science, just like any other human endeavour, is influenced by social factors. It is therefore natural that it should be the object of sociological and philosophical analysis. The emphasis on the social and cultural relativism of science is a distinctive feature of the post-modernist movement that arose during the philosophical debate at the end of the 1970s with the work *La Condition Post-Moderne*, by Jean-François Lyotard (Lyotard, 1979). Some authors currently defend the extreme point of view that scientific theories are just social constructs that result from the process of building a consensus within the scientific community. Consequently, science has no more claims than other socially constructed systems of ideas to uphold that its results constitute objective truths. In this perspective, the status of science is on the same footing as a 'social myth', just one among many others (Hesse, 1992).

The central idea of the post-modernist vision of science is to deny that its methodology is privileged or unique among organized ways of obtaining knowledge. Furthermore, it questions the capacity of science to indicate a path for improving the well-being of humanity. It also contests the idea that science necessarily leads to what can be generically considered as human progress. The roots of the movement can be traced back to Romanticism, and it therefore represents the contemporary antithesis to the descendants of the Enlightenment.

Paul Gross and Norman Levitt (Gross, 1994) have developed one of the most extensive and exhaustive criticisms of the post-modernist line, stressing that the reflections of sociologists and philosophers often reveal a substantial ignorance of the methodology and practice of science. This difficulty of communication became very apparent in the parody article *Transgressing the boundaries: Toward a transformative hermeneutics of quantum gravity*, published by the theoretical physicist Alan Sokal in the social sciences journal *Social Text* in 1966 (Sokal, 1966). After receiving a successful peer review report from the journal and following publication, Sokal revealed that the article was a hoax, a parody of post-modern thought on science. For Sokal, the statements in the article sounded clearly false and should make one laugh, but the peer reviewers and editors of *Social Text*, considered it to be totally plausible, since they more highly valued the literary form and the structure of the text than its substantive contents.

This episode clearly illustrates the fact that scientists tend to apply the written language in their papers in a specific and distinct way from other users. They seek to write in an unambiguous and concise way, describing all the steps that led them to their conclusions with the greatest possible clarity. This concern results from the need to guarantee that the final results can be reproduced, a factor that is of crucial importance to the scientific method. A scientific paper, whether of an experimental, observational, or theoretical nature, should allow other scientists that read it to arrive independently at the same conclusions, following the logical path proposed or an alternative route. If it turns out to be impossible, the authors are likely to fall into discredit. Because of this constraint, the written expression of science, as actually practised by scientists, must be precise, concise, unambiguous, emotionally neutral, and suited to the understanding of its peers. That is the reason why it is generally inaccessible to non-specialists. All such requirements imposed on a scientific paper would guarantee the complete failure of a literary piece.

The article by Sokal intensified and widened the debate about the sociology of science, also known by the rather sensational and inappropriate title of *Science Wars* (Ashman, 2001). Of course, they are not wars, but just a lively reflection, sometimes passionate, of a debate that will continue, as it raises important and contentious questions of growing complexity, such as the cultural relativism of scientific knowledge and the role of science and technology in society, nowadays and in the future.

1.13 Relativity, Disbelief, and Opposition to Science and Technology

All over the world today, but especially in the developed countries, there are many lines of thought that share an opposition to science and technology, focussing on its cultural hegemony and its ambition to become an irreplaceable support and inspiration for human progress. They do not constitute a coherent movement, but three distinct tendencies can be identified that frequently coexist.

A first group, more intellectual, concentrates on philosophical or social questions related to science, and emphasizes the vain epistemological pretensions of science. This group claims that the belief in science, or even science itself, is coming to an end. A second, more openly contentious group advocates an alternative, romantic, and post-modern vision of science. Finally, the third group points out the negative impacts on society and the environment of certain technologies supported by science. They all share the conviction that science must be transformed, bringing to an end the form we know and practise today.

The positions of those who defend the first line of thought, represented mainly by sociologists and philosophers of science, have already been presented, and are those which naturally pose a more substantial intellectual challenge. There is a wide range of tendencies within this group, but the most extreme, contained for example in the texts of Bruno Latour and Steve Woolgar, consider science to be a mere social practice that produces narratives and myths, and that the latter have no greater claim to validity than those produced by other social practices (Latour, 1979; Latour, 1987). They also claim that science manifests a hegemonic tendency of man relative to woman, and of the West relative to other cultures.

The defenders of the second group, called *Dionysians* by Gerald Holton (Holton, 1993), trace their roots to the Romanticism of the 19th century and the protest movements of the 1960s. These advocate a new science that integrates other cultures apart from the dominant West, and they are particularly fond of Eastern mysticism.

Finally, the third group has a more extensive and diversified sociological basis. Its defenders manifest a range of concerns that tend to strengthen and amplify, in a general way, the opposition to science and technology. It becomes ever more evident in the developed countries, but gradually even in the developing countries, that science and technology constitute an essential mainstay of modern life in practically all areas of human activity, including the civilian and military spheres. This group is especially concerned with the risks to the environment, health, and human life that result now and will result in the future from such dependence. The current rates of consumption of natural resources, of environmental pollution, and of global population growth can only be maintained by the permanent intervention of science and technology. The question is whether these rates are compatible with the ecological equilibrium of the planet, and what would be the consequences of an eventual disequilibrium. They are concerned with the risks involved in the growing use of technologies that inevitably degrade, pollute, or destroy the various components of the environment: atmosphere, oceans, coastal zones, water resources, soils, forests, and biodiversity.

The technological risks have their origins in the development and use of the most diverse ancient and modern technologies. They are associated with the production of an increasing variety of chemicals, with the operation of industrial and energy infrastructures, with toxic and dangerous wastes, with various forms of pollution of the air, oceans, coastal zones, rivers, aquifers, and soils. Featuring high among these are contamination with heavy metals, use of pesticides in agriculture, various forms of air and water pollution, radioactive fallout, and emission of greenhouse gases, especially carbon dioxide (CO_2) emitted by burning fossil fuels, which is the main cause of climate change. Besides these civilian technological risks, there are also those of a military nature. The permanent quest for the technological improvement of weapons, in particular chemical, biological, and nuclear weapons of mass destruction, creates a persistent and growing peril for populations, especially those that live in areas of instability, or are victims of armed conflicts. In this third group, awareness of this increasing multiplicity of perils generates a reaction of mistrust toward science and technology. Those who belong to this group tend to make science and technology accountable for the unacceptable risks associated with our present model of development, and they advocate a search for less risky ways to promote development.

The emergence of the current movements against science and technology overlaps notoriously in time with successive vocational crises for these areas among students. Furthermore, one notes a deep scientific illiteracy in most of the more developed societies. The simultaneity of these phenomena looks contradictory and even paradoxical if we think of the remarkable discoveries of contemporary science and the spectacular advances in the quality of life, which were made possible by the applications of modern science and technology. People are thrilled by the contemporary technological wonders: they use them eagerly and take them for granted, almost immediately. In the face of so much success, what reason is there for such clear signs of hostility towards science and technology?

The tremendous development of scientific and technological innovation in our time fails to attract the younger generations to these areas, and there is a growing lack of scientific culture, especially in the developed countries. In part this may be due to the relative difficulty involved in the process of becoming a professional scientist or engineer, which is usually not rewarded by a highly paid job. The temptation is then to choose professions that are financially more attractive. It is nowadays frequent for scientists, mathematicians, and engineers to move to the financial sector in order to get better salaries. According to a 2011 report by the Kauffman Foundation, this movement of scientists to Wall Street may have stifled entrepreneurship in the USA. The benefits that science and technology bring to society are often looked upon as being guaranteed indefinitely and there is no conscious realization that they will require a permanent investment of human and material resources.

It is not unusual in contemporary societies for people to simultaneously hold both positive and negative attitudes towards science and technology. This duality appears in many forms and has diverse origins. Among the group of factors that tend to generate positive attitudes is the awareness that continued economic growth and improvement in the quality of life are more and more dependent on the development of science and technology. On the other hand, there is a growing awareness of the risk of a progressive environmental degradation, and the associated threat to human health and life, associated with that dependence.

The lack of scientific culture can also generate contradictory views and attitudes. Some defend a reduction in the role of science and technology, as if this could be compatible with the actual levels of well-being and security that we enjoy in our ever more complex societies, particularly in the more industrialized countries. Others have an almost unlimited and naive confidence in the power of science and technology to solve all the varied and intricate social, economic, and environmental problems of contemporary societies. They argue that if modern science and technology were repeatedly able to turn something that was almost impossible to imagine into a triviality, then in the future there are no bounds for finding solutions to the problems created by our development growth paradigm.

There are many signs of a changing environment as regards public perception of the value of science. Two particularly revealing examples are the denial of scientific knowledge as regards Darwin's evolutionary theory of life and the denial of anthropogenic climate change. For the general public, the arguments for denial are very often attributed the same value as those that result from the methodology of science. The debate is viewed as the expression of two competing discourses. This trend is intellectually damaging and also, in some cases, economically harmful. It means that there is a tendency for a broader and more varied assault on policies based on scientific evidence rather than prejudice. Again a good example is the climate policy to prevent a dangerous anthropogenic interference in the world's climate, which is clearly a science-based policy.

The scarcity of scientific culture around the world makes it more difficult to establish a distinction between science and pseudo-science. Indeed, the media are often unable to distinguish these themselves, and may sometimes favour pseudo-science because the results tend to be more sensational. Furthermore, the internet is completely neutral as regards its contents and therefore gives equal value to all arguments and propositions, including those that are soundly based on science and all the others that are not. Everything points to a need for scientists to make a growing effort to communicate, inform, and explain what they are doing to the public in general, and above all to elucidate and defend the methodology of science.

The number of science sceptics is clearly growing. Their preferred methodology is to exploit dissent among scientists and to emphasize presumed weaknesses in the scientific theories. This movement is in line with the post-modernist vision of science as a socially constructed system of ideas that competes with other social constructs.

1.14 Science and Religion. A Dialogue in Permanent Evolution

It is important to emphasize that science and technology have a marginal capacity for intervention in society compared to the much more powerful forces of economics, politics, religion, and culture. Religion is particularly relevant, because it affects the heart of consciousness, the most intimate and personal values, while simultaneously establishing a relationship between private and public life. Its firm entrenchment in society is ongoing, and increasing in some countries, including developed countries. Some say that religion is on the return, and statistics apparently point to an increase in the number of believers (Testot, 2008). Regardless of what religion they belong to, the faithful benefit from a gratifying human experience that deeply affects their view of the world.

It is especially interesting to compare the influence of science and religion in contemporary societies from the historical perspective of its contentious relationship, sometimes strong, but today greatly softened. Both offer a view of the world and a roadmap for bettering it. However, they are of quite distinct nature. Religion is a complex combination of feelings of faith, acts of worship, and rituals that establish the communication between man, the sacred, and the divine. The subject of science is neither the sacred nor the divine. Science merely seeks knowledge by way of a very specific and limited methodology that involves observation, experimentation, and the construction of theories that are potentially falsifiable by further observation and experimentation.

Modern science had its origins in the 16th century, at the heart of a Christian Europe. Until the 18th century, the vast majority of those who practised it were believers. It was only midway through the 18th century that atheism began to emerge as a powerful doctrinal current. Its first systematic defenders were Paul Henri Thiry, Baron d'Hollbach (1723–1789) and Claude Adrien Helvétius (1715–1771). D'Hollbach, in his book *Système de la Nature*, laid the ideological foundations for a rigorously materialistic and atheistic concept of nature, man, and society. The main objective was to release man from religion, which he considered to be a form of superstition, in order to allow him to become completely free and his sole master. In parallel, he endeavoured to exclude any supernatural cause to explain the phenomena observed in nature. These positions made him the target of great hostility, even among contemporary philosophers of the Enlightenment, including Voltaire who was a deist. Only Diderot, a convinced atheist, remained among his few loyal friends.

The epistemological clash between science and religion deepened with the positivism of Auguste Comte (1798–1857). He argued the existence of an evolution in human knowledge, starting from theological and metaphysical states and leading up to the scientific state. For him, true human knowledge is the positive knowledge that results from the practice of science and is free from a priori suppositions and prejudices. He organized the sciences into a hierarchy of decreasing abstraction and increasing complexity, from mathematics and astronomy to physics, chemistry, and biology. Furthermore, he coined the word 'sociology' to designate the empirical science of social phenomena, divided into a static part, which studies the social order, and a dynamic one that studies social progress. According to Comte, nothing lies outside the range of the scientific method, a clear example of scientism.

In the second half of the 19th century Darwin's theory of the evolution of species gave a new and powerful impulse to positivistic ideas. Herbert Spencer (1820–1903), one of the most distinguished of the English positivists, went further, and defended the existence of a universal law of evolution from the simple to the complex and from the non-organic to the organic, applicable to all realms from cosmology to biology, sociology, and history. Although he admitted that it is possible to know the laws of evolution, he considered that the explanation of their nature and primordial causes lies beyond the scope of human knowledge and understanding. This limitation is a form of agnosticism, a term first used by the English positivistic biologist Thomas H. Huxley (1825–1895) to describe the attitude of those who restrain from pronouncing on absolute principles and unsolved problems, considered to be beyond the reach of the scientific methodology. From this viewpoint, God may exist but if he does, his existence is beyond human knowledge.

During the 20th century, various forms of atheism and agnosticism became more visible and widespread. They lost the negative character of a deviant and reproachable way of thinking that had been attributed to them by theists in previous times. For theists, the spread of atheism is mainly due to an erroneous extension of the positivistic vision of scientific knowledge to the religious domain.

On the other hand, the influence of religion on society worldwide is stable or growing in some countries. In the group of developed countries, the USA stands out, possibly due to its relatively recent independence and the fact that its political history was profoundly tied to religion and the freedom of religion. In recent polls, around one third of all adults declare that they have had some form of a religious rebirth in Christianity and therefore consider themselves to be born-again believers. Sixty percent of these state that they believe in the literal existence of hell, as the destiny for those who do not behave well on Earth.

Sometimes scientific and theological views of the world clash in a very visible way. An illuminating example is Darwin's theory of evolution, rejected by about 40% of the population of the USA, especially among the protestants of the Southern States, a far higher value than that found in the more secular societies of Europe. They insist on the truth of the creation of the world as told in the Bible. Consequently, they consider that Darwin's theory is false, since it does not reflect that truth. The defence of creationism has turned into a strong political force in the USA, with a considerable influence on the way biology is taught in high schools. In the 1920s, John T. Scopes, a high school teacher in Tennessee, was sentenced in court because he insisted on teaching Darwin's theory of evolution to his students, overruling a law that prohibited it. From the 1960s onwards, the conflict between Darwinists and creationists evolved and forced the latter to defend the hybrid concept of creation science as opposed to evolution science. The aim was to get round the prohibition of the teaching of religion in state run schools, as specified in the American constitution. With the new designation, it was possible to teach the biblical version of creation just as it is written in the *Book of Genesis*. This attempt to defend convictions of a religious nature under the guise of science has been vigorously denounced by scientists, and especially biologists, in the USA.

The debate continues to evolve, as do the terms in which it is expressed. Now, instead of creation science, the preferred term is 'intelligent design' (Ruse, 2008). The argument is based on the belief that life is too complex to be explained just by natural causes, whence some intelligent force, most probably of a divine nature, must have created the primordial forms of life on Earth and guided their evolution. However, there is no way of testing the existence of this presumed intelligent force by observation or experimentation. According to the methodology of science, it must be outside its reach.

However, the pressure to teach both the creationist and the Darwinist viewpoints persists. One of the latest rounds in this controversy is centred on how evolution should be taught in the high schools of Texas (Herald Tribune, 2009). The creationists insist that the curriculum should include views about the scientific strengths and weaknesses of Darwin's theory, although the examples of weaknesses are clear misrepresentations of science.

Meanwhile, the debate between the creationist and evolutionist views is becoming global. Members of the hierarchy of various religions frequently reject, in a more or less explicit way, the idea of evolution by natural selection that led from the most primitive and simple forms of life to *Homo sapiens*. They defend the idea that Darwin's theory of evolution cannot be proved conclusively, and that the way life evolved indicates the presence of a divine force, inaccessible to the methodology of science. The legitimacy and capacity of science to conclude that the origin and evolution of life can be fully explained without resorting to any form of divine intervention is simply denied. At a pinch, the descriptive content of Darwin's theory of evolution can be accepted within the religious viewpoint, but it is not considered to be sufficient to fully explain what has happened. This is a particularly interesting example of an irreducible distinction between science and religion.

During the 20th century and especially since its end, we have witnessed an enormous effort to bring science and religion closer together and to seek a coherent synthesis that is compatible to both. In the USA and in other more industrialized countries, conferences, discussions, and university courses about science and religion are ever more frequent, especially those regarding the problems of their evolving relationship. Most of these initiatives are organized by practicing Christians, many of them scientists, and they consider almost exclusively the Christian religion. They claim that the Universe as described and understood through science can also be recognized as a divine creation. They argue that the meaning conferred by Christian theology on the Universe is compatible with the view of the same Universe given by science, although many scientists consider it to be devoid of any special meaning. They search for signs of God's intervention in the context of the description of the Universe obtained with the methodology of modern science. Finally, they look for ways to reconcile dogmas of faith, such as the incarnation of God and the resurrection of Jesus Christ, with science.

The attempts to achieve these objectives are wide-ranging. Some assume that certain religious teachings, in particular within the Christian faith, such as the creation of the Universe, Earth, and man, are incompatible with the knowledge provided on these subjects by modern science. However, they consider that this is no impediment to searching for a much needed closer bond between science and religion. Christian de Duve, Nobel Prize winner in medicine, puts forward as a possible solution the idea that religions may provide a reinterpretation of certain aspects of their scriptures, such as the week of creation described in the *Genesis*, in such a way as to be more compatible with modern science, but with no change in the essential doctrine (Duve, 2002). But he nevertheless recognizes the enormous difficulties that this proposal would involve.

Others claim that the Universe, revealed in all its detail and complexity by science, is not only compatible with a divine creation, but also that it constitutes one of the proofs of God's existence. An example of this type of argument is based on the anthropic principle, proposed and named by the mathematician Brandon Carter in 1974 and later extended by the physicists John D. Barrow and Frank J. Tipler (Barrow, 1986). According to this principle, the fundamental physical laws and constants that we find in the Universe are the only ones that could have led to the existence of intelligent life. If, for example, we imagine a different law of gravity, where the gravitational force is stronger or weaker, it would not be possible to form planetary systems with planets in stable orbits around a star. In the same way, laws of electromagnetism distinct from the ones we observe in nature would prevent the stability of atoms. Everything seems to work together in order to make it possible to form atoms, molecules, planets, living creatures, and finally human beings. This precise and somewhat improbable concatenation is interpreted by various authors as the result of a deliberate act of a divine being who planned and created the Universe in such a way that we would inevitably emerge.

According to John Polkinghorne (Polkinghorne, 1988), the very existence and nature of the Universe attest to its divine origin, and consequently to its purpose and final cause. In this interpretation the narration of the Universe provided by science constitutes a new form of theology. By following this approach, he suggests that God may act, for example, through the stochastic processes of quantum mechanics and the chaotic natural processes that lead to apparently unpredictable results.

However, a detailed analysis strictly based on the scientific methodology reveals the contradictions of the preceding theses. The anthropic principle becomes questionable by establishing itself on the assumption that an undetermined set of universes characterised by different fundamental laws and constants may exist. This supposition of a Platonic nature is just a conjecture, which is completely unsusceptible to verification by observation and experimentation, because the observable universe is the only one we have available to us. We are subject to one set of physical laws that characterize our Universe, and therefore unable to observe and compare other distinct physical laws. Consequently, the possibility that a set of alternative universes with physical laws expressed by different mathematical formulae exists, although crucial to the argument, lies outside the scope of contemporary science.

Some scientists have developed new theories in which our universe is not unique, but only one in an infinite population of universes, which the British astronomer Martin Rees has dubbed the 'multiverse'. Others, like the American cosmologist Alan Guth, assert that it is possible to create other universes, each with its own space and time. We are far from being able to test these theories with data from observation or experimentation, so they are no more than intellectually stimulating conjectures. The possibility of an immense multiplicity of universes and the trivialisation of its creation has destroyed the apparent singularity of our own universe. We may assume that the universes of the multiverse have different physical laws or by-laws, and we may even imagine being able to 'squeeze' ourselves from one to the other. But since in each of them we would be subject to the local by-laws, could we survive?

The fact is that there are many different opinions about the meaning of everything that science has allowed us to find out about the Universe. For the physicist Steven Weinberg, the deeper and more extensive our scientific knowledge of the Universe, the more meaningless it becomes. For others, such as the physicist Freeman Dyson, the Universe revealed by science has a meaning given by what is interpreted as the inevitable appearance of man. Anyone may have or may develop his own opinion on this subject. In this type of debate, we leave the realm of science. We go beyond rational analysis, based on the methodology of science, of the facts in themselves, stripped of any purpose or finality, and seek to address their significance to us. We thus enter the realm of Pascal's *esprit de finesse* which is dominated by feelings, prejudice, belief, religious faith, and life experience.

There have been other attempts to launch the foundation of a coherent synthesis of science and religion. J. Wentzel van Huyssteen (Huyssteen, 1998) seeks to achieve it through post-modernism. On the one hand he strives to transcend the religious dogmatism of the past by including the great diversity of religions, and on the other he stresses that science is not purely objective, since it is a socially constructed system of ideas. This common form of relativism brings them closer, but is unable to conceal the fact that, in their essence, they are two very distinct forms of human representation and expression.

There is yet another relationship between religion, science, and technology which is characteristic of the contemporary world, although with distant origins. Religions form an important part of the identity of a given community, nation, or civilization, and are also a most relevant component of their history, their collective memory, and their solidarity. Science and technology, on the other hand, are by nature universal and reject any kind of link with a religious or cultural identity. However, they influence religion and culture in a more or less direct and explicit way by constituting an essential support for today's phenomenon of globalization. This process tends to establish an economic and cultural uniformity across all countries and all peoples around the world. Some religions tend to react to this standardising pressure by enhancing their identity and by isolating themselves. In certain cases, the reassertion of identity assumes such extreme forms of fundamentalism that it may lead to fanaticism and violence. There is no reason to establish a direct causal relationship between this extremism and modern science and technology. However, there are deeply marked interactions in the triangle formed by globalization, science and technology, and religion. More to the point, the concern for establishing a synthesis at all costs between religion and science in the context of the Christian faith can be interpreted as a manifestation of the hegemony and hubris of Christianity.

1.15 The Great Challenges of the Twenty-First Century

For many centuries, human activities constituted an insignificant force in the dynamics of global terrestrial systems. Today the situation is very different. Humanity has begun to perturb and change the biosphere, the atmosphere, and other systems on a spatial and temporal scale never reached before. The spatial scale is global as it covers the whole of the Earth's surface. The swiftness of these anthropogenic changes has no parallel in the past, given that they happen over time intervals of decades to a few centuries, instead of the many centuries to millennia which are characteristic of natural changes resulting from the dynamics of terrestrial systems. It is very likely that the new correlation of forces between humanity and its planet will affect it in a very deep and long-lasting way. This is the reason why Paul Crutzen and Eugene Stoermer consider that we have entered a new era, characterized by significant human interference in the terrestrial systems, which they have called the *Anthropocene period*.

Since the end of World War II, we have participated in and witnessed something truly remarkable which has been called the Great Acceleration (Hibbard, 2007). The last 60 years have been a period of rapid social and economic development worldwide, characterized by a very significant growth in world population, a strengthening trend toward economic and cultural globalization, the integration of major countries, such as China, India, and Brazil, into world trade and the world economy, a growing global use of energy per capita (where the energy is drawn mainly from fossil fuels), an increasing and unsustainable consumption of natural resources, the extraordinary development of information technologies, and a considerable increase in human mobility, communication, and the flow of information worldwide. The Great Acceleration has lifted hundreds of millions of people out of poverty and improved the quality of life of many more. On the downside as regards the environment, over the past 60 years, humans have very rapidly and extensively modified many ecosystems in order to meet demands for food, water, timber, fibres, and energy. We are witnessing a substantial and irreversible loss of biodiversity and a degradation of the environment on a global scale. And of course, the Great Acceleration was made possible essentially by the development and dissemination of science and technology and by substantial worldwide improvements in political and economic structures and institutions.

One of the main consequences of this accelerated development has been an increase in the uncertainty and the risk associated with the future. It is clear that the pace of growth over the last 60 years cannot be maintained forever. An increasing number of people are trying to find and implement more harmonious relations with the environment, and new forms of governance capable of leading us to sustainable development. Some of these movements look upon scientists and engineers with scepticism, since they consider them entirely committed to the contemporary development paradigms and therefore unable to contribute effectively to the quest for sustainability. Nevertheless science and technology will have a vital role in ensuring a smooth transition to more sustainable development paradigms.

The coexistence of contradictory values associated with the currently dominant model of development based on growth tends to generate uncertainty and some anxiety. The future is, and always will be, uncertain. But today the uncertainty involves highly complex risks that arise from severe and increasing inequalities between and within countries, fast-growing security concerns, increasing violence, widespread and long-lasting conflicts, and dangerous environmental problems at the local, regional, and global scales. We live in a world of increasing complexity, permeated by growing uncertainties and risks.

Until very recently it was widely accepted that the best way to fast-track into development was the worldwide dissemination of neoliberal forms of capitalism. It was thought that the straightforward application of this model to developing nations would eradicate poverty and turn them into developed countries. Globalization, and in particular the outward orientation of formerly closed economies, has brought innumerable benefits to hundreds of millions of people. However, this export-led growth model is coming under stress, and has in many cases seriously aggravated inequalities. There were signs in the past that the exportation of neoliberal policies to developing countries, sometimes without any adaptation, constituted a serious threat to millions of people who lived on the margin of the global economy. Gigantic multinational companies seeking to dominate agriculture in developing countries tend to create innumerable risks which are particularly serious to rural populations, to the sustainability of natural resource management, and to the environment. Supra-national institutions like the International Monetary Fund, the World Bank, the World Trade Organization, and the Organization for Economic Cooperation and

Development have, for many years, influenced the economic policies of a large number of developing countries, but with very limited success. Although it may be said that without their intervention economic crises would have been more frequent and damaging, the fact remains that in most cases they have been unable to diminish, in a significant way, the inequalities of development and quality of life between rich and poor countries.

At the geopolitical level we are still at a time when the USA dominates the world economically and militarily and is therefore the only superpower. However, we are also at the beginning of a transition period characterized by the increasing power of emerging economies such as China, India, Russia, and Brazil. The most significant aspect of the present situation is that these countries are now big and vitally important players in the global economy. They are also responsible for a fast-increasing share in what is becoming an unsustainable consumption of natural resources. Another crucial aspect is that inequality in the world is increasing rather than decreasing. Abject poverty and hunger can be found in some of the emerging economies, side by side with shiny new factories and an increasing number of luxury buildings, shops, and cars.

The barbarous attacks in New York on 11 September 2001 have stressed quite clearly that we constitute, in an increasingly integrated way, one global community that shares very similar goals, ambitions, vulnerabilities, and risks. The existence, development, and well-being of a community presume the respect for a set of rules that make living together possible. These rules should be just, obtained by democratic processes, and respected by all. The reality is, however, very different.

There are many failed or dysfunctional states, and many countries have autocratic regimes and suffer from widespread and increasing corruption. Terrorism and violent conflicts are persistent problems. Furthermore, there are frequent examples of tension between the West, championed by its dominant values and models of liberalism, freedom of expression and democracy, and Islam, championed by a stronger integration of religion and society. This deep fissure is reinforced by territorial issues that feed the terrorism of various Islamic extremists against the West and its global power. However, the political uprisings in North Africa and the Middle East that started in January 2011 have clearly shown that the majority of people in Islamic countries aspire to greater respect for human rights, freedom of expression, democracy, and better social and economic well-being, just as people do in the West. But Western nations did not fully recognize this convergence and long supported corrupt and dictatorial political regimes in these countries.

It is widely recognized (Stiglitz, 2002) that in our globalized economy everybody should benefit from social justice and human rights, and have access to the essential goods and services of the modern world. This is considered essential to ensure the dignity of human life as it is presently understood in the industrialized countries. Here, too, the reality is very different. Millions do not benefit from basic human rights and freedom. There are profound social and economic inequities and inequalities, often coupled with weak or dysfunctional governance, corruption, and religious and cultural tension and hostility. They tend to generate instability, extremism, violent confrontation, and terrorism, a growing risk on all scales.

A very good example of the uncertainties regarding the future is provided by the financial and economic crisis that started and developed in the West in 2008–2009. Very few considered it possible at all, and the uncertainty about the evolution of the global economy is still great in 2011. Capitalism is known to have periodic crises. However, the present global downturn is different from previous ones, since it occurs at a time of increasing scarcity of natural resources, mainly due to the growing demand from the emerging economies. The prices of staple foods and other commodities rose sharply between the beginning of the century and 2008. Food security became precarious in several countries after the price of cereals — wheat, rice, and corn — increased in 2007 by a factor of more than two. The price of oil reached 147 US dollars per barrel on 11 July 2008, and then fell by more than 65% in the following six months. Since 2003, the world GDP has increased annually by more than 2.5%, and in 2007 it reached 5.17% (IMF, 2010). This period of strong growth was supported by a world financial system that was opaque and unsustainable as regards its procedures and structures. Access to credit in the West was encouraged beyond sustainability, particularly in the housing market. The subprime financial crisis in the USA had a cascade effect throughout the world, and led to a near meltdown of the global banking system. Confidence in the markets and among consumers is still low, especially in the more industrialized countries.

One of the most remarkable aspects of the Western financial and economic crisis of 2008–2009 is that it was not predicted or even thought possible by the overwhelming majority of economists. Since it happened in an emerging scenario of increasing scarcity of natural resources and global environmental problems, such as climate change, we may conclude that unsuspected critical situations are likely to recur, and also that uncertainties regarding the future are likely to grow. Furthermore, it shows that our increasingly globalised model of development is riddled with deep and far-reaching instabilities and unsustainabilities.

There are many open questions in the world that are crucial for the future. How can we attenuate the inequalities between developing and developed countries, and also within countries? How can we bridge the huge gap between the severe poverty in some developing countries and the extreme wealth of the more developed ones? How can we significantly reduce the number of hungry people in the world, which has oscillated between 800 and 1000 million in this century and was 925 million in 2010, according to the FAO? How can we create or reconstitute good government in failed, dysfunctional, or weak states? How can we disseminate and implement democratic governmental practices worldwide, to effectively fight corruption, ensure respect for the law, enforce human rights and social justice, and invest in education? How can we moderate religious and cultural tensions? How can we control the exodus from rural to urban areas and avoid the latter becoming dominated by poverty, drugs, pollution, insecurity, and crime? How can we achieve the sustainable management of natural resources — especially water, soils, terrestrial fresh water, and marine ecosystems? How can we stop the rising loss in biodiversity? How can we achieve the sustainability of energy systems to ensure security of supply, affordability, and environmental compatibility? How can we create paradigms of social and economic development and patterns of consumption that reduce inequities, improve security, and remain sustainable as regards the availability of natural resources and the environment? How can we reduce the number of violent conflicts and wars, and the risk of future hostilities? How can we stop the continuing arms race, and especially the development and proliferation of nuclear weapons? How can we reduce the risk of financial, economic, and environmental crises? How can we achieve sustainable development? What would be the outcome if we fail to address and solve most of these challenges?

In the current situation where levels of risk are gradually increasing, these questions should have a very high priority and polarize the political agenda of governments worldwide during the 21st century. However, states have other more pressing concerns associated with the preservation and assertion of different forms of identity, in all areas of human activity. The most important are the preservation and advancement of economic and military power and associated hegemonies at the national, regional, and global levels, and the continuous enhancement, particularly in the more industrialized countries, of patterns of consumption and lifestyles which serve as a role model for the whole world, although impossible to sustain globally.

We are facing an increasing number of challenges that require new forms of governance with global legitimacy. However, the world power structure is profoundly fragmented and hierarchical. Countries are dominated by the need to defend their national interests, aspirations, and dreams, and this posture is not favourable to addressing and implementing adequate solutions to global problems, especially those that involve a timeframe of a few decades or more. The total aggregate of all the national agendas is increasingly divergent from an effective response to the main global challenges.

There are of course various future development paths for humanity. At the end of the 21st century, we may have a more strained, more conflicting, and more fragmented world, with larger social and economic inequalities, an increasing global population, and little concern for the sustainable use of natural resources and for environmental preservation. Another possible scenario might be a world with a stabilized global population that has significantly advanced towards social and economic convergence, the sustainable use of resources, and the preservation of the environment. These future socio-economic scenarios are nowadays frequently used in various types of study. The Intergovernmental Panel on Climate Change (IPCC, 2000), established in 1988 by two United Nations organizations, first developed them with the purpose of constructing scenarios on greenhouse gas emissions up to the end of the century. Each scenario is a condensed 'story' of a possible future characterized by specific demographic, social, economic, technological, and environmental developments. The scenarios must be internally coherent, but have no probabilities attributed to them. They provide distinct visions of the future and are therefore a powerful tool for longer-term planning.

The question we wish to address here is the role of science and technology in the multiplicity of decisions at an individual and collective level that will determine the future development path of humanity. Throughout the 21st century, will the application of the methodology of science, based on rationalism, on the appreciation, analysis, and interpretation of concrete facts, and on free debate, intensify and extend to meet the main challenges posed by the search for sustainable development? Or will it become weaker, with a reduced field of action?

The systematic application of science and technology to seek sustainable solutions for development will demand new conceptual structures, new methodologies, and a new willingness by scientists of the natural and social sciences and engineers to integrate the natural and social sciences. Science will have to handle the functional complexity of the systems that unite human communities and ecosystems, called socio-ecological systems by Jeffrey Sayer and Bruce Campbell (Sayer, 2003). These systems are characterized by multiple and wide-ranging spatial and temporal dimensions of mutual interaction and response to internal and external stimuli. Furthermore, responses are difficult to unravel, simulate, and model because they are more or less delayed, frequently non-linear, and commonly affected by significant uncertainties.

Science must also contribute to finding development paradigms capable of integrating phenomena with enormous differences in their spatial and temporal dimensions. One example of this type of challenge is how to integrate the globalization of the economy with agricultural practices at a local and regional level in developing countries. Another example is how to model the diversity of social and economic driving forces at those levels in order to project their impact globally. Science should be more alert to the possibility of exploring new realms, in which knowledge is useful both to the advancement of science itself and to society.

The main characteristic of a science more committed to sustainability will be its ability to face the challenges of the future. Only science can help us develop scenarios of future evolution of the socio-ecological systems, including global changes of both anthropogenic and natural origins. These scenarios afford us a view of the future, admittedly with many uncertainties and inaccuracies. If we believe in the methodology of science, this vision has the greatest probability of actually resembling the future. Our ability to project into the future will be ever more important as regards management, control, and the quest to minimize social, economic, technological, and environmental risks that are likely to increase during the 21st century. The question is whether each one of us, at the private and institutional level, and especially politicians, will make the best of the decision-making opportunities available to us to choose those options that most contribute to a sustainable development in the social, economic, and environmental domains. In order to choose thoughtfully, it is crucial to be well informed and to recognize the possible consequences of the various options in the short, medium, and long term. It is essential not to sacrifice the medium and long term for the short term. It is necessary to be able to project the future scenarios that may result from various priorities where goals and behaviours are concerned, both individually and collectively. Finally, one must accept the inevitable uncertainty inherent in all attempts to make projections into the future.

1.16 From the Past to the Future

To be able to meet future challenges, it is essential to know the long history of our
past and the conditioning factors of the present. A very significant part of our common future is inevitably constrained by the biological characteristics of *Homo sapiens*, but probably a larger part depends on our capacity for cultural evolution. We
are part of the history of the Universe and there is no way we can divorce ourselves
from it, no way of escaping from its fundamental physical laws. It is important to
know and reflect upon that history from the most remote beginnings through to the
emergence and evolution of life on Earth and the appearance of *Homo sapiens*. This
knowledge is essential to understand the present and to address questions about the
future. There is an inescapable continuity between the past and the future. With
science we have acquired the wonderful ability to reconstruct the past, often with
great detail: our own past as a biological species, the past history of life on Earth,
that of Earth itself, and even that of the Universe as a whole. We can now unravel
with sufficient reliability the way this adventure has run from the very first instant
when the Universe was only a singularity that generated time and space. Furthermore, this same scientific knowledge also provides a way of revealing the future.
The point is that the Universe, our own galaxy, the Milky Way, our Solar System,
and our planet Earth will all evolve into the future according to the fundamental
laws of nature, which we now know with impressive accuracy. We can thus project
the evolution of those systems into the future with a high degree of reliability. In this
process the presence of humans on Earth is merely a rather remarkable epiphenomenon. How long and in what manner will it last? What are the main driving forces
that will determine its duration? What are the leading uncertainties when trying to
answer these questions? From the outset, there is much more uncertainty regarding
the future of humans in the short and medium term of 10 to 100 years, than there
is in the future of our Earth in the long term of hundreds of millions years. Before
addressing such uncertainties, let us explore our own history on the planet Earth,
starting right from the very beginning.

The next chapter is a brief overview of this adventure, precisely up to the 16th
century when modern science began to flourish. It is an amazing story. From then
on, science began to influence the history of civilization in a profound way, and
among a great variety of benefits, gave us the remarkable ability to travel backwards
and forwards in time. However, to reach out towards and project into the future, one
must first seek to understand the current situation in all its immense complexity.
We are immersed in an increasingly globalised civilization, ever more complex, and
subject to a fast rhythm of transformation. We must analyse it well if we are to be
able to address and scrutinize its future. That is the aim of the following chapter.

Chapter 2
A Very Brief History
of the Universe, Earth, and Life

2.1 From the Big Bang to the Formation of the Galaxies

Let us begin at the beginning. About 13.6 billion years ago (Spergel, 2003) the
Universe we inhabit came into existence in a phenomenon called the Big Bang, in
which space, time, energy, and the laws of nature emerged. The initial singularity
evolved to states of extremely high energy density and temperature causing a very
rapid expansion of space through non-space at a speed far greater than that of light.
The prodigious expansion of space at that time, termed inflation, corresponds to the
inflationary period of the evolution of the Universe. Since then, the universe has
cooled from initial temperatures of the order of 10^{32} K (kelvin thermodynamic temperature scale), down to its current temperature of 2.73 K, or $-270.42°C$, observed
in the cosmic background radiation.

There are other astrophysical observations that support the theory of the Big
Bang. Space continues to expand and this can be observed by the motion of galaxies relative to the Milky Way (where the Solar System is located) with a velocity
proportional to their distance. This relationship, called Hubble's law, was first proposed by the American astronomer Edwin Powell Hubble in 1929 after long and
persevering observations of distant galaxies with the telescope at the Mount Wilson
Observatory in California. Using the fundamental laws of physics to describe the
early history of the universe, it is possible to reconstruct the formation of protons
and neutrons out of a plasma of elementary particles called quarks and gluons. Then
follows the primordial nucleosynthesis, a process in which the first atomic nuclei
were formed from the neutrons and protons. These first two steps lasted only about
15 minutes. The resulting relative abundances of the elements produced in the primordial nucleosynthesis agree with current astrophysical observations. About three
quarters of the total mass of the atomic nuclei present in the Universe are hydrogen
nuclei, in other words protons. The other quarter is made up of helium nuclei, formed in the nucleosynthesis. The atomic nuclei of elements with greater mass, such
as carbon, oxygen, and iron (all of which are essential for life), were formed much

later in the central cores of stars. They constitute a very small percentage, on the order of 1%, of the total mass of atomic nuclei in our current Universe.

After the primordial nucleosynthesis, there were no stable neutral atoms, only atomic nuclei and free electrons. The temperature was still too high for the atoms not to be immediately ionised. When it dropped below 3 000 K and the Universe was about 400 000 years old, the electrons were captured by the protons and other nuclei to form neutral atoms. In this way, the electric charge was neutralized and the Universe became transparent to light or, in other words, to the photons of the electromagnetic radiation. These photons began propagating freely in space, preserving to this day a picture in the cosmic background radiation of the distribution of matter at that moment when matter and radiation decoupled.

Observation of the anisotropies in the cosmic background radiation confirms that the expansion of the Universe amplified small initial quantum fluctuations in the spatial distribution of matter. Gravity pulled these slight fluctuations and collapsed them into objects where the first stars were born. A filamentous distribution was formed, with local accumulations of mass, which were to be the origin of the clusters of galaxies. The first stars to be formed by gravitational collapse, called Population III stars, brought back radiation to the visible part of the universe's electromagnetic spectrum.

This was the end of the dark age of the Universe that lasted for about one billion years, from the time the Big Bang flashed up to the ignition of the first stars. The latter had an enormous mass, of the order of three hundred times the mass of the Sun ($M_{Sun} = 2 \times 10^{30}$ kg). Initially, they were made up exclusively of hydrogen and helium. The first nuclei of elements from carbon to iron formed in their central cores through nuclear reactions. At the end of their lifetimes, they transformed into supernovas, exploding violently and fertilizing their neighbourhood with heavier elements. The interstellar material, enriched in this way, continued to condense and collapse giving rise to a second generation of stars, called Population II, with a broader mass distribution. In this population the stars of larger mass, which implies a shorter lifetime, continued to produce supernovas that spread heavy elements throughout the galaxies. Interstellar matter, now much richer in metals, continued to condense, forming the Population I stars. The gravitational collapse of a small part of a giant molecular cloud made up of gas and dust gave rise to the Solar System, in which the Sun is a Population I star. The enrichment with heavy elements was essential for the formation of the terrestrial planets, in particular the Earth.

2.2 The Four Fundamental Forces of Physics

It is important to realise that this remarkable evolution since the Big Bang was determined precisely by the characteristics and properties of the four fundamental forces that science has identified in the Universe — gravitational, electromagnetic, weak nuclear, and strong nuclear. It all took place like a grand play in a theatre, where each force played a role that intervened decisively in various scenes. The strong nuclear force had a crucial role in the transition phase from the quark–gluon plasma into protons and neutrons and at the beginning of the primordial nucleosynthesis. The weak nuclear force determined the relative abundance of protons and neutrons and decoupled the primordial neutrinos (elementary particles that travel almost at the speed of light, have a very small mass, and interact mainly through the weak nuclear force). These particles continue to wander around the Universe with a temperature very close to the photons of the cosmic background radiation. The electromagnetic force, through the value of the ionization energy of hydrogen and of other atoms, fixed the moment when the first atoms formed and the primordial photons became decoupled from matter. The gravitational force determined the large-scale structure of the Universe, from the dynamics of large clusters of galaxies to the dynamics of stellar clusters inside galaxies and the internal dynamics of planetary systems.

However, the role of this last character in the play reveals something mysterious in the Universe. The explanation of the relative motions of galaxies in galaxy clusters and of stars in rotating galaxies requires the presence of another form of matter, different from the one we know, formed by protons, neutrons, and electrons. This matter of an unknown nature is called dark matter.

Furthermore, the outward motion of the more distant galaxies at the edge of the universe reveals that some other form of energy, also of an unknown nature, is accelerating the expansion of space rather than slowing it down. This new energy is called dark energy. Only 4% of the total energy of the universe is in the form of normal matter, made up of the nuclei, electrons, atoms, and molecules that constitute stars, planets, and living beings. The much larger remaining part comprises about 23% dark matter and 73% dark energy. Surprisingly enough, science has not yet been able to completely decode the nature or origin of these two forms of energy.

2.3 And Before the Big Bang?

When speaking about the Big Bang, it is natural to question what was happening before it took place. Just 10^{-44} seconds after the Big Bang singularity, the Universe was already an enormous concentration of energy in a very small volume. Under these extreme conditions, the theories of general relativity and quantum mechanics imply that space and time are completely intertwined and therefore inseparable. Time and space are no longer autonomous, so that any talk of before and after loses its usual meaning.

In spite of this limitation, we can continue to develop theories to describe what happened in the very early universe, on the basis of the fundamental laws of physics. Quantum mechanics allows us to conjecture that the Big Bang resulted from a quantum fluctuation in the quantum vacuum that led to the inflationary process of expansion. For what reason did the fluctuation occur? There is an inherent uncertainty in quantum phenomena that makes it impossible to know more than the probability distribution for the occurrence of such phenomena. Let us consider an example. In a sample of a radioactive substance, we cannot know which of the nu-

clei, in a future time interval, will disintegrate and which will not. It is only possible
to determine a probability law that gives the number of disintegrations as a function of time. All this leads us to suppose that quantum vacuum fluctuations capable
of producing a Big Bang may reoccur or even be provoked by the creation of appropriate conditions. According to Alan Guth, who first proposed the inflationary
theory, it is not dangerous to create a universe in our homes! It would not occupy or
dislocate any part of the space of our Universe, although it would grow in an explosive manner, creating its own space until it reached cosmic dimensions (Brockman,
1995). Theories about other universes, real or virtual, are fascinating; however, they
are far beyond our present capacity to test them. Not being potentially falsifiable
through observation or experimentation, they remain in the limbo of conjectures.

2.4 How Did the Solar System Form
and What Is the Source of the Sun's Energy?

Spiral galaxies were born from the gravitational collapse of gigantic clouds of gas
and dust. The stars and planets also resulted from the gravitational collapse of clouds
of gas and dust of lesser dimensions, lodged within spiral and irregular galaxies. No-
tice, however, that these gravitational processes were not straightforward, as they
had to overcome powerful obstacles, such as the differential rotation of clouds, tur-
bulence, and strong magnetic fields. Our own spiral galaxy, the Milky Way, resulted
from the contraction of a giant cloud, nearly spherical and with an estimated mass
of 10^{11} solar masses. Its present shape is approximately that of a disk with a diameter of 160 000 light-years (one light-year is the distance light travels in one year,
equal to 9.46×10^{15} metres) and a thickness of 2 000 light-years. At its centre it has
a powerful black hole with a mass equivalent to 2.6 million solar masses. We know
about its existence because we observe stars that circle the invisible object at very
high speeds.

Inside the Milky Way, there are molecular clouds made up essentially of hydrogen molecules, H_2, and helium atoms. The other common molecular species are
OH, H_2O, CO, and NH_3. With radio telescopes, it is possible to detect the presence
of more complex molecules, some frequently found in living organisms, such as
nitriles, aldehydes, alcohols, ethers, ketones, and amides. Although more than 99%
of the mass of these clouds is in the form of gas, they also contain dust which, by
scattering the stellar radiation and cooling through infrared radiation, maintains the
temperature at the centre of the cloud as low as 5 to 10 K.

Low temperatures and relatively high densities favour the gravitational collapse
of the molecular cloud, generally triggered by the passage of a density wave of a
galactic spiral arm or by the material ejected in the explosion of a neighbouring
supernova. Our Solar System formed in a molecular cloud, where a progressive process of star formation gave rise to an open cluster, or in other words, a group of stars
initially close together but moving apart as time went by. The gravitational collapse
of a part of the molecular cloud, which was probably triggered by a neighbouring

supernova explosion, produced a protostar, surrounded by a rotating accretion disk, whose matter nourished the future Sun. It is in the interior of these accretion disks, currently observed in extra-solar planetary systems, that planets form. The flattened shape of the accretion disk results from the fact that the collapse is much easier parallel to the rotation axis than perpendicular to it, due to the requirement of angular momentum conservation.

Once the central region had reached temperatures on the order of 10^7 K, thermonuclear reactions began to cause the fusion of hydrogen into helium, and the protostar became a star. Those reactions turn mass into energy in accordance with the equivalence law of relativity $E = mc^2$ ($c = 3 \times 10^8$ ms^{-1} is the speed of light in a vacuum). They produce enormous amounts of energy that end up being emitted as electromagnetic radiation from the surface of the star. In the case of the Sun, the surface temperature is close to 5 800 K, corresponding to a luminosity of 3.9×10^{26} W.

The luminosity of a star is fairly stable as long as there is sufficient hydrogen in its central region to supply the nuclear fusion reactions that transmute hydrogen into helium. The duration of this stability period, corresponding to the time the star spends in what is technically termed the main sequence of the Hertzsprung–Russell diagram, depends on the initial mass of the star. The larger the mass, the shorter the stability period, and the shorter the lifetime of the star will be. In the case of the Sun, a star of spectral class G2, it will last approximately 1.1×10^{10} years. A much more luminous star of the spectral class O5, with a surface temperature on the order of 41 000 K and a mass of 40 solar masses, will remain in the main sequence for only a million years. On the other hand, a much less luminous star of the spectral class M5, with a surface temperature of 3 500 K and a mass of 0.5 solar masses, will remain in the main sequence during a far longer period, of the order of 30×10^{10} years.

When the hydrogen in the central region runs out, the star is forced to seek other ways to produce energy in order to avoid collapse. At this stage, the fusion of helium into carbon starts with a violent expansion that transforms the star into a red giant. The Sun is approaching the halfway mark in its period of stability in the main sequence. The fusion of hydrogen into helium started about 4.6×10^9 years ago, implying that there are still 6.4×10^9 years before it becomes a red giant with a luminosity 3 000 times greater than at present.

Despite the stability of the hydrogen fusion process, the luminosity of the Sun is not exactly constant. It increases with time in an approximately linear way until it becomes a red giant. Initially, 4.6×10^9 years ago, the luminosity of the Sun was about 70% of what it is today, and in the next 10^9 and 5×10^9 years, it will increase by 10% and 40%, respectively. As we shall see, this is going to have dire consequences for life on Earth.

All these predictions are based on the application of the fundamental laws of physics as regards the production of energy in the interior of stars. The secret of the extraordinary stability of the Sun's luminosity during a period of about 10^{10} years resides in the properties of the weak nuclear force. The first reaction in the cycle of nuclear reactions that transform hydrogen into helium consists of the fusion of two protons, one of which becomes a neutron to produce a deuteron, the nucleus of the hydrogen isotope called deuterium. The reaction happens through the weak nuclear interaction that transforms a proton into a neutron. For that to happen, the two initial protons must be very close to one another. However, the protons, due to their positive electric charge, repel one another, making the probability of this reaction very low. The probability increases with temperature. Nevertheless, in the central region of the Sun, where the temperature reaches 1.5×10^7 K, the probability of one proton fusing with another reaches a value of 50% only after a time interval of 10^{10} years! The extremely slow process of nuclear fusion guarantees the prolonged stability of the energy generation process, and consequently, of the luminosity of the Sun. This property has been essential in allowing life to evolve from microbial forms to the more complex plants and animals living today.

As for the continuous but very slow increase in the Sun's luminosity, the explanation is very simple. It stems from the equation of state of a perfect gas, which describes relatively well what happens in the interior of the Sun. The equation relates pressure, temperature, and the number of particles per unit volume. When the number of particles diminishes in a given volume of gas, the temperature has to increase in order to maintain pressure. That is the situation in the interior of the Sun, as fusion transforms hydrogen into helium. The nuclear fusion of four protons into one nucleus of helium reduces the number of particles in the ratio of four to one. For the pressure in the internal region to remain constant, in order to prevent the Sun's gravitational collapse, the temperature must necessarily increase. Consequently, the luminosity of the Sun also increases. Although it is a relatively slow process, it will have dramatic consequences for the biosphere, and maybe also for human civilization if it still exists at that time, long before the Sun becomes a red giant.

2.5 How Did the Planets of the Solar System Form?

Let us go back again to the molecular cloud whose gravitational collapse gave rise to the protostar of the Sun and the accretion disk surrounding it, called the solar nebula. In accordance with Kepler's laws, the rotational movement of the solar nebula is faster in the region closest to the protostar than it is further away. The differential rotation produces frictional forces that cause matter to migrate from the accretion disk to the protostar, while some remaining matter forms planets (Lissauer, 1993). Curiously, when it radiates out into interstellar space, the infrared radiation resulting mainly from the heat caused by friction allows the observation of accretion discs in extra-solar systems. In this way, we can now observe the evolution of planet-forming nebulas similar to our own primordial solar nebula.

During the collapse of the molecular cloud into an accretion disk, gas and large quantities of dust particles accumulate, due to electromagnetic forces and interactions between the gas and dust. In this process, grain-like conglomerates begin to form, and gradually increase in size. Computer simulations of the accretion mechanism indicate that, after about 1 000 years, the grains agglomerate to form bodies with a diameter of 10 m. After 10 000 years, these planetesimals have a diameter of 10 km, and within 100 000 years, they can reach 1 000 km. Following this phase, the mass of the planetesimals is sufficiently large for the gravitational force they generate to begin to play an important role. They tend to collide, fragmenting or making larger bodies, which will eventually constitute the protoplanets. The heat generated by such collisions and by the disintegration of radioactive isotopes from elements such as potassium, thorium, and uranium, cause a migration of materials of higher density to the central region of the protoplanets.

The chemical composition of the protoplanets is determined by the temperature of the planet-forming nebula in the region where they are born. Near the Sun, in the relatively hot regions where the four terrestrial planets — Mercury, Venus, Earth, and Mars — formed, the protoplanets are mainly composed of silicates and compounds of iron and nickel. Substances such as water, methane, carbon monoxide, carbon dioxide, and ammonia would have been in a gaseous state, incorporated in small amounts in the planetesimal grains. Therefore the terrestrial protoplanets contained relatively little carbon and water. Further away from the Sun, the situation was very different. Various compounds such as water, methane, ammonia, and others rich in carbon and nitrogen solidified and formed ices when the temperature dropped to values below 150 K. These temperatures are reached beyond what is known as the ice-forming boundary, situated at a distance of approximately 5 AU from the Sun (1 AU $= 1.5 \times 10^{11}$ m is the astronomical unit of distance, equal to the average distance between the Earth and the Sun). On the other side of that frontier, ice grains with a comparatively small relative velocity favoured the formation of planetesimals of large mass by attracting and accumulating enormous amounts of molecular hydrogen and helium. This rapid process of mass accretion produced the four Jovian planets — Jupiter, Saturn, Uranus, and Neptune — which are extremely large, and formed mainly of hydrogen and helium.

After their formation, the planets were still surrounded by the remains of the solar nebula and by the accretion disk. The dissipation of the remaining interplanetary gas and dust resulted from a new phenomenon caused by the young Sun. When the beginning of the nuclear fusion reactions approaches, the protostar enters the T-Tauri phase, first observed in a star of the Taurus constellation. Very strong magnetic fields produce powerful jets of gas that emerge from the nebula in the direction of its rotation axis, as well as stellar winds, which eject into space large quantities of matter from the outer layers of the protostar. In the final part of the Sun's T-Tauri period, the accretion diminished and the jets weakened, while the solar wind intensified. Its propagation out through the nebula would have gradually dissipated and carried away the remaining gas and dust from the young Solar System, thus ending the initial accretion process. Only the planets, a few planetesimals, and the bodies that gave rise to the comets would have remained. Approximately 10^8 years after the initial gravitational collapse of the solar nebula, planet Earth was formed (Wetherill, 1990).

2.6 The Importance of the Comets and the Moon in the History of Life

Comets are composed of water ice and ices of other substances rich in carbon, like carbon monoxide and methane, mixed with dust. They come from two reservoirs at the outskirts of the Solar System. One, called the Kuiper Belt, lies at a distance of 50 to 500 AU from the Sun. The other, the Oort Cloud, forms a spherical layer at the outer limits of the Solar System, at a distance of 5 000 to 100 000 AU from the Sun. Throughout the history of the Solar System, some cometary bodies have been dislodged from their reservoirs to become comets, through the action of gravitational forces of various origins. After their formation, the terrestrial planets were bombarded with an intense shower of comets that resulted from the tidal forces exerted on the Kuiper Belt by the Jovian planets and the nearby stars of the open cluster to which the Sun belongs.

Surprisingly, this shower of comets was crucial for the appearance of life on Earth. Remember that, due to their formation in a relatively hot region of the solar nebula, the composition of the terrestrial planets would have been deficient in oxygen and carbon. Water and carbon monoxide and other carbon compounds are volatile gases at those temperatures and would not therefore have formed grains that could be incorporated in the planetesimals. A large part of the water that formed Earth's oceans, and of the carbon in all living organisms, came from comets that collided with Earth in the early stages of the Solar System. Nowadays, the formation of comets is extremely rare. Only the passage of stars beyond 200 000 AU, the distance that separates us from the nearest star Alpha Centauri, would be able to dislodge cometary material from the Oort cloud, thereby sending comets into the inner Solar System.

In the early history of the Solar System, the frequency of impacts on planets by large objects was much higher than it is today. One of these collisions with Earth was also decisive for its future history: about 4.5×10^9 years ago, a planetesimal, about the size of Mars, with a radius approximately half the radius of Earth, collided with the Earth protoplanet. In this gigantic crash, the two metallic cores merged and the less dense material of the mantle, made up mostly of silicates, was thrown into space. It gave rise to a disk of planetesimals that eventually agglomerated to form the Moon. Today, this consensual theory is able to explain why the Moon does not have a metallic core and has only half the density of Earth.

The presence of a relatively large satellite like the Moon had various implications for Earth. The tidal forces exerted by the Moon on Earth cause a friction that slows its rotation rate. However, the total angular momentum of the Earth–Moon system is constant in time. Thus, according to the fundamental laws of physics, the increase in duration of an Earth day, initially only 5 hours, increases the distance between the Moon and the Earth. After its formation, the Moon was very close, being only 3 to 5 Earth radii away from the Earth's centre, while today it is on average 60 Earth radii away.

The constancy of the total angular momentum has another remarkable consequence that has direct implications for the history of Earth. The seasons of the year and the contrast between the climates that characterises them result from the inclination angle or obliquity of the Earth's rotational axis. Its present value is 23.44°. If, for example, the angle were 97.9°, as in the case of Uranus, the contrast between summer and winter, particularly in high latitudes, would be brutal. However, the direction of the rotational axis of a planet can suffer violent variations due to collisions and also due to the gravitational forces exerted by other planets. In the case of Earth, the presence of the Moon, just as in a gyroscope, has held the obliquity steady and has thereby guaranteed the stability of the four seasons of the year over billions of years.

For about 600 million years, the young Earth continued to suffer the impacts of planetesimals of large mass, some with a diameter of over 100 km. On the Moon's surface we can still see the marks left from such collisions, produced by planetesimals with different sizes. One can observe craters with various diameters, but the smaller ones predominate. These sizes follow a power law, according to which there are about eight times more craters every time the diameter is halved. On Earth, most craters have disappeared due to tectonic movements, volcanic activity, and erosion by the atmospheric weather.

The bombardment of Earth by planetesimals had yet another important consequence. The energy released by the impact of these bodies of huge dimensions was sufficient to vaporise the oceans and a large part of the Earth's crust to a depth of several miles. The heat resulting from these collisions and the disintegration of radioactive isotopes in the Earth's interior led to a differentiated structure. Due to the high temperatures, the iron melted and migrated to the centre of the planet where it now forms a dense core. This warm metal nucleus is the furnace that provides the energy needed to maintain the convective movements in the mantle. They are responsible for the volcanic activity and the displacements of the tectonic plates that lead to the formation of mountains. They are also essential for the stability of the carbon cycle, as will be discussed later. After this period of large and frequent impacts, the surface of the Earth was made of liquid rock with a temperature in the range of $1\,500°C$. The atmosphere contained large quantities of water vapour, carbon dioxide, and methane, arising mostly from the fusion of comets and planetesimals coming from the region where the Jovian planets formed. The last period of catastrophic impacts ended about 3.9×10^9 years ago (Cohen, 2000). The subsequent period of relative calm allowed for the restoration of the oceans and the formation of a solid rocky crust. The abundance of water coming mostly from random collisions with comets and planetesimals has determined the relative proportion between land masses and oceans on the Earth's surface. A larger amount of water could have submerged the continents completely, and that would have had profound consequences for the evolution of life.

2.7 The Last Great Collision with Earth

The stabilization of the Solar System made impacts with Earth progressively less frequent. This frequency decrease was particularly significant between 4.1 and 3.7×10^9 years ago. However, nowadays, the collision probability of objects with large dimensions, although considerably smaller, is not zero. Collisions with meteorites or comets with diameters of 1 and 10 km have, in our time, an average return period of 300 000 and one million years, respectively. The last collision with an object, probably a comet, with a diameter between 9 and 17 km happened about 65 million years ago at the end of the Cretaceous period. The remains of the gigantic crater created by the catastrophic impact were discovered in Chicxulub, in the Yucatán peninsula of Mexico. Estimates of the energy released in the collision lie between 3.8×10^8 and 3×10^9 megatons of TNT. In order to make this value less abstract, note that 1 megaton of TNT is equivalent to 50 times the energy released by the atomic bomb dropped on Hiroshima.

The impact generated very high temperatures that vaporized a large volume of ocean waters, corresponding to a 30 cm surface layer, and ignited widespread forest fires. Enormous amounts of ash and dust ejected into the atmosphere blocked out the solar radiation and produced a continuous night that lasted for many months. The impact also generated a gigantic tsunami with a height of around 1 km that deposited detritus and sand in a circular region with a diameter of about 3 000 km, from Alabama to Guatemala. Recent estimates indicate that between 40 and 75% of all species became extinct, including the dinosaurs. The detailed study of the records left by the Chicxulub collision can be used to reconstruct and understand the initial period of the Earth's life when comets and planetesimals of all sizes bombarded it incessantly.

2.8 The Atmosphere, the Oceans, and the Appearance of Life

The primordial atmosphere of the Earth formed from the gases released by the constituent planetesimals and by those resulting from the volatilisation of colliding comets and planetesimals. The main components were carbon dioxide (CO_2), methane (CH_4), molecular nitrogen (N_2), water (H_2O), ammonia (NH_3), and molecular hydrogen (H_2). Gradually, as the Earth cooled, the water vapour condensed and formed oceans of freshwater that became salty by reacting with the Earth's crust. The behaviour of the atmosphere of a planet throughout its history depends essentially on the gravitational acceleration at its surface, the atmospheric temperature, and the molecular mass of the gases that make it up. The reason for this dependence results from the fact that the constituent molecules have a velocity distribution which is determined by the temperature and the mass of the molecule. The higher the temperature or the smaller the molecular mass, the greater the average speed of the molecules. This relationship is important because it characterizes the conditions under which a planet will lose some or all of its atmospheric gases. The speed that a body must reach in order to break free from the gravitational field of a planet, called the escape velocity, increases with the mass of the planet. A planet only retains those gases whose average molecular speed is much lower than the escape velocity.

Earth's escape velocity is 11.2 km s^{-1}. This value is much higher than the average speed of the molecules that make up its present atmosphere. Nevertheless, the Earth was unable to retain the hydrogen and helium that were there originally, because of their relatively small molecular masses. Mercury could not retain any atmosphere at all, due to its small mass and the high temperatures that result from the planet's proximity to the Sun. The Moon, because of its very small mass, has no atmosphere either. Venus has a longer and more complex history. Initially the atmospheres of Venus and Earth were very similar. However, in the case of Venus, the higher temperatures did not allow the formation of oceans. The ultraviolet solar radiation dissociated the water vapour into hydrogen that ended up escaping from the planet, and oxygen, which was later consumed in chemical oxidation reactions on the surfaces of rocks. In the end only CO_2 remained. It actually constitutes approximately 97% of the Venusian atmosphere, and is responsible for an extremely intense greenhouse effect, which increases the globally averaged atmospheric temperature by about 480°C.

At the end of the period of frequent catastrophic impacts, the Earth's environment was rich in water in the liquid state, dominated by intense volcanic activity, chemically aggressive and subject to a strong flux of ultraviolet radiation. All these conditions were favourable to the synthesis of organic compounds. Probably the first living organisms formed on the edges of the oceans or in submarine hydrothermal vents. It could be that life first appeared at the time of the great collisions, and had to survive or reinitiate its genesis several times.

In any case, the most ancient records of life have ages of approximately 3.8×10^9 years and were found in carbon globules encrusted in metamorphic rocks in the south west of Greenland (Mojzsis 1996). It is well known that photosynthesis reduces the ratio of isotopic abundances $^{13}C/^{12}C$ by 2.5% by preferentially processing ^{12}C rather than ^{13}C. Measurements of that ratio in the Greenland carbon globules reveal similar depletions in ^{13}C, which indicates the presence of photosynthesizing bacteria. According to some authors (Schopf, 1999), the oldest fossils that have been found are colonies of algae-like single-celled prokaryotic organisms, that is, each comprising a single cell with no nucleus, but already using photosynthesis. Another example is the layered fossils of stromatolites, which are colonies of photosynthesizing bacteria that still exist today. The oldest fossils date from about 3.5×10^9 years ago. From then on, fossilized remains of living organisms became progressively more abundant and diversified.

Let us now consider the consequences of the use of photosynthesis, first by prokaryotic bacteria and later by more complex organisms, on the Earth environment. This is a metabolic process, which converts atmospheric carbon dioxide into organic compounds, especially sugars, using the radiant energy from sunlight, and releases molecular oxygen (O_2) into the atmosphere.

The presence of significant amounts of oxygen in the atmosphere, apart from being a direct consequence of the presence of life, contributed greatly to its evolution. A remarkable example is the formation of the stratospheric ozone layer, which, by absorbing a large part of the ultraviolet radiation from the Sun, protects living organisms from its harmful effects. Oxygen atoms, resulting from the dissociation of the O_2 molecule by the Sun's ultraviolet radiation, react with molecular oxygen and form ozone (O_3). The O_3, in its turn, dissociates under ultraviolet radiation to produce atomic and molecular oxygen which, by repeating the same mechanism, produce ozone again. The dynamics of this cyclic process forms a layer of ozone in the stratosphere, at an altitude of around 30 to 50 km, which allowed life's conquest of land by preventing most of the Sun's ultraviolet radiation from reaching the Earth's surface. While atmospheric O_2 and O_3 are essential to sustain life as we know it, the CO_2 also played a crucial regulatory role in its history.

2.9 The Greenhouse Effect and the Regulatory Role of Atmospheric CO_2

First, it is important to remember that CO_2 is a greenhouse gas, which means that it has the property of absorbing and emitting infrared radiation. Second, it should be noted that there are various greenhouse gases in the Earth's atmosphere, the most important being H_2O, CO_2, and CH_4. They reside mainly in the troposphere, the lower part of the atmosphere up to a height of about 10 km, a region where the temperature falls with increasing altitude. The greenhouse gases contribute decisively to the energy balance in the atmosphere because they absorb the infrared radiation emitted by the Earth's surface, but also by the atmosphere itself. This absorption heats the troposphere. Greenhouse gases are responsible for the Earth's greenhouse effect, which increases the average global temperature of the troposphere at the Earth's surface by about $33°C$ from $-18°C$ to $15°C$. An increase in the concentration of greenhouse gases leads to an increase in the average global temperature of the troposphere. What is the reason for this?

When the concentration of greenhouse gases in the troposphere increases, the infrared radiation they emit will originate on average from a higher and colder level than before. According to Planck's law of radiation, this means that since the gases are cooler they emit less radiation into space. More energy would be coming in from the Sun than going out. To restore the energy balance, the temperature of the troposphere must necessarily increase so that the total amount of radiation emitted into space is equal to that received from the Sun.

Venus gives a good example of the consequences arising from an intensification of the greenhouse effect. Temperatures on Venus reach about $700°C$ because the atmosphere is 97% CO_2, although initially it was similar to that of the Earth. The Earth avoided a similar fate because of its capacity to retain water in the liquid state, which formed the oceans.

Since the terrestrial atmosphere contains large amounts of water vapour, the CO_2 reacts with H_2O to form carbonic acid (H_2CO_3). This acid weathers silicate rock by forming calcium carbonate ($CaCO_3$), which is found in calcite and is the main component of limestone. $CaCO_3$ ends up deposited on the ocean floor after transportation by streams and rivers. This process, called weathering of silicate rock, removes CO_2 from the atmosphere and fixes the carbon in the $CaCO_3$ molecules. One might think that it would remove all CO_2 from the atmosphere, but this does not happen, because the ocean floor and the limestone deposits penetrate the mantle in subduction movements of the Earth's lithosphere plates. Due to the very high temperatures in the mantle, the carbon in the calcium carbonate ends up forming CO_2 again that escapes into the atmosphere in volcanic eruptions and in other forms of volcanic activity. This closes the carbonate–silicate cycle, which has a thermostatic function in the atmosphere.

The flux of carbon in the cycle depends on temperature because the efficiency of the weathering chemical reactions increases with temperature. When the climate becomes colder, there is less water vapour in the atmosphere, weathering is less efficient, and the CO_2 concentration increases. Thus, the greenhouse effect intensifies and the global average temperature tends to rise. If the climate becomes warmer, weathering increases and the temperature therefore tends to decrease. Here we have an ingenious natural process of climate regulation controlled by CO_2.

CO_2 participates in yet one more cycle of vital importance. Remember that photosynthesis sequesters CO_2 from the atmosphere and produces O_2. On the other hand, cellular respiration in plants and animals, and decomposition and burning of organic materials, releases CO_2 back into the atmosphere. Both cycles integrate into the carbon cycle whose equilibrium is essential for the permanence and evolution of life on Earth.

2.10 Is There Life Outside the Solar System? The Discovery and Study of Extra-Solar Planets

The history of planet Earth from its origins, briefly addressed in the preceding pages, is very probably still incomplete and affected by considerable uncertainty. Note, however, that the recent observation of extra-solar planetary systems provides new insights and can be used to test models and simulations. It thus widens and deepens our understanding of the formation and evolution of planetary systems and their planets.

Michel Mayor and Didier Queloz made the first indirect observation of an extra-solar planet in 1995. Very accurate measurements of the radial velocity of a star similar to the Sun in the constellation of Pegasus, named 51 Peg and situated at a distance of 94 light-years, revealed the presence of a planet with a mass of about half that of Jupiter. The method used is conceptually very simple. It consists in measuring the radial velocity, using the Doppler Effect, of the motion of a star around the system's center of mass produced by the gravitational tug of an orbiting planet. Applying this idea, the mass of the planet and the average distance to the star can both be estimated. Unexpectedly, the planet orbiting the star 51 Peg is found at a distance of only 0.005 AU from the star, in spite of being a Jovian type planet.

This is 7.8 times smaller than the distance between Mercury and the Sun. Due to the small radius it completes one revolution in just four days. The presence of a massive planet so close to the star was indeed a surprise, since it contradicts existing models for the evolution of the Solar System. One possibility is that it may have migrated from its formation site on the other side of the ice-forming boundary until it stabilised in an orbit very close to the star. The discovery aroused enormous interest and stimulated the search for new planetary systems. Since that time, the number of extra-solar planetary systems has continued to grow steadily. In mid-2007 the number of detected extra-solar planets was 249, while at the beginning of 2009 more than 340 were known, and by November 2010 there were 504 (IEPC, 2010).

One of the main objectives of the search for extra-solar planets is to find out more about the origin and evolution of the Solar System and about life beyond it. We are also deeply curious about the possibility of not being alone in the Universe. And so we search for signs of life and above all intelligent life in extra-solar planetary systems. Current research is focused on planets similar to Earth, since they probably have the most favourable conditions for supporting the emergence and evolution of complex living organisms. There are good reasons for supposing that carbon forms the basis for hypothetical extraterrestrial life forms, because it generates an immense variety of chemical compounds — the organic compounds. The explanation for this remarkable property resides in the electronic structure of the carbon atom. It has four electrons in the outer shell, available to form four chemical bonds that lead to the formation of extensive and complex three-dimensional molecular structures. Atoms of other elements, such as oxygen and nitrogen, tend to produce planar or linear molecules. At present, we know more than 10 million organic compounds, but only around 200 000 inorganic compounds formed by all the other elements.

We could imagine extraterrestrial life with other types of chemistry based on different elements such as silicon. However, the higher abundance of carbon in the Universe increases the probability that life in other planetary systems also uses organic chemistry. Environmental conditions must allow water to be in the liquid state, since that is the most likely solvent for life based on organic compounds. Life as we know it on Earth is mostly composed of the elements carbon, hydrogen, nitrogen, oxygen, sulphur, and phosphorus, and the fact that it does not use other elements more abundantly is something of a mystery, since some of them could in theory serve the same functions. Recently, it has been shown for the first time that a bacterium isolated from Mono Lake in California can eat and grow on a diet of arsenic in place of phosphorus (Wolfe-Simon, 2010).

We have no reason to suppose that the success of life on Earth would not take place on another planet with similar conditions to our own. It is this assumption that leads us to define the habitable zone of a planetary system as the region where we may expect life based on organic chemistry to be theoretically possible. On Earth, we find micro-organisms, called extremophiles, adapted to physically or geochemically extreme conditions. Some live in very hot environments where the temperature can reach 120°C, such as hydrothermal vents on the ocean floor or deep inside the Earth, at depths that can reach more than 3 km.

Regions on Earth where the average annual temperature is lower than $-10°C$ or greater than $30°C$ have no life, or only very simple life forms. That range of temperatures corresponds to a variation in the average solar radiation flux of at most a factor of two. In the Solar System, it is safe to admit that the Earth provides ideal conditions for life. Thus, it is usually assumed that the solar habitable zone corresponds to a region in our planetary system where the solar radiation flux varies between half and twice its value at Earth's average distance from the Sun. This value, called the solar constant, is equal to $1366 \ Wm^{-2}$. In conclusion, the solar radiation flux varies by a factor of four within the solar habitable zone. But how far from the Sun are the boundaries of that zone? To answer that question we must first note that the radiation flux decreases with the square of the distance from the Sun. Therefore the solar habitable zone lies between distances of $\sqrt{0.5}$ and $\sqrt{2}$, which correspond, respectively, to 0.7 AU and 1.4 AU. Curiously, both Venus, at a distance of 0.723 AU, and Mars, at a distance of 1.524 AU, are close to the boundaries of the solar habitable zone.

The position of the habitable zone of a planetary system varies with the spectral type of the star, which is closely related to its luminosity. The very bright and massive stars have the disadvantage of having a very short lifetime that could prevent the evolution of life up to complex and possibly intelligent forms. At the other end of the scale, the much less massive dim stars, although abundant, have the disadvantage of producing a habitable zone very close to the star. Due to this closeness, the tidal forces exerted by the star decelerate the rotational movement of the planet until it is synchronised with the orbital rotation, as has happened with the Moon relative to the Earth. Then the planet always has the same hemisphere facing the star, leading to extreme values of the surface temperature on both hemispheres. It would not be a benign environment for the evolution of life.

In conclusion, we have well-founded reasons for supposing that the most favourable conditions for life correspond to planets of terrestrial type situated in the habitable zone of planetary systems of stars of spectral type F, G, or K. It is thus natural to focus our attention on the detection of extra-solar planets with masses close to the Earth mass, and at distances approximately between 0.5 AU and 1.5 AU from stars of spectral type G, like the Sun. Planets with small mass or with large orbital radius produce smaller radial velocities in the star. Recent technological advances allow the measurement of velocity variations of the order of 1 to 3 ms^{-1}. This accuracy allows only the detection of planets with a much larger mass than the terrestrial planets and with an orbit relatively close to the star. In order to detect terrestrial planets in the habitable zone, measurement accuracies must be improved to $0.01 \ ms^{-1}$. Meanwhile studies using the radial velocity method show that about 5% of stars similar to the Sun have orbiting planets of large mass and small orbital radius. A similar percentage of terrestrial mass planets may exist around stars similar to the Sun.

There other ways to detect extra-solar planets, such as the transit, astrometric, gravitational microlensing, and pulsar timing methods, but the most promising is direct imaging. The problem is that planets are extremely faint light sources compared to the stars they orbit. Let us consider the example of Earth: its luminosity, resulting from the reflection of nearly 30% of the Sun's radiation, is about 2×10^9 times smaller than the luminosity of the Sun. Young Jovian planets glow quite brightly in the infrared and that glow makes imaging much easier. Furthermore, photometric and spectral analyses promise to reveal much about them. Recently, two research groups recorded for the first time images of Jovian planets orbiting stars in our neighbourhood (Marois, 2008; Kalas, 2008). One of the planets orbits the very well known star Fomalhaut, one of the brightest stars in the night sky. It is a young star, only 100 to 300 million years old, still surrounded by part of its planetary disk of dust, and relatively close to us at a distance of 25 light-years.

An interesting development was the launching of the Kepler spacecraft in March 2009, carrying a powerful telescope designed to discover Earth-like planets in the habitable zone of their planetary systems, using the transit method. It will measure the brightness of about 100 000 stars, looking for tiny periodic dips when a planet crosses in front of its star, a phenomenon known as a transit. Once Kepler finds a planet, it will be able to calculate its size, mass, orbital period, distance from the star, and surface temperature. It will thus be able to identify Earth-like planets in Earth-like positions. This is a very important first step towards finding planets on which life is likely to exist. On 2 February 2011, the Kepler mission announced that it had found 1235 planetary candidates circling 997 host stars, which is more than twice the number of previously known exoplanets. Probably only 90% of these planetary candidates will turn out to be real planets. Kepler's latest haul includes 68 planetary candidates of Earth-like size and 54 in the habitable zone of their star.

We are in a golden age of planetary science. Within the next 20 to 30 years we will probably get the first image of a pale blue Earth-like planet orbiting a distant star. There are several ambitious projects aiming to do this, such as NASA's Space Interferometry Mission, scheduled for launch in 2015–2016, which will use the astrometric method. The NASA Terrestrial Planet Finder probe will also attempt to image Earth-like planets, but the launch date has not yet been determined. These projects consist in putting into orbit infrared telescopes that cancel out the radiation from the star using interferometry, thereby allowing direct observation of planets, a method known as nulling interferometry. But then how will we know whether those planets support living organisms? The answer is that, by analysing the spectral image of the planet's atmosphere, we may discover its composition. The presence of absorption bands in the infrared part of the spectrum characteristic of H_2O, O_3, and CH_4 would be a very encouraging sign that life may exist on the planet.

Before the end of the century many Earth-like planets in the habitable zone of their star are likely to have been detected. Furthermore, we will probably know a considerable amount about the composition of their atmospheres. The next challenge will be to try to communicate with those planets that have an atmosphere similar to ours.

2.11 The Origin of Life
and the Search for the Oldest Common Ancestor

We now return to the question of life and its origins on Earth. The first essential question one must address concerns the very nature of life itself. From the phenomenological point of view, a living organism can be characterised by the basic properties of metabolism, growth, use of energy, individuality, preservation of identity, procreation, and mutation, that is, changing hereditary information. Furthermore, these properties must manifest themselves in a completely abiotic environment. They exclude viruses, which need to associate themselves with other biological organisms in order to survive and reproduce. Although the frontier of life may be ambiguous at the level of virus, this is not so at levels of complexity higher than bacteria.

To characterise life and address the question of its origins on Earth, it is very important to realise that all known living organisms function by means of internal biochemical processes coordinated by DNA, or deoxyribonucleic acid. DNA molecules are the ultimate long-term repository of genetic information, transmitted from generation to generation. They are long molecules with a double-helical structure, known as the famous double helix, formed by two chains of nucleotides joined by pairs of four bases — adenine, cytosine, guanine, and thymine. In each pair the bases are always joined in the same way, i.e., cytosine with guanine, and thymine with adenine. The order of the bases in each of the spirals of DNA constitutes the language of life. Written there is a sequence of the 20 amino acids which join up in chains in various ways to constitute the proteins. In the DNA each group of three successive nucleotides, called a codon, constitutes a code that identifies a specific amino acid. The table of correspondences between the codons and the amino acids is the genetic code. The production of the proteins essential to life, and in particular to sustain the reproductive processes, is carried out by copying the codes encrypted in the DNA molecules.

To direct the assembly of proteins, life turns to another type of molecule, the RNA or ribonucleic acid, which acts as a messenger. It has only one spiral of nucleotides and the same bases as DNA, except that uracil replaces thymine. Most RNA serves as a messenger sent by the DNA with its genetic information to the sites of protein formation inside the cells. There the specific RNA molecule provides the information needed to synthesize a given protein. Three types of processes participate in the circulation of genetic information:

- Replication, whereby DNA molecules give rise to other DNA molecules, preserving their genetic content.
- Transcription, in which genetic information is transferred from DNA to RNA.
- Translation, where information is transferred between RNA and proteins.

Essentially, these processes involve only the transfer of information. The synthesis of DNA, RNA, and proteins results from chemical processes catalyzed by specific enzymes and supplied energetically by adenosine triphosphate or ATP.

The total hereditary genetic information encoded in the DNA of an organism is its genome. It subdivides into units, called genes, each of which contains, in its sequences of bases, a code for a distinct protein chain, except for a few genes coding for functional RNA molecules. The latter, called ribozymes rather than messengers, carry out catalytic functions. The human genome, whose complete sequence was deciphered in 2001 (IHGSC, 2001), contains approximately three billion base pairs in roughly two metres of DNA, rolled up in 23 chromosomes and contained in a sphere one hundredth of a millimetre in diameter, the cell nucleus. All the instructions that specify a given human being from birth until death are contained in the person's genome. It is a marvel of miniaturisation at the molecular scale.

The fact that all living organisms found on Earth use a biochemistry based on DNA, RNA, the genetic code, and proteins built from 20 amino acids is a strong argument to support the idea that life had a common ancestor that used similar structures and processes. This ancestor was probably a prokaryotic bacterium, that is, a unicellular organism in which the cell has no nucleus. The question is: how did it form and evolve? Surprisingly, as regards evolution, the genetic code of current living organisms contains most of its history.

With the very powerful and sophisticated technologies of molecular biology, it is nowadays possible to compare sequences in homologous molecules containing genetic information — DNA, RNA, and proteins — taken from different living organisms. Here, 'homologous' means molecules that carry out analogous functions in different organisms. Let us consider, for example, the sequence of amino acids in a given protein that can be found in the cells of various living organisms, such as, bacteria, fungi, invertebrates, birds, reptiles, and mammals. The sequences corresponding to the different groups of species, although very similar, are not exactly the same. The differences result from mutations that occurred during the evolutionary process, and they reveal evolutionary relationships between those groups. Two sequences that differ by a large number of mutations indicate a common ancestor in the remote past. If, however, the number of mutations is small, the common ancestor is closer in time. The fact that the mutation rate in a given molecule is nearly constant in time allows us to establish a molecular clock for this molecule and to construct a phylogenetic tree, or tree of life, for all living organisms. The number of different mutations between two tree branches measures the time elapsed since the two branches diverged from their most recent common ancestor.

Using comparative sequencing of ribosomal RNA in a large number of living organisms, including various types of bacteria, the geneticist Carl R. Woese (Woese, 1987; Woese, 1998) was able to reconstitute the tree of life. He classified all organisms into three domains that he called *Bacteria*, *Archaea*, and *Eucarya*. The first domain includes the bacteria, which are prokaryotic unicellular organisms, both pathogenic and non-pathogenic from the point of view of bacteriology. The organisms in the second domain are also unicellular and include those that produce methane, found in environments poor in oxygen, and the extremophiles, adapted to habitats that are very acidic, very salty, or very hot. Finally, the third domain is the one where the first eukaryotic cells formed, that is, cells with a nucleus and cytoplasm. It was from this branch of the tree of life that all the multicellular organisms, including plants, fungi, animals, and man, eventually developed.

There are clear indications that the most recent common ancestor of the three domains established by Woese already used DNA. What was this organism and how did it form? We still do not have the answers to these questions, but only conjectures. In times prior to the most recent common ancestor, the tree of life could have had other branches that disappeared without trace. We can also admit that other trees of life may have evolved in parallel, but that only life based on DNA survived.

In any case, what is the origin of DNA, RNA, and the proteins, that is, how did life as we know it arise? According to Christian de Duve (Duve, 2002) the answer to this question is simple: we do not know and we may not know for a long time to come. The fascination of this topic and the immense challenges it places on science have encouraged intensive research activity. There are various ideas and theories still lacking robust observational and experimental support. In the evolution of life, the complex mechanisms of archiving genetic information in DNA, and of transcribing it to RNA, were probably preceded by a period during which only RNA existed. This corresponds to what Walter Gilbert (Gilbert, 1986) called the RNA world. According to this hypothesis, for a while RNA played the role of carrying the genetic information now transmitted by DNA, leading later to the formation of DNA and proteins. As for RNA, it could have arisen in aquatic environments on the Earth's surface, through metabolic processes catalysed by abiotic molecules. Other recent theories propose that life had its origin near the hot hydrothermal vents on the deep ocean floor, in particular on the surface of solids, such as pyrite (FeS_2), which controlled and supplied energy to metabolic processes (Wachtershauser, 1998).

2.12 The Oxygen Catastrophe

The high oxygen concentration in today's atmosphere results from enormous and sustained emissions of O_2 produced by photosynthesis during Earth's history. These emissions started at least 3.5×10^9 years ago, which is the age of the oldest fossils of prokaryotic bacteria. Using the energy from solar radiation, the bacteria transformed carbon dioxide and water into oxygen and, for example, formaldehyde. However, geochemical records allow us to conclude that the concentration of atmospheric oxygen began to increase significantly only 2.3×10^9 years ago. The reason for this delay was that the oxygen released by photosynthesis produced the oxidation of the abundant iron dissolved in the oceans, giving rise to iron oxides (Fe_2O_3). The oxides deposited and formed well-known geological structures rich in iron, called banded iron formations, with ages between 3.5 and 2.0×10^9 years. Currently, these deposits constitute about 90% of the world's iron ore reserves, and are essential to sustain our industrial societies, both now and in the future.

Only after the saturation of this sink did the oxygen start to accumulate in the atmosphere. The change in the atmosphere's composition created an enormous challenge for life, one of the most difficult to overcome in the history of its evolution. Living organisms transformed so that they could survive in an atmosphere rich in oxygen, a remarkable example of life's capacity to adapt to changes in the environment, including those it had induced itself, as in this case. Oxygen is deadly poisonous for anaerobic bacteria because, by entering into contact with certain biological substances, it generates highly toxic compounds such as the hydroxyl radical (OH) and hydrogen peroxide (H_2O_2). To survive being poisoned in this 'oxygen catastrophe', the anaerobic organisms found refuge in oxygen-free environments, such as the deep seas, or developed the necessary means of defence. Initially, these means involved associations with enzymes capable of neutralizing the toxic compounds produced by free oxygen.

2.13 The Formation of the First Eukaryotic Cells

The accumulation of oxygen in the atmosphere was a crucial factor for the development process leading to eukaryotic cells, whose evolutionary branch on the tree of life parted from the other two prokaryotic branches about 3.5×10^9 years ago. Starting at that time, it took about 1.2×10^9 years to form the first eukaryotic cell that constituted the common ancestor of all eukaryotic cells and of all multicellular organisms. These primordial cells with a cytoplasm and a nucleus were already adapted to survive in an oxygen-rich atmosphere, which meant that they could not have been around for more than 2.3×10^9 years. Analysis of fossil data suggests that the need to adapt to a profoundly toxic environment led life to discover a new model for the cell, more structured and more complex, and with remarkable evolutionary capabilities. The impasse created by oxygen accumulation in the atmosphere took more than one thousand million years to overcome, but produced extraordinary results.

Eukaryotic unicellular organisms constitute a highly heterogeneous group, called protists, which diversified in a spectacular way, giving rise to three groups of multicellular organisms: fungi, plants, and animals. They have essentially the same basic properties as the prokaryotic cells: similar chemical composition, metabolic processes, and genetic mechanisms. However, their organisation and internal structure is so much more complex that it seems improbable that they evolved from prokaryotic bacteria. Furthermore, the dimensions of the two kinds of cells are very different. A typical eukaryotic cell occupies a similar volume to tens of thousands of prokaryotic cells. The secret of the success of the eukaryotic cells resides in their functional structure, characterised by a cytoplasm where most of the metabolic processes take place and a nucleus where the genes are stored and which centralizes the majority of the genetic operations, including DNA replication, when the cell is preparing to divide. It is in the cytoplasm that we find the organelles or small organs responsible for the metabolism of oxygen, such as the peroxisomes, mitochondria, and chloroplasts. Curiously, the last two were initially free aerobic bacteria but were later adopted by host cells and ended up integrating into their structure.

While the prokaryotic cells tend to multiply exponentially, when environmental conditions are favourable, eukaryotic cells are subject to complex internal control processes, being able to go for long periods of time, or even indefinitely, without dividing. Specific mechanisms wake the cell from its dormant phase and stimulate the replication of DNA in the nucleus, leading to a doubling of the number of chromosomes. The envelope of the nucleus ruptures and, by a structurally complex process called mitosis, the cell divides in two, each with the same complete set of chromosomes.

2.14 The Invention of Sexual Reproduction

One of the most outstanding discoveries in the evolutionary process of eukaryotic cells was sexual reproduction, later adopted almost universally by multicellular organisms, fungi, plants, and animals. With rare exceptions, multicellular organisms have two parents, that is, they result from the conjugation of two germ cells. This process permits the combination of the hereditary material of both parents. Sexuality implies that each individual has two complete genomes, while the germ cells, called haploids, have only one complete genome formed by one single set of chromosomes. The ovule or sperm cells are haploid. Diploids are the cells with two genomes that result from the conjugation of two haploids, that is, the fertilization of the ovule.

To form gametes from cells that have inherited two sets of chromosomes from the fertilized ovule, one of the two sets of chromosomes must be dropped. This reduction in the number of chromosomes happens through a process called meiosis, in which the cells divide twice, but the chromosomes duplicate only once. Four haploid cells result, each one a bearer of a full set of chromosomes. However, contrary to what one would naturally expect, each of these sets is not a faithful copy of one of the two maternal or paternal sets. During meiosis the chromosomes of each pair align one against the other and randomly exchange homologous segments of DNA. This reshuffling process, called crossing over, implies that each of the gamete cells resulting from meiosis has a different version of the characteristic genome of the species. The organism born from the conjugation of two gamete cells is necessarily unique. We are like our parents but we are not identical to either of them.

Sexual reproduction has the remarkable advantage of promoting genetic diversification within the same species, allowing it to adapt more rapidly and efficiently to environmental changes. The conjugation of two gamete cells belonging to different organisms in meiosis is constantly shuffling the genes, thereby diversifying the individuals and conferring upon the species an increased capacity to adapt to the variable external conditions. Since the genes in a diploid cell result from an imperfect duplication process, sexual reproduction allows experimentation with a large number of genetic variations without compromising the identity of the species.

The cloning of organisms that reproduce sexually is the result of biotechnological processes capable of producing genetic copies, which is clearly contrary to the natural tendency for genetic diversification (Rideout III, 2001). In cloning, the nucleus of an embryonic cell or a somatic cell from the cloned organism replaces the nucleus of the ovular cell. This new cell contains two sets of chromosomes, just like the cell that results from the sexual fertilisation of an ovule. However, instead of having their origin in two gamete cells from different organisms, the two sets of chromosomes belong to the same organism. Thus, the organism that results from the nuclear replacement is a genetic copy of the organism that provided the nucleus. The first adult cloned animal was a frog *Xenopus laevis* obtained from an embryonic cell in 1958. Later, in 1996, Ian Wilmut's research team produced the first mammal successfully cloned from an adult somatic cell, a ewe named Dolly. The cloning of mammals by nuclear transfer has a very low probability of success and the cloned animals often show abnormal development and early senescence. Human reproductive cloning is a controversial issue that would very likely lead to severely disabled children. In spite of a non-binding United Nations Declaration on Human Cloning, it is probably inevitable that research groups will try to clone human beings in the future.

The discovery of sexual reproduction in the history of life also had important implications regarding the consequences of mutations. In unicellular organisms, each mutation creates a new lineage. In multicellular organisms, each mutation creates new genes in the gene pool of the species that is in permanent transformation through sexual reproduction.

Considering this capacity to reshuffle and transform, the question arises as to how a new species originates, a process called speciation. Our knowledge in this field is still very sketchy and incomplete. Speciation frequently results from a separate evolution as regards the environment, which ends up inhibiting fertilisation with individuals from the unchanged species. The environment, however, is only one of several factors. There are groups of organisms which, for no apparent ecological reason, speciate rarely or very slowly. These include the so-called living fossils. On the other hand, at the other extreme, there are examples of species which, in a relatively limited region and over just a few thousand years, produce hundreds of new species. An example is the more than 400 endemic species of cichlids (a fish from the Cichlidae family) in Lake Victoria in Africa, which did not exist as recently as 12 000 years ago (Mayr, 2001).

The evolution of species, from the first unicellular eukaryotic organisms, corresponds to a functional and structural increase in complexity, which implies more information stored in the genome, and therefore a larger number of genes. These increasing trends became very significant with the specialization of cells into organs in multicellular organisms and with sexual reproduction. Over about 2.5 billion years the number of genes increased by a factor of approximately 12 from 500 in the first prokaryotic unicellular organisms to 6000 before the first eukaryotic protists. With the invention of multicellularity, cell specialization, and sexuality the number of genes increased to about 35 000 in man, in only one billion years. However, the increase in complexity is not the only evolutionary way that living organisms have to win the struggle for survival. Some unicellular organisms with a small genome, such as bacteria, have developed strategies that have allowed them to survive up to our time and to compete with much more complex life forms. The reason for their success lies in their fast reproduction rate, combined with a high mutation rate. These characteristics increase the probability of finding successful adaptations to an environment that is continuously changing. The ability of some pathogenic microorganisms to become rapidly resistant to new antibiotics and drugs constitutes a remarkable example of their adaptation capacity, and it creates a tremendous challenge to modern medicine and pharmacology.

2.15 The Invention of Multicellularity and the Loss of Virtual Immortality

Nearly one thousand million years slipped by between the appearance of the first eukaryotic cells and the oldest traces of fossil evidence for multicellular organisms. Later, about 550 million years ago, at the beginning of the Cambrian period, there was an immense and relatively rapid surge of new life forms, known as the Cambrian explosion. It is therefore probable that the eukaryotic organisms remained unicellular for many hundreds of millions of years. The main difficulty in tracing the path of evolution during the Precambrian is that the organisms were probably soft-bodied and did not leave fossil remains, in contrast to the Cambrian multicellular organisms, some of which had skeletons. Note that, well before the emergence of the first eukaryotic cells, unicellular organisms organized themselves into colonies where different species specialize in complementary functions that support the colony. Colonies of cyanobacteria called stromatolites, which occur widely in the fossil record of the Precambrian but are rare today, are a good example.

However, the great innovation, achieved through genetic mechanisms, was the development of cellular associations arising from one initial cell, in which all cells share the same genome. This landmark discovery opened the way for the remarkable diversity of life forms. The fundamental mechanism that allows the formation of multicellular organisms is cellular differentiation, that is, the coherent functional specialization within the same organic structure of cells derived from the same parental cell. This division of labour led to the rapid development of specific external functions of protection, support, and mobility, which would be impossible at the unicellular level.

How can cells with the same genome have such different functions and structure? Cellular differentiation became possible due to the development of regulatory genes or super-genes, responsible for controlling a system of genetic switches that switch on and off the expression of certain genes. These super-genes direct and control the differentiated development of the cells of multicellular organisms, beginning from the fertilized ovular cell, in such a way as to permit the formation and harmonious functioning of its different organs. After fertilization, the ovular cell divides into two cells, which similarly divide to produce four, eight, and so forth. Very soon, the cells cease to be identical and start to differentiate according to their position in the body structure under construction, through the action of the regulatory genes.

The emergence of multicellular organisms had a rather extraordinary consequence. It created the concept of life span for a living being. The unicellular organisms, which reproduce by simple cell division, are theoretically immortal. Their disappearance or death does not result from old age but from some form of destruction, such as being eaten by other organisms. Reproduction is simply a duplication of the complete organism just as it is. In multicellular organisms, time has another value. Cells have their death programmed in the genome. This mechanism, called apoptosis, controls the ageing process and imposes a maximum life span. In certain very simple multicellular organisms, such as the volvox, a freshwater alga with the form of a blastula, it is possible to abolish apoptosis. Thus death is programmed and not a necessity. Research has shown that a mutation in a single gene obliterates the mechanism of apoptosis, freeing the volvox from the ageing process and rendering it immortal. The same type of intervention becomes progressively more difficult with the increasing complexity of multicellular organisms, due to the large number of regulatory genes. The reason why multicellular organisms have vastly different life spans is still not fully understood. Programmed death is the price life has to pay for its extraordinary diversity and the development of intelligence.

What is the reason for the sudden Cambrian explosion that followed a period of more than 1.4×10^9 years of slow and inexpressive evolution? The great variety of families of organisms that slowly differentiated themselves over hundreds of millions of years suddenly produced an astonishing multiplicity of new species, some larger, stronger, and more complex, others fragile and with strange and surprising forms. Stephen Jay Gould masterfully describes the creatures of this incredibly diverse fauna in his book *Wonderful Life* (Gould, 1989).

The sudden acceleration of life's evolution probably resulted from the increase of atmospheric oxygen to values of 20 to 30% of today's concentration. At that time, cells developed mechanisms for the use of oxygen by means of oxidization processes coupled to the production of ATP molecules, which constitute the main energy source in the cell. The greater abundance of oxygen in the atmosphere and in the oceans, by dissolution of atmospheric oxygen, increased the energy available to the cells. This allowed evolution to develop organisms of greater size, with greater mobility and sophisticated protection and competition strategies. Some of these may have been efficient predators that contributed to the development of rigid body protection structures. One should also note that the Cambrian was a period with high global average temperature, about 7°C above present levels, and with a high CO_2 concentration, probably greater than modern values by a factor of more than ten. Nevertheless, the simplest and most plausible theory to explain the Cambrian explosion is to admit that life, by producing oxygen, changed the environment in a way that favoured its own development.

2.16 Life Starts to Seize Land

The evolution that took place during the Cambrian period created the conditions for life to come out of the oceans and conquer a new development arena — terra firma. Various mechanisms and processes had to be invented to overcome the challenge of adapting life forms to the conditions of the new environment. The force of gravity is much more significant in the atmosphere than in the oceans, where the buoyancy force resulting from the greater density of the aquatic medium facilitates flotation. There are other adverse factors in the atmosphere, such as the danger of desiccation, the higher intensity of solar radiation, strong winds, and severe temperature variations.

The transition from the ocean to land was relatively less difficult for animals than for plants (Kenrick, 1997). While 10 of the 25 sub-kingdoms or phyla of the animal kingdom phylogenetic tree managed to adapt to the land environment, only one phylum from the plant kingdom managed to adapt successfully. This difficulty promoted a greater creativity in the transition process of plants. Animals such as spiders, insects, and land vertebrates have organic and functional structures relatively similar to their marine ancestors. Plants, however, had to develop a series of new, more complex and diversified organic structures, bearing very little resemblance to those of their ancestors.

But why did plants and animals adapt to the inhospitable land environment, abandoning the simpler strategy of evolution in the comfort of aquatic surroundings? The question is especially significant because natural selection operates in the immediacy of the here and now. To be successful, mutations have to bring some benefit. The main reasons and the detailed history of life's transfer from ocean to land are still not well understood. Presumably it happened because the plants and animals that lived in shallow internal waters, close to the oceans, were systematically faced with periods of drought, when the water tended to disappear, thereby destroying their habitat. Plants facing up to this challenge, had a period of rapid growth in the wet season and started to produce spores in the dry season. These spores would have been highly resistant to desiccation and heat, with a good capacity for survival, and capable of dispersion to more favourable sites. As regards animals, the drought periods in the wetlands would have forced them to develop new means of breathing and locomotion.

Probably the most primitive forms of terrestrial flora did not leave any trace. The oldest fossilised remnants of plants with any complexity are spores dated around 470 million years ago from the Ordovician geological period. Plants successfully solved the major problems of holding the body upright against gravity and supplying above-ground cells with nutrients and water. A dual vascular system was developed that conducts mineral sap drawn from the soil by the roots to the leaves and conducts organic sap formed through photosynthesis in the leaves to the roots and other non-photosynthesizing parts. The invention of a particularly resistant substance, lignin, allowed the development of the first treelike plants. Trees became robust, some very tall, measuring more than 30 meters, and abundant, forming luxuriant marshy forests in the Carboniferous period between 360 and 299 million years ago. These forests sequestered huge amounts of CO_2 from the atmosphere. Their fossil remains produced the immense coal deposits that fuelled the industrial revolution and still sustain a large part of our energy needs. They are known as the coal forests.

In a relatively short time, plants would have profoundly changed the landscape of the islands and continents, and decisively influenced the evolution of species by interacting with the other kingdoms of life. There are root fossils that reveal a symbiotic relationship with fungi. These make nutrients in the soil more easily assimilated by plants, while fungi benefit in turn from the energetic compounds produced by the plant metabolism. These symbioses, which are still very frequent, were probably essential to the initial development of terrestrial plants of increasing complexity. The sustainability of the forest is only possible if there is a mechanism that transforms the wood and all the other residues that result from the death of trees, avoiding accumulation. Fungi and bacteria solve the problem by digesting cellulose and lignin, recycling residues into organic molecules that make up the nutrients in the forest's ecosystem.

Photosynthesis in these very large carboniferous forests removed huge quantities of CO_2 from the atmosphere. This process led to a lower atmospheric concentration of carbon dioxide, which reduced the greenhouse effect and probably contributed to the onset of a long ice age that started in the Carboniferous. It was the longest in the last 500 million years and ended only in the Permian period, which lasted from 299 to 251 millions years ago.

2.17 The Major Permian–Triassic Extinction of Species

The geological era from the beginning of the Cambrian to the present, called the Phanerozoic, divides into three major periods: the Paleozoic (old animal life), Mesozoic (middle animal life), and Cenozoic (recent animal life). The Paleozoic ended in a dramatic way with the most severe extinction of species, called the Permian–Triassic extinction, at the transition between the Permian and Triassic geological periods, 251 millions years ago. This was the third of five major extinction events identified in the fossil record (Raup, 1982). About 57% of all families and 83% of all genera became extinct. Terrestrial vertebrates lost about 75% of their families. It is the only known event of mass extinction of insects, which is a particularly robust class of animals.

The Permian–Triassic event was not instantaneous and probably occurred in several phases lasting millions of years. This biological catastrophe happened during a maximum of continental compression into one supercontinent called Pangaea, and in an age of very intense volcanic activity, marked by flood basalt eruptions, which covered huge areas with lava. These eruptions, which formed extensive areas of basalt in Siberia, called the Siberian Traps, were very destructive for the flora and fauna, and probably changed the concentration of CO_2 and the amount of aerosols in the atmosphere. The lava flows in shallow seas caused the dissociation of methane hydrates in the sea floor, releasing very large amounts of methane into the atmosphere. The climate became warmer and arid in the vast continental interior of Pangaea. With the resulting global warming, the temperature gradient between the equator and the poles decreased, slowing down or even stopping the thermohaline circulation. This slowdown or collapse reduced the mixing of oxygen in the ocean, which became anoxic and therefore less suitable for marine life. There are also indications of significant marine regression, with the average sea level falling by about 250 meters, and of dramatic consequences for marine and coastal ecosystems. Furthermore, some authors (Becker, 2004) have suggested that these profound global environmental changes were compounded by the impact of a large comet or asteroid. A similar event, which happened in the transition from the Mesozoic to the Cenozoic, at the end of the Cretaceous period, was responsible for the fifth major extinction. The full history of the Permian–Triassic extinction is still unknown, but it is likely that several convergent causes aggravated the outcome.

2.18 The Importance of Seeds, Flowers, and Fruits

About 380 million years ago, plants developed a revolutionary innovation: seeds. Fertilization takes place in the plant, inside the ovary, which later produces seeds that leave the plant, to be carried by wind or water and lie dormant in the soil until conditions are favourable for germination. With the invention of pollen and ovules, the sperm cells of seed plants no longer needed the liquid water environment to swim and wriggle their way to the egg cell. The new reproduction mechanism immensely facilitated land colonisation and proved to be very robust, surviving the Permian–Triassic extinction that destroyed a large part of the Carboniferous forests. The conifers, which appeared in the Carboniferous period, constitute one of the more representative families of plants that carry bare seeds, called gymnosperms (from the Greek word *gymnos*, meaning naked).

In the Cretaceous period, more than 250 million years after the appearance of seeds, plants developed a mechanism for producing flowers and fruits. A new group of land plants emerged called the angiosperms. The name derives from the Greek word *angeion*, which means small vessel or envelope, since the seeds were contained in fruits. The angiosperms diversified rapidly and contributed in a significant way to the magnificent increase in biodiversity in both the plant and animal kingdoms. The fruits that enclose and protect seeds are frequently edible. In this way, animals became agents of seed dissemination, just like wind and water. Birds, bats, squirrels, and monkeys transport the seeds of tasty and nutritious fruits.

Another property of angiosperms, probably even more important, is that animals, insects, birds, and bats can also pollinate flowers. The animal pollinators promote gene transfer between widely separated plants of the same species, which would be unlikely or even impossible through wind pollination. This form of pollen transfer, the only one available for conifers and other non-flowering seed plants, would not have developed the diversity found in angiosperm forests. Nowadays vascular plants number around 300 000 species, of which about 250 000 are angiosperms.

The appearance of the angiosperms led to the development of a very productive symbiotic relationship with animals, which receive food in exchange for pollination and seed dissemination. The forest ecosystems created by the angiosperms, with their pollinating animals, the herbivores, the predators, and the parasites, influenced the evolution of life in a crucial way for the development of primates and eventually human beings.

There are many examples of co-evolution in the symbiotic relationship between flowering plants and the associated pollinating animals. The remarkable variety and exuberance of flowers results largely from the development of interaction strategies with animals. Each one of the organisms exerts a specific pressure of natural selection over the other, and this in turn results in a process of co-evolution. Both the plant and the pollinator tend to evolve together in a way that is favourable to the success of pollination. Co-evolution is also common between predator and prey, and host and parasite.

In the pollination of orchids, there are striking examples of co-evolution, described in the pioneering work of Charles Darwin (Darwin, 1862). One remarkable example is the epiphytic orchid *Angraecum sesquipedale*, from Madagascar, which has a very long spur, 20 to 35 cm long with nectar at its tip. Darwin predicted that there must be a pollinator moth with a sucking organ or proboscis long enough to reach the nectar at the end of the spur. The moth's efforts to reach the nectar would get the pollinarium that holds the pollen grains attached to its body. These grains would later pollinate another plant visited by the insect. Although the process was well known, a moth with a 35 cm long proboscis was unbelievable. However, in 1903, such a moth was discovered in Madagascar and called *Xanthopan morganii praedicta* in Darwin's honour.

The important point is that the orchid spur is adapted to a proboscis of comparable length of a specific moth that collects the nectar in flight while pollinating the flower. The increase in the length of the orchid spur and moth proboscis happened by mutual pressure between the two, over thousands of generations, induced by the need to assure the best conditions to pollinate the flower and collect its nectar. The orchid seeks the contact of the moth to favour pollination; the latter seeks to avoid contact as it endangers its flight near the flower.

There are other surprising strategies. Some orchids, such as *Ophrys speculum* from South West Europe and North Africa, use a new principle of flower pollination known as pseudo-copulation. The flowers almost exclusively host the males of a specific insect that performs copulation movements during its visits. They imitate the corresponding females and the confused insects, while carrying out their sexual practices, get the pollinarium stuck to their bodies, whereafter they deliver pollen to other flowers on future love-making expeditions. In the case of *Ophrys speculum*, the sexual mimicry imitates the female of a hymenopterous stinging insect called *Dasyscolia ciliata*. Besides this visual trickery, other orchids produce aromatic chemical substances with the some odour as the pheromones produced by the females to attract the males (Schiestl, 1999).

2.19 Origin and Evolution of Mammals

The first animals originated in the oceans from extremely simple heterotrophic organisms. They progressively developed various organs coordinated by a network of nerve cells, which enabled them to move, feed, digest, breathe the oxygen dissolved in the water, excrete, and reproduce. A decisive step to accelerate and diversify the evolution of animals was the process of segmentation that resulted from the duplication of certain regulatory genes. The organism transformed into a chain of semi-repetitive and initially semi-independent segments, almost identical with one another, as in the primitive annelids of the Cambrian. The fossil record of that period has another important phylum of segmented animals, namely, the arthropods, which includes among others, the insects, the arachnids, and the crustaceans. Evolution toward more complex animals proceeded with the invention of a cord-like central fiber called the chorda, which characterizes the phylum of chordates, and became the precursor of the spinal cord in vertebrates.

The oldest fossils of terrestrial animals are predatory arthropods from the Silurian, about 420 million years old. These animals may have lived in coastal terrestrial ecosystems and fed on algae and on plant and animal marine organisms that lived close to the seashore. The fish that were the ancestors of land vertebrates developed primitive forms of lungs in order to adapt to muddy and oxygen-poor waters. Apart from solving the problem of breathing in the atmosphere, they also had to solve the problems of support against gravity, and mobility. The four lateral fins strengthened and transformed into the four legs of early amphibians, about 370 million years ago. This is the reason why all land vertebrates have four limbs.

To solve the problem of desiccation, the first amphibians used the water environment to lay their eggs and also for developing their offspring. A very important invention was the amniotic egg that could survive out of the water with protective membranes and a porous calcified shell, which allowed the embryo to breathe. This development allowed animals to conquer land far from the sea. The amniotes evolved into two lineages, one leading to mammals and the other to reptiles and birds. The first birds appeared in the Jurassic period as descendents of theropod dinosaurs. Mammals appeared earlier, around 225 million years ago in the Triassic. The thermal insulation of the body by means of feathers and fur was an important innovation for birds and mammals, compared to their reptilian ancestors. It allowed these animals to become warm-blooded, and this favoured an accelerated metabolism in which biochemical and sensory reactions operate at optimal levels. Maintaining a relatively high body temperature is much more important in small nocturnal animals than in large diurnal animals, for which the ratio between surface area and volume is smaller, as in dinosaurs. There are indications that the emergence of fast-moving carnivorous dinosaurs with a powerful vision forced contemporary mammals to become nocturnal. The analysis of their fossilised remains also reveals that they developed better hearing and a keener sense of smell. Such improvements required a larger brain to process more information and a greater energy supply to keep it functioning. These conditions could be satisfied due to an increased metabolism in the earlier mammals.

Another crucial breakthrough in the evolutionary path of mammals was the placenta, an interface device between mother and foetus, through which the two circulatory systems transfer oxygen, nutrients, and detritus without rejection by the mother's immune system. The complete development of the foetus inside the maternal body, in contrast to what happens in the marsupial mammals, provides more time and better conditions for the genesis of more diversified evolutionary forms, including larger brains.

2.20 Mammals Benefited from the Catastrophe at Chicxulub

During the first 150 million years of mammalian evolution, the volume of the brain did not grow significantly. Fossilised remains indicate a clear tendency for a brain size increase relative to the body only during the last 50 million years, well after the disappearance of the dinosaurs. These facts suggest that the pressures of natural selection in the direction of larger brains appeared only in the Cenozoic. What were these pressures, so important for the later development of the primates?

The fossil analysis from the end of the Cretaceous, 65 million years ago, reveals the extinction of many species of plants and animals, including dinosaurs, and the appearance of others. Mammals proliferated and diversified, probably due to dinosaur extinction, and in the plant kingdom, the angiosperms became dominant. The main cause for this widespread extinction was very probably the impact of a 10 km wide asteroid on the Yucatán Peninsula in Mexico, which produced a crater with a radius of 180 km known as the Chicxulub crater. The catastrophe left many traces, now well identified. Some palaeontologists consider that the extinction of species that marked the beginning of the Cenozoic had other causes besides the collision. They claim that the dinosaurs had already begun their decline before the impact, due to climate changes induced by a changing configuration of continents and sea currents associated with plate tectonics. The asteroid collision was only the last straw in this process, precipitating and amplifying the extinction. In any case, there can be no doubt that it had profound consequences for the later evolution of the flora and fauna.

Very likely, the dinosaurs became extinct in the colder world that resulted from the huge emissions of dust particles and smoke, owing to the lack of thermal insulation of their bare skin. Furthermore, the new climate and reduced photosynthesis destroyed a significant part of the plant cover, making survival much more difficult. However, birds and small mammals, precisely due to their smaller sizes and the thermal protection of their bodies, managed to survive by feeding on insects, worms, larvae, roots, and other vegetable products. In the sea, the blocking of the solar radiation caused the extinction of many forms of phytoplankton and various animals at all levels in the food chain, including the great marine reptiles.

At the beginning of the Cenozoic there were no very large animals, and the amphibians, reptiles, mammals, and birds lived in an environment with a very diverse flora and a great abundance of insects. After rapid diversification, some mammalian evolutionary lineages gave rise to predator and prey pairs, with clear signs of a tendency for higher rates of increase of the encephalization quotient, or brain volume relative to body size. But larger brains consume much more energy, which means larger nutritional requirements. A mammal requires five to ten times more food than a reptile of the same weight (Allman, 1999). There must have been strong natural selection pressures to increase the size of such a high energy-consuming organ. However, we are still unable to reconstruct all the details of what actually happened. We know only the sequence of the more important evolutionary transformations.

2.21 Primates Develop in the Angiosperm Forests and Come Down to the Ground

Primates evolved and branched off from their ground-dwelling insectivorous ancestors in the upper Cretaceous period. They began to follow insects up the trees where they could also find an abundance of leaves, fruits, and nuts. Primates — apes, monkeys, lemurs, lorises, and tarsiers — since their differentiation, tended to have larger brains than other mammals, relative to their body weight, and this tendency intensified during the Tertiary period. The majority adapted to a daylight life in the trees, in which eyesight was more important than a sense of smell. It may be that the initial impulse toward higher encephalization resulted from the change to a diurnal lifestyle and to the adaptation to the tropical forest habitat (Sussman, 1991). These animals had to dominate the three-dimensional structure of the forest, and find out how to locate and remember the location of trees with edible fruits or nuts. To jump from branch to branch at the top of the trees to avoid the predators prowling around at ground level demands high visual acuity. Over millions of years, natural selection shortened the muzzle and separated the eyes, leading to a stereoscopic frontal vision that allows better distance perception by processing the images from the two visual fields.

Flowering trees are clearly a key factor at the origin of primates. The arboreal lifestyle and a more intense social interaction in their habitat contributed decisively to the greater encephalization of primates relative to other mammals. It also led to the development of longer arms and prehensile feet to help support the body while the arms were looking for fruit, nuts, and insects. Meanwhile the arms and shoulders developed a greater flexibility, and the hands adapted to be able to pick up, hold, and examine food. Carrying food to the mouth by hand, after a careful examination, is a characteristic of all primates. In the Oligocene, about 25 million years ago, the evolutionary branch of the lesser apes, which includes today's gibbons, separated from the branch that led to the Hominidae or Hominids. This branch includes the orangutans and the subfamily of the Homininae that includes gorillas, chimpanzees, and all members of the human lineage.

The Proconsul, which lived about 20 million years ago in the humid forests of Africa and Eurasia, was a remarkable precursor for the future evolutionary characteristics of the Hominids. It had a larger brain and was heavier than its ancestors, with a weight of about 50 kg. It had no tail, and its legs were strengthened to support a more upright posture. The Proconsul was not adapted to living in the treetops like its ancestors. It preferred the lower part of the tree where the branches were stronger, and probably made frequent visits to the ground. This tendency for a more active life on the ground was part of the evolutionary process that led to the Hominids. The forest of flowering trees had ended its role as a propitious habitat for the initial evolution of primates. Endowed with relatively powerful brains, the Hominid ancestors were now ready to address the new challenges of evolution on the ground.

2.22 The First Hominids

The Hominids are a taxonomic family that includes the orangutans, gorillas, chimpanzees, and the human lineage. Orangutans separated from the evolutionary lineage that led to humans about 14 million years ago, followed by the gorillas about 10 million years ago. Finally, about 9 to 6 million years ago, the branch of the chimpanzees, man's closest living relatives with only 1 to 2% difference in genetic heritage, separated as well. Our knowledge about the timing of this evolutionary divergence and the beginnings of the Homininae lineage is still very incomplete.

The *Australopithecines* were probably among our earliest direct ancestors. These include the *Australopithecus*, meaning chimpanzees from the south. Raymond Dart (Dart, 1925) first discovered their fossilised remains in South Africa in 1924. They lived on the savannah and walked upright, but their encephalization was not much superior to that of the gorillas or chimpanzees. Their activities were mostly on the ground, but they continued to be good tree climbers, and would probably have taken refuge in the trees in times of danger, and also slept there.

According to some authors (Coppens, 2004), the ramification that separated our ancestors, whose habitat was preferentially the savannah, from the gorillas and chimpanzees, whose preferred habitat was the forest, resulted in part from environmental changes. Tectonic movements of extension about six million years ago produced the Great African Rift, stretching south from the Dead Sea in Israel and Jordan through the Red Sea and Ethiopia to Mozambique. The formation of the Rift produced a succession of relatively high plateaus along East Africa, where the drier climate favoured the development of the savannah, inhabited by the *Australopithecines*. Furthermore, this was a time of global change toward colder and drier climates, which led to the break-up of the great rainforests and to more open woodland and grassland. It seems likely that the great diversity of East African regional climates produced by the African Rift would have contributed to the strong species diversification that fostered the appearance of *Homo sapiens* (Trauth, 2005).

One of the most famous fossil remains of an *Australopithecus* is 'Lucy', a skeleton of an *Australopithecus afarensis* specimen found in 1974 by a team led by Maurice Taieb, Donald Johanson, and Yves Coppens (Johanson, 1981) in the Afar depression, which is part of the Great African Rift in Ethiopia. The majority of Homininae fossils consist of a few pieces of disconnected and often deformed bones, but in the case of Lucy, the researchers were able to collect 40% of the skeleton. Lucy was a youngster, around 20 years old and one metre high, who lived about 3.3 million years ago. Her pelvis and backbone clearly indicate that she was bipedal and her feet resembled ours, although they still bore the traits of a more apelike ancestry. However, the volume of the brain, of the order of $400\,\text{cm}^3$, was still closer to that of a chimpanzee.

A remarkable proof of bipedalism is the fossilised footprint remains of *Australopithecus afarensis* discovered by Mary D. Leakey (Leakey, 1979) in Laetoli in northern Tanzania. These are only a few hundred thousand years older than Lucy. The form and the sequence of the footprints preserved in moist ash from a volcanic eruption that later dried clearly indicate that they walked upright.

The trend toward walking on two feet probably resulted from natural selection, induced by adaptation to a dryer climate and to the environment of the savannah, where it was relatively more difficult to find food. Standing upright increases security, since it broadens the field of view above grass and shrubs and makes it easier to survey the surroundings. It also had the great advantage of freeing the hands to transport offspring and food, to make and use tools, and perhaps even to develop communication through gestures. Lucy's hand anatomy would already have allowed the use of sticks and stones for protection from predators. Life on the ground created situations of great danger, especially at night, when continuous vigilance was necessary and some means of protection had to be ensured. Cohesive social groups probably formed with strong interactions between their members for the protection of life, in order to overcome this vulnerability. This would have been especially relevant for the newborn, who were defenceless, and whose survival lay completely in the hands of the mother, assisted by the group.

Lucy belonged to a successful group of species whose lineage lasted about a million years before originating a bifurcation into two new lineages. One of these led to the robust *Australopithecines* or *Paranthropines* (*Australopithecus robustus* and *Australopithecus boisei*), which had large skulls with big jaws and huge molars, probably resulting from adaptation to a vegetarian diet including fibrous plants and harder fruits and seeds. However, it became a failed evolutionary path that went extinct about 1.4 million years ago. On the other branch, which gave rise to the genus Homo, the jaw and the molars remained small, indicating a more varied diet. As well as being omnivores, they were probably scavengers and efficient hunters of the savannah's abundant fauna. However, the crucial reason for their future success was their adaptation to the environment through increasing encephalization.

The evolutionary path that led to *Homo sapiens* from the common ancestor with the chimpanzees is still far from being well understood, remaining full of uncertainties. The great difficulty in the reconstruction of this part of our past evolution results from the fact that fossils are very rare, mostly fragmented, distorted, and geographically dispersed. Nevertheless, fossil remains do allow one to conclude that, about 2.5 million years ago, the genus Homo existed in the form of *Homo rudolfensis* and *Homo habilis*. Their evolution led, about two million years ago, to the appearance of *Homo erectus*. This was the first member of the genus Homo to migrate outside of Africa, about 1.8 million years ago. Fossils found in Java in 1891, initially called *Pithecanthropus erectus*, and known in China as Peking Man, are in fact specimens of *Homo erectus*.

Today we have a better knowledge of *Homo erectus*, following the extraordinary discovery of the Turkana Boy, a fossil found by Richard Leakey and his collaborators in Lake Turkana, Kenya, in the middle of the 1980s (Walker, 1993). The discoverers were able to retrieve more than 80% of the boy's skeleton. He lived 1.6 million years ago and was between 11 and 15 years old when he died. He had longer legs than his ancestors, which helped him to cover longer distances, and his height, had he achieved adulthood, would have been about 1.8 m. The most remarkable characteristic, estimated from the study of his skull, is that the volume of his adult brain would have been 900 cm^3, a significant increase over the 600 cm^3 of *Homo habilis* and the 400 to 450 cm^3 of the *Australopithecines*. In little more than two million years, the brain volume of our direct ancestors had doubled! The fossil record reveals that about 600 000 years ago the *Homo erectus* lineage evolved in Africa and possibly Europe into a larger-brained form called *Homo heidelbergensis*. This new species had a brain with a volume estimated between 1 100 and 1 200 cm^3, a value that is very close to the average of 1 350 cm^3 of today's *Homo sapiens*. About 200 000 years ago, a new speciation event in Africa gave rise to *Homo sapiens*, which eventually dispersed to Europe and Asia, and later to Australia and the Americas.

2.23 What Were the Reasons for the Remarkable Encephalization in the Homo Lineage?

The increase in the enchephalization quotient by a factor of more than 2.5, from *Australopithecus afarensis* up to *Homo sapiens*, which took place over about 3.5 million years, is the most significant and surprising characteristic of our ancestral lineage, and is very likely the main reason for our great success in the realm of the evolution of species. During that period of time, the volume of the brain increased by a factor of about 3.4, while the body mass only increased by 30%.

What were the selective pressures behind that evolution? The sustained increase in the volume of the brain would not have been possible without the permanent action of a natural selection mechanism that guaranteed greater reproductive success. As already mentioned, the brain is a very demanding organ in terms of energy consumption. With only 2% of our weight, the brain consumes 20% of the energy required for life. The selection pressures leading to our amazing high-energy evolutionary path must have been very strong. What was the nature of that natural selection?

To find an answer it is important to analyse the dominant evolutionary specializations of the Homo lineage. One of the most distinctive is tool use and tool making. The oldest known stone artefacts, consisting of very crude choppers and flakes, were discovered in Ethiopia, and date from 2.5 to 2.6 million years ago (Ambrose, 2001), the time when the Homo lineage probably diverged from the *Australopithecus* evolutionary branch. To make them, our ancestors would strike stones together in a way that demanded a skill and dexterity of the arms and hands far superior to that of non-human primates today. The new tools served to chop the meat of their prey, or to scrape and break the bones to obtain the marrow and to open fruits and nuts that would otherwise have been inedible. These early tools were known as the Oldowan industry, named after the Olduvai Gorge in the northern Tanzania region of the East African Rift. They were common and underwent little refinement for roughly a million years. Meanwhile the diet became richer in animal and plant proteins, favouring the process of encephalization. About 1.6 million years ago, *Homo erectus* developed a more sophisticated stone tool technology, known as the Acheulean industry, which lasted until 100 000 years ago and produced much more diversified tools as regards manufacture and use. It was during this time, probably between one and a half and one million years ago, that some populations began to use fire systematically (Brain, 1988). With this development, the diet was not exclusively raw food, and this had implications for the evolution of the digestive tract and for the way of life, especially as regards comfort and protection against the cold.

The other distinctive specialization in the evolution of the genus Homo was language. While the brain was developing a considerable capacity to construct a perceptual world, the ability to communicate these perceptions to others was also fast progressing. This was undoubtedly one of the most remarkable adaptive innovations of this period of our prehistory. It opened the way to cultural evolution and later to the development of civilizations. The ability to speak played a decisive role in the evolution of social relationships, facilitating communication and social cohesion within the groups. Vocal communication requires the capacity to produce sequences of different sounds organized into separate syllables.

The ability to speak resulted from an anatomical adaptation in the human lineage, consisting in the lowering of the larynx and the formation of a longer pharynx. The larger volume of air in the pharynx allows the mouth, lips, and tongue to produce a great variety of sounds. But these developments came at a price. A lowered larynx can be deadly since it becomes much easier to lodge food in the windpipe. Furthermore, it becomes impossible to breathe and swallow at the same time. In a human baby, as in the great apes, the larynx is high up on the throat, allowing him to suckle and breathe at the same time. As the baby grows, the larynx moves rapidly downwards, enabling the pronunciation of the first articulated words. Then the child starts to speak and to learn a language during a period of time that is one of the most extraordinary and magical of our lives. Human language is a generative system that makes it possible to construct an almost infinite number of sentences, each with a different meaning, from the 50 000 to 100 000 words that make up the average vocabulary of an adult. This served as an immensely powerful framework for the future development of cultural evolution.

At the level of the brain, communication by means of language corresponds to the coordinated use of cortex motor areas that control the larynx and the mouth. Furthermore, we know that the use of a language requires the integrity of the Broca area in the left part of the brain. The morphology of the fossil craniums of *Homo habilis* and *Homo erectus* indicates that they could accommodate a brain empowered by a Broca area. We may therefore infer that they could use some form of vocal communication, but it is not possible to reconstruct the evolution of language from its most rudimentary and primitive forms.

The development of a language demands a more complex and larger brain to allow for the communication of a great variety of perceptions and abstract concepts. A correlation between the gradual development of the capacity for vocal communication and the encephalization process is therefore likely to have existed over at least the last two million years (Jerison, 1991). Nevertheless, language is not identifiable with a natural selection pressure. It is a remarkable faculty, likely to be part of the answer to the selection pressures that led to the increase in brain volume, but not its cause.

Tool use and tool making is not a natural selection mechanism either, and it could not therefore be the driving force behind brain expansion, although some correlation between them is likely to have existed. As already noted the technology of stone tool making did not change significantly for a long period of about one million years in the last 2.5 million years, while encephalization was a continuous process right through that period.

Environmental change was most probably an important factor in the early development of the human lineage. The Earth's climate in the Tertiary period was relatively warm until about 35 to 30 million years ago, when an ice age started with the gradual formation of ice caps in the polar regions. First, ice sheets formed in East Antarctica, then in West Antarctica, and finally, about 5 million years ago, in the Arctic, beginning in the south of Greenland. The ice age was probably the result of climate forcing induced by changes in ocean currents and by mountain building events arising from the separation of the continents, which followed the break-up of Pangaea. About 2 million years ago, the climate became considerably colder and extensive permanent ice sheets formed in the Arctic, around the North Pole, in Greenland, and in the northern parts of Europe and North America. Since that time, there are records of relatively rapid alternations between cold glacial periods and shorter, warmer interglacial periods. These cyclic climate changes, with a period of the order of one hundred thousand years, have their origin in small variations of the parameters that characterize the orbital and rotational movements of the Earth. Glacial periods have a typical duration of 80 to 100 thousand years, while the interglacial periods are shorter, of the order of 10 to 20 thousand years. Clearly, these cycles would have caused pronounced global climate changes during the evolution of the Homo lineage up to *Homo sapiens*, which lasted about 2.5 million years. The resulting environmental changes created a variable ecological context that could limit or enhance the encephalization process.

The savannah appeared in East Africa very early on, approximately 6 million years ago. On the other hand, due to its latitude and orographic characteristics, tropical Africa presents a wide range of climates and habitats, some of which could serve as a refuge in times of global cooling. The situation is very different at higher latitudes, for example in Europe, where the glacial periods are much more severe and cause profound changes in the flora and fauna. We know that during the period of greater encephalization, our human ancestors inhabited regions where, due to the colder and drier climate, the savannahs were losing their trees and becoming predominantly shrubby. In this habitat it was more difficult to find protection from the carnivorous predators, and it was probably necessary to develop new strategies for survival. In the savannah, lions, leopards, hyenas, and wild dogs, all faster runners than our ancestors, would have been a permanent threat. On the other hand, it is very likely that in the mosaic of local environments of East Africa, the populations of the genus Homo could always have found amenable living conditions. Surviving the ice age in this part of the world was probably one of the minor challenges facing them. It therefore seems unlikely that the increased encephalization resulted primarily from climate and environmental changes in Africa, as some authors have suggested (Calvin 2002; Stanley, 1998).

So what were the selection pressures that powered brain expansion in the human evolutionary lineage? They probably stem from the survival and reproductive advantages associated with a social life of increasing complexity. The development of a more interactive and complex social life based on alliance building and collective responsibilities in the group for activities such as hunting, food sharing, protection, and intergroup fighting, would have improved the capacity for survival and mating opportunities. Those that are more adept at building and maintaining alliances, and also at manipulation, will tend to be more reproductively successful. To predict and manipulate the behaviour of other individuals, to establish networks of friendship, and finally to become consummate social tacticians requires a higher level of mental functions, such as thinking, planning, and consciousness, which can only be achieved through the development of a larger brain.

The process eventually became irreversible because, by reaching a certain level of social complexity, new survival pressures within the groups would have worked in such a way as to increase complexity even further. Our lineage trapped itself in an evolutionary ratchet from which there was no going back. Our main allies and opponents became members of our own communities. Thus, the Homo lineage forged an evolutionary path in which nature was no longer the leading adversary. This development had profound consequences as regards our relationship with the environment, and would continue to do so in the future. The progressive divorce from nature of the selective evolutionary pressures since that time has been the main reason for the immense success of the Homo lineage in the biosphere, but at the same time, it is probably the main cause for the troubles that await us in the medium and long-term future.

The increase in brain volume had profound consequences for the characteristics of the newborn and their mothers. In evolutionary terms, it was not possible to widen the pelvic canal to accommodate larger craniums, while at the same time adapting the pelvis to bipedalism. Consequently, a large part of brain growth had to be gradually postponed to the period after birth. Nowadays the volume of a child's brain nearly doubles in its first year of life. The disadvantage with this solution is that the newborn has almost no mobility, being defenceless and completely dependent upon its mother. Human babies have a degree of survival independence comparable to a newborn chimpanzee only when they reach the age of 17 months. The very slow development of our ancestors' babies demanded far more intensive childcare and a closer bond between the parents. To fulfil this requirement, the mutual sexual attraction of the parents increased through anatomical changes such as large breasts, permanent sexual responsiveness, and an intense and deep relationship at the mental level between the partners.

All primates are highly sociable animals, but the species of the Homo genus developed the social interactions in the groups in which they lived, typically between 20 and 100 individuals, to an extraordinary degree. The intimate cooperation between the members of the group, in a frequently hostile environment, usually with scarce resources, promoted competition and aggression towards other groups of the same species. The capacity to cooperate within the group is likely to have evolved at the same time as the fighting strategies between groups. The more the intergroup domination and warfare capacities developed, the more important it became to ensure protection, cohesion, and cooperation within the group. The need to develop all these increasingly complex social interactions and strategies, with their clear reproductive advantages, was very probably the main selective pressure in the process of brain expansion.

It is important to realize that this evolutionary process also had a dark side. Our Homo ancestors probably lived in small, stable, and socially very cohesive groups, characterised by strong bonds and complicity between the males. Membership of a group implied a deep mistrust and hostility towards those belonging to other groups. Some Asmat tribes of New Guinea, where members of an adversary group are identified as 'the edible ones' (Watson, 1995), provide an extreme contemporary example of this type of aggressiveness. According to Jacques Monod (Monod, 1971), our Homo ancestors, after dominating the environment, had no serious adversary other than their own kind. The members of the opposing social groups became the most dangerous individuals in their world, excluding parasites and pathogenic agents. Intergroup fighting, often mortal strife, became one of the principal factors of selection of our Homo ancestors. To kill a member of one's own group was equivalent to murder, but to kill a male of another group was probably the best way to gain social prestige within the group. As Ernst Mayr (Mayr, 2001) writes: "There is no doubt that hominid history is a history of genocide."

Various archaeological findings support this thesis. Recent excavations in a cave in Spain revealed the presence of bones of specimens of the Homo genus and animal bones mixed together (Fernández-Jalvo, 1996). Both kinds of bone exhibit signs of dismemberment and scraping by the Homo species that lived in the cave about 300 000 years ago. Very likely, they served the same end — food. On the other hand, carnivorous predators had chewed only about 4% of the bones found, a sign that these were mostly hunters and not scavengers. Homo bones found in other caves also show clear signs of scraping with stone tools, indicating that the meat very probably served as food. It seems highly likely that the groups competed with each other for food, especially when hunting, and for territory, while some of them, under certain circumstances, were cannibals.

The heritage that resulted from our ancestors having been highly social beings living in small cohesive groups accustomed to strong intergroup competition and warfare left us with a profoundly dichotomous morality based on the tension between 'them' and 'us'. The history of civilizations shows that we have a double capacity for the most abnegated altruism and the most violent and abject aggression. The way altruism first emerged in the evolution of the social behaviour of *Homo sapiens* is still a subject of much debate. A recent study (Bowles, 2009) supports the idea that lethal intergroup competition and conflict may have contributed significantly to the proliferation of a genetic and cultural predisposition to behave altruistically towards the other members of one's own group. According to this point of view, altruism, in its origins, is a form of response and adaptation to the rigours of warfare.

In conclusion, the main reasons for the extraordinary increase in encephalization in the last 3.5 million years were very probably the pressures of natural selection that resulted from the growing complexity of a social life characterised by fighting between small, tightly knit groups, to gain resources and territory, and by defence and protection activities within the groups. By adopting this evolutionary path, it became impossible to regress, because it would imply becoming a victim of the increasing complexity of the social environment. We became our own prisoners, in a self-sustainable and one-dimensional process determined mainly by internal mechanisms of social behaviour rather than by the external natural environment. Nature always had a deep influence on the living conditions of our ancestors, but from about 3.5 million years ago, it probably ceased to be the main driving force in their evolution. In the evolutionary process of the Homo lineage, a trend was initiated that led to the domination of nature. Could it be that this subtle divergence leads inevitably to a relation between man, nature, and the environment that implies an unsustainable use of natural resources, environmental degradation, and aggression of nature, with irreversible destructive consequences, such as we are increasingly witnessing today? To answer this question, one must analyse the history of the *Homo sapiens* species, from its emergence until it became the sole representative of the Homo genus.

2.24 How Did *Homo sapiens* Arise?

Paleoanthropology began as a scientific discipline with the discovery in Neanderthal, Germany, in 1856, of the fossil remains of a specimen of a Homo genus species later called *Homo neanderthaliensis*. Since that time, the Neanderthals have been, and in many respects continue to be, an enigma. Morphologically they are close to *Homo erectus*, with prominent brow ridges and large upper and lower jaws. However, they also have rather unusual features, such as a very robust stature, strong heavy bones, and a cranium that is broad and rounded when viewed from behind, with a large capacity, averaging some 1400 cm^3. They lived in Europe and western Asia from about 350 000 years ago and survived in the very harsh climate of these regions during four glacial periods. But then, about 28 000 years ago, they mysteriously went extinct.

Some time earlier, between 70 000 to 40 000 years ago, a new species had appeared in Europe. This was *Homo sapiens*, to which today's humans belong. However, the oldest fossilised remains of *Homo sapiens* were found in sub-Saharan Africa and date from as early as 200 000 to 150 000 years. Recent genomic studies indicate that *Homo sapiens* and *Homo neanderthalensis* share a more recent common ancestor, about 700 000 years old, and that their ancestral populations split around 370 000 years ago (Noonan, 2006). In comparison with the Neanderthals, the skeleton of *Homo sapiens* is slender and less robust, the cranium has a flatter face, the brow ridge is not so prominent, and the jaw is less massive, usually with a well-differentiated and protruding chin and smaller teeth. Curiously, the average cranial capacity of fossil *Homo sapiens* is only slightly larger than that of the Neanderthals. The most distinctive feature of the latter is their heavier bone structure and stronger muscles, which probably resulted from adaptation to the cold climates of Eurasia, especially during the glacial periods.

For a long time it was assumed that *Homo sapiens* spread out of Africa to Eurasia between 80 000 and 60 000 years ago and replaced the archaic hominids it encountered without mingling with them. There are now strong reasons to believe that it did not happen like that. Recent comparisons of the Neanderthal genome to the genomes of present-day humans from different parts of the world show that the sequences are equally close to all present-day humans, even to those from Papua New Guinea and China where Neanderthal specimens have never turned up (Green, 2010). This result points to interbreeding between Neanderthals and *Homo sapiens*, probably in the Middle East and parts of Europe between 80 000 and 45 000 years ago, before the latter fanned out into Eurasia and split into different groups. There was probably also genetic mixing in various parts of Eurasia, between *Homo sapiens* and locally evolved descendants of *Homo erectus* from previous out-of-Africa waves.

There are clear indications that the climate changes that resulted from the alternating glacial and interglacial periods affected the evolution of the Homo, particularly *Homo neanderthaliensis* and *Homo sapiens*. Furthermore, genetic studies reveal that, in the very cold period at the end of the penultimate glacial maximum, about 140 000 years ago, the populations of *Homo sapiens* went through a severe population contraction or bottleneck, which was probably caused by the decrease in the global average temperature and the increased aridity (Lewin, 2004). There is also evidence of a more recent bottleneck caused by the very large volcanic eruption of Lake Toba in Sumatra that occurred around 71 000 years ago. The eruption plunged the Earth into a volcanic winter that lasted about six years. The genetic evidence so far gathered suggests that the human populations already living in Europe and eastern parts of Asia became extinct by the direct and indirect consequences of the large drop in the average global temperature, and that only about 10 000 adults survived the catastrophe in some regions of equatorial Africa (Ambrose, 1998). There are also indications that the Toba eruption caused genetic bottlenecks in other mammals, such as the orangutan, the Central Indian macaque, the East African chimpanzee, and the tiger. These episodes remind us of the random nature of evolution and the fragility of our common past and future.

Homo sapiens arrived in Europe during a glacial period, but brought with him a more diverse and sophisticated stone technology, including stones to grind flour, utensils made from bone and horn, such as needles, probably to sew skins, sharp spearheads, and fishing hooks. These artefacts were apparently more evolved and sophisticated than those used by the Neanderthals in the same period. There are records of *Homo sapiens* having inhabited the Middle East and eastern Mediterranean around 100 000 years ago, but at that time, he did not yet possess the cultural attributes, characterised by a more sophisticated technology, that he used when the population expanded into Western Europe. The early dispersal around Africa and into the Middle East probably took place during the last interglacial period, which peaked 120 000 years ago, when the climate was warmer and wetter. It was only after about 60 000 years that he migrated west into Europe, and in particular to the south of France. The fossilized remains of these early humans in Europe, first found in a cave in Cro-Magnon, near Les Eyzies on the Dordogne, in 1868, correspond to the so-called Cro-Magnon Man. By then humans, meaning *Homo sapiens*, had a much more highly developed culture. They had a powerful capacity for representation which manifested itself in the remarkable cave paintings in southern France and northern Spain, such as Lascaux and Altamira.

The interpretation of such facts indicates the emergence of a new cultural evolutionary trend in the Homo genus. It is important to emphasize that the development of a more sophisticated culture among the *Homo sapiens* inhabiting western Eurasia about 45 000 years ago took place with essentially the same brain as the earlier *Homo sapiens* that lived in Africa about 200 000 years ago and in the Middle East 100 000 years ago. These ancestors had the same brain, but a less developed culture. There was a decisive turning point after which the determining factors of survival and expansion came to be of a cultural nature. We do not know the reasons for the emergence of this cultural evolution, characterized by an increased symbolic and technological complexity. However, a recent study (Powell, 2009) has shown that the first signs of a distinctive *Homo sapiens* culture, such as shell beads for necklaces, and the use of pigments and sophisticated tools, like bone harpoons, are associated with relatively high population densities. Probably the opportunities to develop these primordial cultural manifestations resulted from the stronger interaction and exchange provided by larger populations living close together. A demographic mechanism may enhance the degree to which human populations are able to develop culturally inherited skills. Interestingly, one finds once more that the decisive human capacity for cultural evolution has a social origin, associated with increased opportunities for more intense and diversified interactions between people.

The brain of *Homo sapiens* is sufficiently powerful to make cultural diversification and development possible. Our brain is a very complex organ which in adulthood contains more than 100 000 million neurons, each connected on average with 10 000 other neurons, and is able to process about 1 000 computational cycles per second. The detailed plan of this gigantic labyrinth of inter-neuronal connections is not stored in the genes, but develops epigenetically by means of successive processes that take place during the development of the brain. Genes only provide a general framework for the cerebral structure. The detailed plan of inter-neuronal connections establishes itself under the influence of stimuli received from the body and from the outside world. During the growth of the brain after birth, the neurons continually establish transient connections between themselves in an essentially random fashion (Edelman, 1987; Changeaux, 1983). Many of these connections break down rapidly, unless outside stimuli lead to repeated use and eventual stabilization. The neuronal network is initially established and subsequently stabilized throughout life by the continuous use of neural interconnections through the various types of mental processes, especially those associated with learning, thinking, and decision-making.

These scientific findings highlight the decisive importance of communication and education in the psychic development of a child. The way in which we deal with and care for a child since birth directly influences the development of the brain and consequently the personality, aptitudes, and even the future capacity to transmit, recreate, and innovate the culture to which the child belongs. The attainment of ethical concepts and practices and advanced forms of altruism, associated with a given culture, requires a continuous process of learning and integration into that culture. Altruism for outsiders to the social group to which one belongs is a relatively recent type of behaviour in the history of man, which is not innate but results from an educational and cultural process. From a wider point of view, the communication, acceptance, and practice of the basic social principles of modern society, such as equality, democracy, tolerance, and human rights, is only possible through a permanent effort of cultural learning.

2.25 Cultural Evolution Becomes Dominant

The fossil record indicates that the *Homo sapiens* brain has not undergone any significant anatomical evolution since the species appeared about 200 000 years ago. The amazing cultural development of man, from the primitive phase of the hunter–gatherer up to the appearance of agriculture and the predominantly urban civilization of today, took place with essentially the same prototype of brain. The volume of the brain ceased to grow in the sole representative of the Homo genus that survived. Apparently, over the last 200 000 years, there have been no natural selection pressures to produce a larger brain, which implies that it would not guarantee reproductive advantages. Evolution occurred on a cultural level, with regard to knowledge, technology, beliefs, faiths, practices, customs, arts, and any other skills and habits of man as a member of the society in which he lives. This evolutionary process operates through selective processes that distinguish between winners and losers, analogous to the processes of biological evolution. Culture gave *Homo sapiens* the remarkable capacity to evolve in a new and more dynamic way that does not depend on the selective propagation of the genetic code. Cultural evolution overshadowed biological evolution, but just like it, proceeded through mechanisms of natural selection that work on the cultural innovations rather than the genetic ones. The analogy is also applicable to the result of the selection that consists in the reproductive success of groups that manage to gain advantages from their cultural innovations. However, from the point of view of time and space, there are profound differences between these two independent evolutionary processes. Cultural evolution is a much more rapid and extensive process than biological evolution. The rules of transmission and inheritance are markedly different since, in cultural evolution, knowledge, ideas, technological innovations, and behavioural patterns go from one individual to another regardless of whether they are parent and offspring. The rate of cultural evolution over the last 30 000 years, and the extent and intensity of its impact on the environment, is incomparably greater than the previous evolutionary processes of natural selection in the Homo lineage over the last 2.5 million years.

The relatively rapid cultural evolution of *Homo sapiens* had profound implications, not only for *Homo sapiens* himself, but also for the other Homo species, and more generally for the environment. In this context, it is particularly interesting to enquire about the relation with *Homo neanderthalensis* and the reasons for the extinction of this species around 28 000 years ago. We do not know them and probably will never know for sure. All we can say is that something in their interaction with the environment, or with *Homo sapiens*, caused them to go extinct. It could have been competition with the more developed culture of modern humans. However, we still have scarce information about the scale and relevance of the cultural difference between the two species (Balter, 2004).

We do know that the present human populations of sub-Saharan Africa have a greater diversity of DNA than all the rest of humanity. The diversity diminishes toward the northeast region of Africa, and overall in other continents, indicating that the *Homo sapiens* ancestors stayed longer in Africa than in the other continents. This interpretation is in agreement with the fossil finds, which reveal a migration of *Homo erectus* out of Africa about 1.8 million years ago, followed much later, probably during the last interglacial period, by an analogous migration of *Homo sapiens*. The fossil analysis also shows that, when *Homo sapiens* was already in Africa and Europe, the populations in Southeast Asia evolved from *Homo erectus*. A remarkable example of the evolutionary complexity of the Homo lineage is the recent discovery in the island of Flores, Indonesia, of fossil remains of what is probably a new species, named *Homo floresiensis* (Brown, 2004). Flores Man, nicknamed the Hobbit, had a very small body, only about one meter tall, and survived until approximately 12 000 years ago. The skull is anatomically closer to *Homo erectus* than to *Homo sapiens*, with a cranial capacity of only 380 cm^3. The diminutive stature of Flores Man may have resulted from insular dwarfism, a form of speciation observed in other mammals living on small islands.

The origins of *Homo sapiens* are not reducible to a single event focused on a well-defined time horizon, but are likely to spread over the last 200 000 years, and include intermittent and geographically scattered trends towards the anatomical features of modern humans. Demography evolved through geographic dispersals and replacements, as well as local adaptations. The important point is that *Homo sapiens* became the only surviving species of the homo genus and its populations rapidly reached and settled on all continents. Cultural evolution, centred on social mechanisms, enabled the species to migrate successfully and to adapt to very diverse climates and habitats. The ecological niche became increasingly broad, leading later to profound and irreversible impacts on the environment.

The evolutionary path of the *Homo sapiens* species in the tree of life since the very beginning has been briefly described here. It is of course possible to carry out the same exercise for any other living species, such as a butterfly, a mammal, or an orchid. Each of them has an ecological niche, which in most cases is very specific and geographically limited. Let us consider the chimpanzees, the closest living evolutionary relatives to humans. They live in various types of forests in western and central Africa. The number of chimpanzees in the wild is steadily decreasing because of hunting, deforestation, and susceptibility to human infectious diseases. Their habitats are disappearing at an alarming rate due to farming and other human activities. At the beginning of the 20th century, there were between one and two million chimpanzees in Africa. Nowadays the total population is likely to be between 100 000 and 300 000. There is fossil evidence that ancestor species of chimpanzees and humans lived side by side in East Africa sharing the same habitat in the middle

Pleistocene (McBrearty, 2005), the first epoch of the Quaternary period, which started 2.6 million years ago and ended 12 000 years ago. Since then our biological and cultural evolution has completely changed the relationship between humans and nature and in particular with our closest living relatives.

2.26 Impacts of the Cultural Development and Territorial Expansion of *Homo sapiens*

About 30 000 years ago, the cultural diversity of *Homo sapiens* began to expand. New and more sophisticated tools were developed for hunting, and humans began to have a significant impact on the environment. One of the most remarkable examples is provided by the consequences of hunting on certain types of fauna in the Americas after the arrival of *Homo sapiens* on these continents. Mitochondrial DNA studies suggest that nomadic *Homo sapiens* hunters migrated more than 30 000 years ago from Siberia onto Beringia, a land bridge roughly 1600 km wide between Asia and North America, which emerged during the last glacial period and was not glaciated at its maximum. Later, about 17 000 years ago, they migrated gradually southward to North America as the glaciers started to melt and the sea level to rise.

Armed with spear-throwers and fluted stone spearheads, they hunted in logistically well-organized groups. The discovery of very similar hunting artefacts in Siberia and North America dating from that period tends to confirm this migration hypothesis (Goebel, 1999). These hunters found a vast continent with an abundant fauna, some of which was incapable of defending itself effectively against attacks by humans. A wave of species extinctions started in northern America and gradually moved all the way south to Patagonia, as man advanced southward through North, Central, and South America. The disappearance of several species of relatively large land animals, especially large mammals, known as the Pleistocene megafauna, took place up to the end of the last glacial period when the Pleistocene terminated. This timing has led some authors to consider that the cause of the extinctions was the changing climate rather than the human hunters. However, the same megafauna had already resisted previous glacial periods, and at the end of a glacial period, survival conditions improve due to a milder climate and the expansion of vegetation.

It is particularly interesting to compare the extinction of species in different continents during the Pleistocene, which affected about 200 genera (Martin, 1984). In the Americas, there are no indications of significant extinctions until about 11 000 years ago, after man had arrived from Asia and established himself across North and South America. The species that disappeared most consistently were the large herbivorous mammals, and the predators which depended on them from the ecological point of view. The mammoth, the giant ground sloth (larger than an elephant, with a length of over six meters), the glyptodont, a species related to living armadillos, some more than four meters long, several species of bear, horse, camel, and antelope, the giant bison, and the sabre toothed cat all disappeared. But the wave of extinctions tended to spare smaller animals. Many of the species that resisted had previously migrated from Asia, where they had probably adapted to the human presence, acquiring strategies of flight and adequate protection. In Europe and Asia, the extinctions in the Pleistocene started earlier, about 30 000 years ago, and lasted until the end of that period. Among many others, mammoths, woolly rhinoceroses, and mastodons disappeared. Nevertheless, the extinction of the megafauna in Eurasia was a less dramatic and slower process than in the Americas, probably because *Homo sapiens* initially had a less well developed hunting culture and technology, and the fauna would have had time to adapt to his presence. In Africa, the extinctions started even earlier, around 150 000 years ago, but the African fauna had more time to adapt to contemporary humans and thus increase its capacity for survival. Compared with Eurasia and the Americas, it was the continent where the smallest number of species became extinct. In Australia, *Homo sapiens* arrived around 60 000 years ago, and after about 30 000 years, the majority of the marsupials, including the giant kangaroo and the large flightless birds, had disappeared. It is very likely that the extinctions occurred due to the ruthless hunting activities carried out by humans (Miller, 1999).

The Pleistocene extinctions between 150 000 and 10 000 years ago have one important feature: unlike previous extinctions, they did not generate evolutionary tendencies for substitution by new species. The ecological niches that they created became empty or were occupied by already existing animals, some later domesticated by man. This absence of evolutionary substitution corresponds to an effective loss of biodiversity and constitutes one more argument in favour of humans being mainly responsible for the extinctions. The disappearance of animal species that were relatively easy to catch and had a significant nutritional value, caused by intensive hunting in the wild territories into which humans migrated, continued, although on a lesser scale, throughout the Holocene. This is the present geological epoch of the Quaternary, beginning 12 000 years ago, which corresponds to an interglacial period in the current ice age. It also important to note that human civilization dates entirely within the Holocene.

Islands such as Madagascar and New Zealand suffered a considerable number of extinctions in historical times, when they began to be colonised by man. In Madagascar, after the arrival of man around 2 000 years ago, 15 species of lemur, including three giant lemurs, flightless elephant birds, giant tortoises, and pigmy hippopotami, all disappeared. The lemurs that survived are mostly tree dwellers and nocturnal. In Madagascar, New Zealand, and Hawaii, hunting caused the extinction of many species of birds, generally large and flightless, because they had developed in an almost complete absence of predators. Introduction of livestock and other invasive species into the ecosystems by humans increased the competition for food and brought diseases, leading to further extinctions.

However, the extinctions caused by humans during the Pleistocene are nothing compared with today's rate of extinction. The extent of the human presence globally, and the increase in world population have reduced the habitats of many species, some of which are in danger of extinction. Nowadays man interferes in the evolutionary process of practically all species through their use, exploitation, artificial selection, change and destruction of habitats, and introduction of non-indigenous and invasive species. These permanent pressures, which are in many cases increasing, lead to a worrying loss of biodiversity. Before analysing the present environmental situation, and in particular the loss of biodiversity, let us return to the evolutionary history of the human culture.

2.27 The Invention of Agriculture

Hunter–gatherer bands lived by foraging wild edible plants, scavenging, and hunting, at the edge of subsistence, in great insecurity and with the permanent risk of dying from hunger. The development of agriculture, which allowed for regular production of food products and their storage in the less productive seasons and years, is one of the most remarkable advances in the history of humanity. It took place around 10 000 years ago, in the transition from the Pleistocene to the Holocene epochs. There are many conjectures about the origin and the way that humans started to domesticate plants and animals. Despite the large uncertainties, it is safe to conclude that the appearance of agricultural practices in small villages happened independently in various regions of the world: the Middle East, China, Southeast Asia, and Central and South America. The more favourable environmental conditions that resulted from the warmer interglacial climate probably encouraged the establishment of small seasonal encampments where the cultivation of plants and the domestication of animals could begin (Harlan, 1995).

About 14 000 years ago, at the end of the last glacial period, in the so-called Fertile Crescent of the Middle East, which extends from Israel through Syria and south west Turkey to the Tigris and Euphrates river basins in Iraq, the climate became warmer and wetter. The increased rainfall encouraged the resurgence of forests and grasslands, rich in biodiversity, in regions previously occupied by sterile steppes. The human populations of hunter–gatherers that inhabited those regions probably increased, but then about 12 500 years ago, a new colder and drier climatic period called Younger Dryas started, lasting around 1 000 years. The abrupt climate change took place in about a century and the return to a warmer and wetter climate in just a few decades. This type of relatively rapid climatic oscillation lasting a few thousand years, called a Dansgaard–Oeschger oscillation, happened frequently in the glacial periods of the Pleistocene. The Younger Dryas oscillation probably resulted from a perturbation in the North Atlantic oceanic currents, which caused an interruption in the Gulf Stream. During the cold spell, food productivity decreased considerably, forcing humans to search for new survival strategies. Some authors think that the advent of agriculture in the Middle East was a way to counteract the difficulties experienced by an emerging culture during the Younger Dryas oscillation (Fagan, 2004).

The first signs of emmer and einkorn wheat, barley, peas, chickpeas, bitter vetch, lentils, and flax cultivation in the Fertile Crescent appear between 12 000 and 9 000 years ago (Lev-Yadun, 2000). Earlier, before the cultivation of plants, some animals were domesticated. In the case of the dog, there are records of domestication of species close to the wolf at least as early as 14 000 years ago, probably for food, fur, hunting, and protection of the social groups. Later, about 8 000 years ago, goats, sheep, oxen, and pigs became part of the human communities in the Middle East. Beside their high nutritional value, pigs also had a special role in the villages, where they served as sanitary agents by eating food waste. Populations started to become sedentary, living in permanent settlements.

Agriculture developed independently in several other regions besides the Middle East, although probably slightly later. In China, in the Yang-tse-Kiang valley, there are records of rice cultivation as early as 10 000 years ago (Zhijun, 1998), and in Mexico, the domestication of the pumpkin began more than 9 000 years ago, and of maize 5 000 years ago (Smith, 1997). In the Andes potato cultivation began around 8 000 years ago (Pearsall, 1992).

Only a small number of plant and animal species had both nutritional value and the necessary features for domestication. Plants had to adapt to cultivation in soils prepared by man. As regards animals, training them to form herds is only possible in highly sociable species that already live in groups with a well-defined hierarchy. These conditions favoured the emergence of agriculture in regions with a large bio-diversity, where the probability of finding species suitable for domestication was greater. The mountains that bordered the Fertile Crescent to the north did indeed have a high diversity of plants and animals. It was from this variety that the first far-mers selected and tested the cultivation of various kinds of grasses, legumes, spices, and fruits, and the domestication of animals.

The discovery of a new development pattern that brings remarkable and lasting advantages is always likely to involve some negative consequences. In the case of agriculture, there are archaeological findings (Larsen, 1996, Porter, 1998) indicating that it caused a decline in the health of the populations. Diets were less diverse. New illnesses resulted from close contact with domestic animals, and disease propagated easily in densely populated areas. Life became less healthy due to the intensive and sedentary daily workload demanded by new professions and lifestyles. Despite these disadvantages, the agricultural revolution succeeded and prospered in a most remarkable way, forming the foundations of the first civilizations.

Agriculture spread gradually throughout the whole world. This expansion is ar-chaeologically well documented in Europe, where, starting from the Middle East, it reached Greece about 7 500 years ago, Italy about 6 500 years ago, France and the Iberian Peninsula about 6 000 years ago, and northern Europe less than 4 000 years ago (Cavalli-Sforza,1995). One of the main consequences of agriculture was the es-tablishment of relatively big villages, with several hundred inhabitants, occupying much larger areas than the encampments of the hunter–gatherer bands. Fixing the populations in extensive settlements allowed for better safety conditions and pro-tection from enemy groups, predators, and natural hazards. It also promoted food security, social development, and professional specialization.

All these new benefits created a greater well-being that led to significant de-mographic growth. With agriculture and pastoralism, the same lands supported many more people than before. The population increased exponentially, with a high growth constant, for about 5 000 years, and later with a less accentuated growth rate until about 2 000 years ago. According to Joel Cohen (Cohen, 1995), the total human population was around two to twenty million before the agricultural revolution and increased to values around 170 to 330 million at the beginning of the Christian era. After that period, the world population grew moderately to values of the order of 650 to 850 million in 1750, the beginning of the Industrial Revolution. Since that time, the annual growth rate has increased significantly again. Its average value from 1750 to 1950 was six to eighteen times higher than in the period from the year 1 to 1750 (Cohen, 1995). The greater capacity to sustain ever more numerous populations resulted essentially from the scientific and technological development that generated and supported the industrial revolution and the colonization of previously uninhabited or scarcely populated regions.

The initial period of plant and animal domestication that started about 10 000 thousand years ago is remarkable as regards the relation between man and the environment. Progressively, humans discovered and cultivated an increasing variety of plants, not only for food, but also for medicinal purposes, or to produce drugs and venoms. They also bred animals that could be used for food or a variety of other purposes. This process continues today in research centres throughout the world. The main emphasis is on the search for new species with nutritional and medicinal value, genetic improvement, and preservation of the biodiversity of agricultural species. The golden age of discovery at the dawn of agriculture has long since faded away. The challenge today is to preserve the biodiversity that was so useful for the emergence of civilizations and is now endangered.

2.28 The First Use of Metals

In the Neolithic, humanity accomplished the transition from hunting and gathering communities to agriculture and settlement. It also developed a more diversified and efficient stone tool technology, in which stone objects were knapped, flaked, pecked, ground, and polished, and initiated the manufacture of pottery. The essential foundations of civilization had been established, but there was still a long way to go. Archaeological remains in the Middle East, between the Black Sea, the Caspian Sea, and the Persian Gulf, some 6 000 years old, reveal a remarkable new technology that profoundly influenced the later development of civilizations — the manufacture of objects made from copper. The discovery of copper is likely to have occurred much earlier. Working with metals requires labour-intensive and well organized activities that start by mining the ore from the ground, followed by separating the metal from oxygen, sulphur, and other elements with which it is combined, in furnaces at very high temperatures. The discovery of copper probably happened while firing ceramic artefacts through accidental smelting of malachite, a copper mineral that is common in Anatolia and frequently used in items of decoration and jewellery.

It is important to recognize that the use of metals by man in the Middle East resulted primarily from the specific geological characteristics of the region. The average mass proportion of copper in the Earth's crust is only 0.0058 percent. This extremely small percentage makes copper exploitation practically impossible, and even more so its discovery. However, some geological formations around the world, and in particular in the Middle East, have high copper concentrations, of the order of two to five percent. This is the case, for example, with the deposits found on the island of Cyprus, whose name in Greek means copper. The origin of ore concentrations with a high level of metals is ultimately due to plate tectonics, and results from convergent boundaries and associated subduction movements. These slow dynamic processes generate high temperatures and pressures that melt the rock, differentiate compounds, and create deposits that are rich in otherwise uncommon elements. The Middle East is tectonically very active because it is located at the triple junction of the Eurasian, Arabian, and African plates. The relative motions of these plates over many hundreds of millions of years have endowed the region with rich mineral deposits, including copper. This abundance is a remarkable regional property, just like the large biodiversity that facilitated the domestication of plants and animals. The coincidence of both features in the Middle East did indeed occur purely by chance, but it was decisive in creating favourable conditions for the development of the first great civilizations of human history.

The copper age did not last long, however, because the metal, although ductile and malleable, is not very robust. About 5 500 years ago, perhaps by accident, metalworkers found that the addition of arsenic to copper produces arsenic bronze, a harder and more useful metal for making tools and weapons. Later arsenic was replaced by tin, this becoming the sole type of bronze in the late third millennium BC. The lower melting point of bronze compared with copper makes it easier to cast into moulds to produce a wide variety of artefacts. The bronze technology spread from its origins in the Middle East to the west, through western Europe, first to the Balkans, then along the Mediterranean coast to the Iberian Peninsula, and to the east, through Asia, to the north of India and China.

Gold and silver were in use before copper to make ritual and ornamental objects and jewellery. There are records of gold exploration in the eastern desert of Egypt and in Nubia since the first dynasty that began in 3100 BC. The oldest known geological map is a diagram of gold mines in Wadi Hammamat, recorded in the Turin papyrus, which dates from 1150 BC and was prepared for a quarrying expedition by Rameses IV (Harrell, 1992).

According to John Collis (Collis, 1984), the first iron artefacts date from about 2400 BC and were found in Alaca Hüyük, a Hittite archaeological site in Anatolia, but the technology only spread throughout the Middle East after 1500 BC. The same technology developed independently in India around 1800 BC. The relatively late discovery of iron probably resulted from the difficulty in extracting it from iron ores. As with copper, the discovery of iron technology may have been the result of fortuitous circumstances. Metalworkers found empirically that, by slowly heating iron in a charcoal fire, a process called carburisation, they could produce a much stronger metal than bronze, which we now call steel. To understand the physical and chemical processes involved and thus optimize the process of steel production required about four millennia after the discovery of iron. The Hittites of Anatolia mastered the iron technology during the late Bronze Age, and it was the superiority of iron over bronze that allowed them to establish their empire. Later, the geographic spread and the improvement of iron technology led to important developments in agriculture, social life, and warfare. This eventually contributed to the emergence of the Greek and Roman civilizations which superseded the Assyrian and Egyptian civilizations.

2.29 The First City-States

The first city-states of Ur, Uruk, Lagash, Eridu, Nippur, and others, arose in the south of Mesopotamia at the end of the fourth millennium, around 3200 BC. Their inhabitants were Sumerians, a people that came from the east, probably from the Indus valley. They lived in a mosaic of city-states, very competitive among themselves, which governed relatively large territories with common borders in an efficient and organized way (Maisels, 2005). Each city-state had its religious and secular leaders and a tutelary god, guardian of the city, to which people beseeched intervention to placate violent and unforeseen natural forces, especially droughts and floods.

The formation of the first city-states was a decisive step along the path of development towards civilizations. Note that the small rural populations lost some of their autonomy by integrating into centralized and demanding regional social organizations. However, the benefits that resulted from belonging to a city-state outweighed the disadvantages. The deciding factor was probably technological. The Sumerian city-states managed to increase the agricultural productivity of their lands through an ingenious irrigation system, using waters from the Tigris and Euphrates rivers. It was thus possible to sustain a higher and denser population, giving them security and access to a wider variety of natural resources and consumer goods. The city-state had a readily available workforce that was highly organized, diversified, and specialized, and this made it possible to plan and undertake the construction of roads, temples, palaces, and engineering structures for irrigation, as well as dealing with issues of defence and security. These infrastructures increased the capacity to survive in times of crisis and to produce valuable goods for trade in times of plenty. The urban populations, especially the elite, had access to more sophisticated and luxurious products and styles of living, which were inaccessible in the small independent villages. But besides the relative prosperity, there was tension, threats, and incessant warring between the city-states, motivated by disputes over land, rights to water usage, trade, or simply the ambition for power. We know of these stories through cuneiform inscriptions left on clay tablets, where the Sumerians tell of diplomatic triumphs, conquests, battles, trade, and more or less obscure dealings between the city-states.

The history of the decline of the Sumerian civilization in southern Mesopotamia gives us one of the first examples of the consequences that result from an overexploitation of natural resources, and from environmental degradation. Deforestation in the hills bordering the Tigris and Euphrates river basins decreased the area's water supply and led to desertification. Furthermore, intensive irrigation caused salt

to build up in the soil, and the Sumerians did not have the artificial drainage technology that could counter that process. Instead, they had to switch to more salt-tolerant
crops, replacing wheat by barley, and reduce the cultivated area.

Soil salinization and desertification of large land areas ended up weakening the
Sumerian city-states, making them vulnerable to attacks from their adversaries. The
population began to migrate. By 2350 BC, Semitic language invaders from the north
conquered the Sumerian cities and established a regional empire centred on the city
of Akkad. The Akkadian Empire, the world's first, subsumed the independent city-
states into a single state. It stretched from the Persian Gulf to the Mediterranean
and to the headwaters of the Euphrates river in present day Turkey, linking rain fed
agricultural fields in northern Mesopotamia with irrigation agriculture in the south.
After about 300 years, the Empire of Akkad suddenly collapsed, almost as fast as it
had developed, ushering in a dark age that lasted around 1000 years.

What were the reasons for this collapse? In part, they resulted from man-made
environmental degradation, especially desertification. However, it is likely that there
was another cause. Severe droughts occurred in the middle of the third millennium.
It was the longest dry spell for 10 000 years, and lasted for about 300 years. The dry
and windy environment probably triggered the collapse of the northern Akkadian
provinces, which depended on rain for agriculture (Kerr, 1998). The combination of
anthropogenic environmental changes and natural climate change was very likely
the main cause for the decline of the first human empire, a conclusion that may have
a highly symbolic connotation.

The Akkadians were succeded by the Assyrians, who, under Ashurbanipal, in
the VI century BC controlled the whole of the Fertile Crescent as well as Egypt. In
612 BC, Nabopolassar rose in revolt against the Assyrians, sacked the magnificent
capital of Nineveh, destroying its remarkable library, palaces, and works of art, and
established the Neo-Babylonian Empire with its capital in Babylonia.

2.30 The Invention of Writing

The Sumerians were also the inventors of writing, one of the most remarkable inventions of humanity. The oldest inscriptions written on clay tablets, found in the
great temple of Uruk, register the number of sacks of grain and heads of cattle (Harris, 1986). The fact that the first writings describe agricultural accounting indicates
that they probably resulted from the satisfaction of very practical needs. It was necessary to invent a durable and trustworthy process for keeping records of essential
agricultural goods, since it was impossible to achieve this by purely oral means. The
medium used consisted of small tablets of fresh clay where scribes drew the inscriptions using a stylus made of reed or wood, with a tip of triangular section. Initially
the writing was pictographic, consisting of simplified and stylised representations.
For example, the pubic triangle with a trace for the vulva represents a woman, and
a stylized cow's head represents a cow. In about 2900 BC, the curves in the picto-

grams started to disappear, probably because it was difficult to produce them in wet clay, and writing evolved towards more linear and abstract forms.

Later, another remarkable development came about when the writing signs became phonetic, representing the sounds of the spoken language. This new invention, made independently by the Sumerians and the Egyptians, still consisted in using a pictogram, but to represent the sounds or one of the sounds in the spoken word instead of the object. This reformulation paved the way for the development of all modern forms of writing. In its origins, writing was only an enduring recording procedure for agricultural accounts. However, it rapidly evolved into a system for registering the spoken word, becoming a privileged form of communication, thought expression, and literary and poetic creativity, as well as an essential medium for the diffusion of knowledge and information.

While the cuneiform characters spread through Mesopotamia from Sumer, other forms of writing arose in Egypt and China. In these civilizations, writing was considered a gift from the gods that allowed man to record the past on stone, clay, or papyrus. In Egypt, Thoth, represented in the form of two animals, the baboon and the ibis, was the god of writing and knowledge, as well as the patron of scribes. The first inscriptions using hieroglyphics, which means 'sacred carvings' in Greek, date from 3200 BC, and were in use until the end of the IV century AD. On the other hand, Chinese writing, invented later, in the second millennium BC, has survived almost unchanged until today.

About 3 000 years ago, another revolution in writing happened, with the invention of phonetic scripts based on an alphabet. Phoenician was the first major phonetic script. Again, we find that the invention of the alphabet probably resulted from the satisfaction of very practical needs. The alphabet contained only about two dozen letters, making it easier for ordinary traders to learn and use. Furthermore, the alphabet was also convenient for writing different languages. These advantages were particularly useful for the seafaring and trading activities of the Phoenicians, who spread the alphabet along the margins of the Mediterranean Sea.

2.31 The First Records of Religious Practices

The invention of writing by the Sumerians brought us the first documented record of religious activities. In these documents, we can read lists of sacrifices made to various gods in the temples of the city-states like Ur, Nippur, and many others. The gods and goddesses, represented and acting like humans, loved, hated, and fought amongst themselves, only with much more splendour and with terrible consequences. They controlled the forces and cycles of nature and of life itself, from the motions of the stars, to floods, droughts, birth, and death. If the gods were not dutifully glorified and adored in the temples through a system of rites and rituals, temple celebrations, and sacrificial festivals, drought, famine, and other calamities were inevitable. The priests and other religious figures took care of the temple activities and, through their power, created a strong rivalry with the government of the city-states.

Religions and religious practices are certainly much older than the Mesopotamian religion of 3500 BC, the first with a written record. The earliest indications of religious manifestations in both Neanderthal man and *Homo sapiens* date from around 50 000 years ago. There is clear evidence that people buried their dead with proper ceremonies, indicating that they believed in some kind of life after death. Gifts are common in many graves. In some cases, the dead bodies were in a contracted position, stained with iron oxide, and adorned with pierced shells, bracelets, and other artefacts.

The oldest known figurative sculptures, the 'Venus' figurines, dating from about 35 000 years ago (Conard, 2009), made in bone, ivory, and stone, appeared in Eurasia, from France to eastern Siberia. They are woman figures, where the parts of the body that serve sexual or childbearing functions — breasts, hips, buttocks, and sex — are enhanced, whereas the head, face, arms, and legs are usually reduced. It is very probable that these small sculptures were a primitive representation of deities controlling fertility, growth, and abundance. The most remarkable artistic creations of this period are the paintings, engravings, and sculptures inside caves. They reached a magnificent level of development later on, between 15 000 and 11 000 BC, mainly in the south of France and northern Spain, in the caves of Lascaux, Chauvet, Pech Merle, Niaux, Les Trois Frères, Montespan, and Altamira, among many others. The greater part of the cave paintings represent animals such as horses, bison, aurochs, bears, mammoths, woolly rhinoceroses, and deer, and later on reindeer, when the glaciers made their last push southward, just before the beginning of the current interglacial period. The most plausible interpretation is that the animal representations had ritual and magical functions. There are many other types of figures, for example, headless animals, human hands, some without fingers, a human figure with the head of a reindeer, bear paws, and a horse's tail, and women dancing around a phallic man. Although it is impossible to understand completely the meaning of such paintings and sculptures, their interpretation indicates that practices of a religious nature in Europe at the end of the Palaeolithic were centred mainly on the relation between man and animal, especially hunting activities, and on sexual fertility.

During the Neolithic, religion was adapted to the development of agriculture, through the new cults of fertility associated with the earth, plants, and domestic animals. Burial practices diversified, both between and within the different world agricultural communities, with increasing social stratification based on the economic wealth of the constituent families. At the end of this period, there are indications of ritual cremations that suggest a more spiritual vision of the afterlife. Various regions of the world — Europe, the Middle East, and China — have remains of temples with altars, inscriptions, ritual scenes paintings, decorated vases, sculptures, and other objects. Large megalithic monuments constructed at the end of the Neolithic are found in Europe and Asia. A remarkable example is the Stonehenge temple in England. One also finds many dolmens, or megalithic tombs, and the menhirs, huge upright standing stones, isolated or in alignments, particularly in western Europe and the

Mediterranean. The Carnac alignment of menhirs in Brittany, France, is a good example. The megalithic monuments probably had the purpose of forecasting the seasons, through procedures based on astronomical observations, apart from being the meeting place for religious cults. The menhirs are more enigmatic, but likely to have been used for rites associated with fertility and the seasonal cycles.

2.32 The Nature of Religiousness

The various religious cults developed in the first great civilizations of Mesopotamia, Egypt, and other regions of the Middle East had a profound influence on the genesis of monotheistic beliefs, which emerged later, especially Judaism and Christianity. It is important to note that the phenomenon of religion is universal, in that it has manifested itself in all world populations, even those that have not been subject to or have not adhered to the mainstream religions of our time. The religious practices of many ethnic groups that still live in isolated and primitive communities in North and South America, Oceania, Asia, and Africa are a clear example. These religions have a primordial nature and share a common belief in a spiritual world inhabited by beings with powers greater than those that humans have. The fundamental conviction is that we are not alone in the universe. Apart from objects, places, and day-to-day events, there is a spiritual power that frequently takes the form of multiple spirits, either benign or malevolent, or even gods that control certain aspects of the world, or the human way of life, such as hunting, agriculture, the production of metals, war, family relations, and fertility.

The study of the first manifestations of religiousness and the evolution and diversification of cults leads us to conclude that religion is the oldest cultural system, with specific functions of human life protection, especially as regards procreation and rearing of children. Special attention is usually paid to sexuality and food, leading to a set of rules about what we should or should not eat, and regarding the control of sexual activity and marriage. The deep and intimate links between religion, sex, and food contributed to making the family the fundamental unit in religious organizations. From a strictly socio-biological point of view, the value of religion results mainly from its efficiency in contributing to the survival and reproduction of the human species.

The emergence and development of religions since the most remote and primitive manifestations was possible because our brain and, more to the point, our genome gives us this capacity. We are naturally predisposed to religiousness. The exercise of this specific capacity and inclination of *Homo sapiens* in a wide variety of physical, geographic, and socio-cultural environments has led to a great diversity of religions. However, they share many common features as regards their narratives, symbols, beliefs, and practices, because they all have the same root in an inherent human predisposition. The practice of religions over millennia has also had consequences in other domains of human activity. Probably the most important is that it contributed to laying the foundations for and promoting the development of various civilizations. Furthermore, it developed our ability for conceptualization and abstract reasoning.

2.33 Religious Cults in the Egyptian Civilization

Shortly after the emergence of the Mesopotamian city-states and the first written records of the Sumerian religion, Upper and Lower Egypt were unified around 3100 BC, under the rule of King Namer, who was probably the legendary Menes, the first pharaoh of Egypt. For the Egyptians, life and prosperity depended essentially on the daily reappearance of the Sun and the annual flooding of the Nile valley. The powers responsible for these cycles and for those associated with human life and nature were gods, who had to be encouraged and adored through prayers and sacrifices. One of the most remarkable characteristics of the Egyptian religion is the representation of gods and goddesses in the form of animals or other beings that share animal and human features. This supernatural and symbiotic hybridisation reveals the profound association and affinity that the Egyptian civilization had with nature. The most important god, the king of the gods, was Ra the Sun god, father of humanity and protector of the pharaohs.

Other gods were associated with the Sun. One of them was Horus, symbolized by a falcon. When it flies high, its open wings projected against the blue sky are like the Sun. Another was Khepri, who steers the solar globe through the heavens, symbolized by a dung beetle that diligently draws along with its back feet the ball of dung in which it will lay its eggs. There were many other gods and goddesses in the fascinating Egyptian pantheon, some of them arranged in family groups, probably by the priests, in an effort to create a coherent and intelligible system for the people to understand. For example Ptah, the god of creation and fertility, Sekhmet the war goddess, and Imhotep the medicine god, formed a triad identified as father, mother, and son. Frequently the gods were associated with a city or a region, which possibly reflects the diversity of religious practices in prehistoric Egyptian communities, later unified politically. A good example is Amun, the local deity of the city of Thebes, transformed into a national god when Thebes became the capital in the eighteenth pharaonic dynasty. Later he became Amun-Ra, the Sun god, and the greatest expression of a transcendental deity in the Egyptian theology. In the hymn to Amun-Ra, dating from the twentieth dynasty at the end of the second millennium BC, he is the lord of truth, father of the gods, maker of men, creator of all animals, lord of things that are, creator of the staff of life, and maker of the herbage that sustains the life of cattle. He is also the maker of things below and things above, lord of the everlasting, and maker of eternity, and he judges the cause of the poor, between the poor and the mighty. This is a remarkable instance of a premonitory form of language pointing to the future development of monotheism.

There are indications that since the beginning of their civilization, the Egyptians believed in life after death. The oldest tombs contain food products, such as cereals, spices, and domestic utensils. Later they exhibit paintings representing life after death, which largely resembles life itself, but is more opulent and grandiose. The persistence through time of the spirit of the deceased was a fundamental belief. However, the body still constituted an essential support for the new life. That is the reason why it had to be preserved through elaborate methods of mummification. The pyramids are clear evidence of the immense power that the pharaohs had, and the deep conviction of the Egyptians with regard to life after death. Even though they recognized that the pharaoh was human, they simultaneously viewed him as a god, since the divine power of kingship was incarnate in him. The pharaoh was associated with several specific deities, in particular with the sun god Ra, regarded as his father in the fourth dynasty. Upon his death, he assumed the identity of Ra and returned to the company of the other god-pharaohs, while his successor took on the terrestrial duties.

The pharaoh Amenhotep IV, who reigned from 1353 to 1336 BC, moved the capital from Thebes to El-Amarna, changed his name to Akhenaton and brought about a radical reform in religion, politics, culture, and art, when he instituted the cult of a dominant god, Aton, the Sun's disk, creator and giver of all life. He or- dered the priests to destroy all the scriptures and images of most of the other gods on public display in the monuments and temples from Syria to Nubia. Aton was elevated to a supreme place in the Egyptian pantheon. During this period, artistic style became freer and more realistic, departing from the strict idealistic formalism of earlier times. Although Egyptians worshipped a few other gods besides Aton, such as Maat, the goddess of truth, order, balance, law, morality, and justice, the Amarna religious period was probably the first attempt at monotheism. However, it was short-lived. The destruction of El-Amarna, as well as all the references to Akhenaton and the god Aton closely followed the death of the pharaoh. His suc- cessor Tutankhaton changed his name to Tutankhamun, and under pressure from the priests, endorsed the return to the traditional religion. He reinstated the full pan- theon of Egyptian gods and promoted once again the cults of Osiris, Isis, Horus, and Amun-Ra, among many others. Worship of Aton became heretical. The extraordi- nary creativity of the Amarna period, especially as regards more abstract religious forms of belief, was lost. Nevertheless, the monotheistic tendency to single out a dominant god continued to survive in the cult of Amun-Ra.

It may be that the worship of Aton carried on in a more or less clandestine way, and later influenced the emergence of the monotheistic conceptions of Judaism. Freud first put forward this hypothesis (Freud, 1939), which has subsequently been considered by other scholars. One of the arguments is the close similarity, both in meaning and form, between the hymn to Aton and Psalm 104 of the Book of Psalms in the Judaic and Christian bibles (Thomas, 1965). However, this hypothetical causal connection cannot be firmly established.

2.34 The Origins of Judaism

To find the origins of Judaism we must go back to the Canaanites, a Semitic people that settled in an area now occupied by Israel and Palestine, around 3000 BC. Of the numerous deities in the Canaanite religion, the supreme representative was El, the wise, benevolent, and merciful creator of all beings, and father of humankind. Following in importance was Baal, the prince, god, and lord of earth and weather, a great fighter and an implacable enemy of Mot, the god of death. Humans were servants to the gods who held the power to prolong or shorten their lives. Apparently, they believed that divine retribution for good deeds occurred during life on earth and not after death. Canaanite cults, practiced in temples and other open-air sites, included food offerings, animal sacrifices, occasionally human sacrifices, and ritual prostitution.

At the end of the second millennium BC, the tribes from which the Jews originated gave up the nomadic way of life and established themselves in Canaan, already occupied by Canaanites, Philistines, and Amorites, and adopted the gods and rituals of the Canaanites. In the 11th century BC, due to the growing hostilities between Philistines and Amorites, the kingdom of Israel was formed as a means of defence, and Saul became its first king. Later, in the 10th century BC, Solomon was king of a powerful and well-organized state, with Jerusalem as its capital. The city had a sumptuous royal palace and a magnificent temple, where the religious rituals and sacrifices of the day took place. The religion was still polytheistic and largely influenced by the Canaanite cults.

However, Solomon's reign was short. In 926 BC, his kingdom divided into two parts, Judah in the south, with Jerusalem as its capital, and Israel in the north, a larger area, with its capital in Samaria. Around 722 BC, the Assyrians invaded Israel and established complete control over Samaria and the remainder of the kingdom. After that time, the kingdom of Judah, despite the bad relations with the new rulers of Israel, becomes a regional force, and a period of remarkable religious and spiritual renewal began. A history of the kingdom of Judah, written mainly for political and territorial reasons (Finkelstein, 2002), appeared in the VII century BC, probably during the reign of King Josiah, from 640 to 609 BC. Its main purpose was to show the unity of the people in Israel and Judah and to emphasize its distinction from the people of Canaan. This is a history of the Jewish people, which describes its heroic adventures and achievements, beginning with Abraham and his family. About the same time, the compilation of the biblical texts created the structural foundation and the instruments of a new religion. In it there is only one God, lord and creator of the world, omnipresent, transcendent, and eternal, who revealed himself and his doctrine to the chosen people through Moses on Mount Sinai, over 3 000 years ago. The doctrine is contained in the Torah, meaning the law, comprising five books — *Genesis*, *Exodus*, *Leviticus*, *Numbers*, and *Deuteronomy* — which also constitute the first five books of the Christian Bible. The final aim was to have one God, a single temple, that of King Solomon in the capital, Jerusalem, of a single kingdom that would result from the unification of the kingdoms of Judah and Israel, and a single new law, that of the Torah. This was the beginning of the idea of monotheism.

The Torah chronicles the life of the Jews from the creation of the world until the death of Moses and is one of the most important and revolutionary texts in the history of religions. According to the *Deuteronomy*, its main objective is to transform Israel into a sacred nation. In *Leviticus*, the ethical principle of reciprocity or golden rule, which plays a major role in Christianity, is clearly defined and defended for the first time in the Western world. One can read in *Leviticus* 19:18, "You shall not take vengeance or bear a grudge against your countrymen. Love your fellow as yourself", and in *Leviticus* 19:34, "The convert that resides with you shall be to you as one of your citizens; you shall love him as yourself, for you were strangers in the land of Egypt". The heroic epic of Moses in the Torah is a superb literary, religious, and political narrative of the Jews that apparently has no grounding in archaeological findings (Finkelstein, 2002). Its political function would have been to promote a reconciliation of the two Jewish kingdoms and impose their unification on the Assyrian, Egyptian, and Mesopotamian regional empires. However, this grand design failed. On 9 April 587 BC, according to the Jewish calendar, the armies of Nebuchadnezzar invaded Jerusalem, destroying the walls of the city, the temple, and the royal palace, burning hundreds of houses and deporting thousands of Jews to Babylonia. This was a most catastrophic event for the Jewish people, initiating the Diaspora (from the Greek for dispersion) which has become their fate throughout history. Notwithstanding, the foundations of the Judeo-Christian ethical and religious principles had been already firmly established.

2.35 The Origins of Christianity

Christianity was born in Palestine at a time when the Roman Empire was at its peak and dominated the whole Mediterranean. From the point of view of religion, the Roman Empire was pluralist. People professed the Greek and Roman religions and the mystery cults, some with their origin in the Middle East. One important common trait in many of these cults consisted in the faith in a God Saviour who dies but resurrects. The believers achieved immortality by sharing the symbolic death and resurrection of their God Saviour, who could be Mithras, Osiris, Attis, Orpheus, or Dionysius.

In all the diverse forms of Judaism practised in Palestine, there was a recurrent preoccupation with avoiding the influence of the Greek and Roman religions. Many Jews opposed Roman dominion, leading uprisings against the instituted power. This resistance led to a final great rebellion that was violently crushed. The legions of Titus entered Jerusalem on 4 August 70 AD and burned the second temple built by Herod. Many tens of thousands of Jews dispersed throughout the Roman Empire in a new Diaspora.

During the first exile in Babylonia, probably due to the influence of Zoroastrianism, the idea was born that God would send a Messiah, a man with extraordinary wisdom and qualities for doing good, with the power to release the Jewish people from oppression, and whose aim would be to destroy the hostile forces. This hope was so strong after the Roman occupation of Palestine in 63 BC that there are several records of self-proclaimed Messiahs with innumerable followers.

Jesus was probably born in 4 BC in the remote town of Nazareth, Galilee, and lived a little more than 30 years. He did not provide a written testimony, leaving the Gospels, the texts of the disciple Paul, and some references by classical writers of the 1st and 2nd centuries AD as the principal sources of information. Not much is known about his life up to the age of 30, during which he probably received a thorough religious education, based on the Old Testament, in the local synagogue. It seems likely that he had ties with or was influenced by the Essenes, a small isolated Jewish community in Qumran on the Dead Sea, which adopted an austere monastic lifestyle, opposed to violence, and waited patiently for the coming of the Messiah who would free the Jews from foreign oppression.

The life of Jesus rose from obscurity when he was about 30 years old. He was one of the people baptised by the stern Jewish ascetic John the Baptist, who encouraged the Jews to return to God, announced the coming of the Messiah, and proclaimed, "Repent for the kingdom of heaven is coming!" After this meeting, Jesus started to preach, attracting an ever-increasing number of followers. According to the Gospels, his activities during the next three years were preaching, teaching, and healing the sick. He and his group of disciples had no fixed residence, moving from place to place on their mission, and depending on the support and hospitality offered to them. His preaching and parables attracted large crowds. He related very easily with all kinds of people, including those from the ostracized classes, breaking the social barriers and prejudices of the time. He made enemies above all with the Scribes and Jewish leaders by criticizing and condemning hypocritical religious practices, in particular some interpretations and types of behaviour related to the Torah. He disappointed some who recognized him as the Messiah for not wanting to lead the fight against Rome. Jesus' mission lasted only two years before he was arrested by the Roman Governor, as John the Baptist had been before him, and condemned, after a summary and obscure trial, to die barbarously on a cross like a common criminal, evil-doer, or insurgent against the occupation.

The events that followed his death were of enormous importance for the development and expansion of the Christian faith. According to the Gospels, Jesus resurrected three days after the crucifixion and was seen on several occasions by his followers who recognised him as the Son of God. It is not clear to what extent Jesus considered himself as the incarnation of God, or whether the divine quality was attributed to him by his followers after his death. The fact is that Christianity started to spread across the Roman and Persian Empires through the work of many missionaries. One of the most important was the apostle Paul, a Roman citizen of Jewish origin from the town of Tarsus who, after persecuting the Christians, converted to Christianity when the resurrected Jesus appeared to him. He developed many of the fundamental concepts of Christian theology, based on the divinity of Christ who assumes the human condition and suffers death on the cross to redeem the sins and the death of men. Christianity absorbed or adapted many contemporary cultural and religious influences, especially Greek, Roman, and Jewish. The scriptures of Judaism were recognized and accepted, though with different interpretations. The Christian way to express ideas, to develop theological arguments, and to proselytise was largely based on the logic and conceptual philosophy of the Greeks.

On the other hand, Christianity created and developed a model of the Church characterized by a well-organized structure with a system of governance that was independent from the state and autonomous. Furthermore, it had a strong authority and a centralized hierarchy, which was deeply influenced by the Roman political culture. The first centuries were crucial for the development of Christianity. Initially, it suffered violent persecutions, but it eventually emerged as the only official Roman religion at the end of the 4th century, through the edict of emperor Theodosius I (379–395) issued around 395. In the next century, in 451, the first schism led to the separation of the Oriental Orthodox church comprising the Christian groups of Persia, Armenia, Syria, Ethiopia, Egypt, and India. Nevertheless, despite the threat of Islam from the 7th century onwards and new schisms in 1054 and 1517, the Christian religion extended to all continents, and is today the dominant religion in Europe, North and South America, Oceania, and some parts of Asia and Africa.

2.36 The Origins of Islam

Islam arose in Arabia in the 6th century in the city of Mecca, whose merchants controlled the profitable trade between the Indian Ocean and the Mediterranean by camel trains along the western coast. Mecca was also an important religious centre where the Arab tribes practised polytheistic cults and venerated the deity Allah in the temple of Kaaba, a cubic building, founded, according to tradition, by Abraham and his son Ishmael. The building contains a sacred stone in the west corner, reddish black in colour, about 30 cm long, identified as a meteorite. In the first civilizations, meteorites caused great bewilderment. People believed that the gods sent them, which made them sacred. Another example is a meteorite supposedly sent by the goddess Artemis or Diana, kept in her famous temple in Ephesus, later destroyed by Herostratus in 356 BC in an act of arson, to obtain fame at any cost.

Not much is known about the early life of Muhammad ibn Abdullah, the founder of the Islamic religion. He was probably born around the year 570 into the Quraysh tribe, which was dominant in Mecca at the time. When he lost both parents at the age of six, his grandfather, the custodian of the Kaaba, and later his uncle Abu Talib, took care of him. He was a shepherd in the region around Mecca, travelled with his uncle to Syria, and participated in the caravan trading until he entered the service of a rich widow called Khadijah bint Khuwaylid, whom he later married.

In his travels, Muhammad had direct daily contact with the Jewish and Christian communities and probably felt dissatisfied from the moral and religious point of view with the behaviour and polytheism of his fellow citizens. The contrast was also relevant in the political and cultural spheres, with Arabia organized in tribes that did not have significant cultural usages for the local idiom, while the neighbouring nations had organized empires, kingdoms, and principalities, which benefited from established languages, a relatively common practice of writing, and well-developed literatures.

When he was about 40 years old, he experienced revelations and received messages from God that reached him through an intermediary identified as the angel Gabriel, later recorded in the Quran. Initially, Muhammad was frightened and suspicious of being the victim of deceiving spirits, but Khadijah encouraged him to accept the revelations. It was at this time that he began to believe firmly that he was the messenger or prophet of the only God, Allah, the Merciful, Creator of the Universe, and Supreme Judge of Men. He rapidly gained followers, and they met frequently to pray and celebrate the new faith. However, the merchants of Mecca and especially the leaders of the Quraysh tribe considered that the monotheistic doctrine of the Quran, which defended social justice and the rights of the most underprivileged and poor, could become dangerous and harm them by subverting the social and economic order of Mecca.

Muhammad was relatively safe from this hostility due to the powerful positions his wife and uncle held in the community. However, they both died in the year 619. Muhammad's position became unsustainable and he was forced to flee to Yathrib, a city in an oasis about 320 km north of Mecca. In fact, he was invited by envoys of conflicting Yathrib clans as a neutral outsider to serve as chief arbitrator, mainly between the Arab and Jewish communities. This migration, called Hijra, took place in 622 and marks the beginning of the Islamic calendar. Muhammad's mission in Yathrib was a great success: he solved the conflicts between tribes, converted the inhabitants to his faith, and organized the social, political, and educational life by promulgating laws in various domains, for example, about fasting, marriage, divorce, inheritance, and the treatment of slaves and prisoners of war. He married several women, possibly to secure his political alliances, and within 10 years had managed to unify, under his command, the political, civil, and religious government of the Yathrib community, transforming it into the prototype of the Islamic state. Yathrib was renamed Medina, an abbreviation of Madinat un-Nabi, which means 'the city of the prophet', in honour of Muhammad. The establishment in Medina of an Islamic community with political and religious unity constituted a remarkable and revolutionary feat, and created a robust model that was well-suited to the contemporary society in Arabia.

For Muslims, religion is inseparable from the totality of the human experience, including the social, economic, and political fields. All life is sacred and develops in the realm of the Ummah, or the community of the faithful, governed by the dictates of the Sharia, or the divine law of the Quran. In its original form, Islamism is not compatible with the concept of religion understood as a cult separated from political and social organizations and practices, of the kind that prevails nowadays in Christianity.

The religious orientations of Muhammad were deeply influenced by Judaism and Christianity, both in Mecca and in Medina. In the Quran, Moses and Jesus share with Muhammad the supreme recognition of having been messenger prophets or rasuls. Allah sent Muhammad to a people that had not yet had its rasul. He adapted and incorporated into his doctrine many elements of the scriptures, the rituals, and the traditions of both preceding monotheistic religions (Katsh, 1954; Bell, 1926). The attempt to embrace different religious cultures resulted in part from the hope of converting to Islam, not only the Arab polytheists, but also the Jewish and Christian communities in and around Medina.

Most of the Jews and Christians that did not convert to Islam were attacked and fled. Others were forced to pay a substantial tribute to allow them freedom of worship. These new successes made Muhammad more aggressive with all those who did not embrace his new faith. He overpowered the disbelieving Quraysh merchants on the Mecca trade route. But he treated the defeated generously, and this favoured conversions to his new doctrine. His next main objective was to conquer Mecca.

Finally, in 630, after several violent engagements over several years, Muhammad, leading a force of around 10 000 men, prepared to attack Mecca. But his principal rival, Abu Sufyam, surrendered and handed over the city with a minimum of casualties. One of his first actions was to go round the Kaaba seven times and to order the destruction of all images of polytheistic idols found there, including paintings of Abraham, the father of Ishmael who, according to the Quran, is the ancestor of the Arab people. The Kaaba became the house of Allah, the focal point of the daily ritual prayers of all Muslims in the world, and the main objective of the annual pilgrimage to Mecca, the Hajj, which is the duty of every Muslim and is currently the largest pilgrimage in the world.

In the following years, the religion led by Muhammad spread across all Arabia, and its tribes unified in the sense that they all belonged to the Ummah and obeyed the Sharia. The sudden death of Muhammad in 632 created a grave leadership crisis. The Sunnis or traditionalists that make up by far the largest part of the world's Muslims, considered that Muhammad had not left a successor, and looked among their own ranks, choosing Abu Bakr as caliph, which means 'successor'. On the other hand, the Shiites or partisans considered that Muhammad had designated his first cousin and son-in-law Ali ibn Abi Talib as his successor. Against their will, the Shiites accepted Abu Bakr and two more caliphs that succeeded him. However, there was considerable confrontation, and the schism between Sunnis and Shiites came in 661 after the assassination of Ali, following his designation as caliph. Nowadays, the tension between Shiites and Sunnis, who make up 85–90% of all Muslims, is still ongoing, with various degrees of violence.

2.37 The Expansion and the Cultural Golden Age of Islam

Islam underwent a remarkable and amazingly rapid expansion. By the end of the Umayyad Dynasty (661–750), only 118 years after the death of Muhammad, Islam had spread to the Atlantic through North Africa, the Iberian Peninsula, and the south of France, and to the borders of China across the whole of the Middle East and India. In the 10th and 11th centuries, a cultural golden age emerged when Islam reached the height of its wealth. Arts, literature, philosophy, mathematics, physical and natural sciences, and medicine all flourished, contributing significantly to the foundations of modern science in the 16th century.

An important impulse to encourage enquiry about life and the world is supposed to have come from a dream of the Abbasid caliph of Baghdad, Al-Mamum (813–833), in which Aristotle appeared and convinced him that there was no contradiction between religion and reason. Free from any misgivings, the caliph ordered the construction in Baghdad of the famous House of Wisdom, an academy of higher education with a vast library, where books from Greek, Syrian, Persian, and Hindu authors were gathered and translated into Arabic.

The search for knowledge by the Muslims was rooted in the conviction, shared by the Greeks before them, that beyond the apparent chaos of the world around them, there is a fundamental order governed by universal laws accessible to human reason. Physicians made great progress, building upon the experience of the Persians, and several remarkable discoveries date to this time, such as the pulmonary blood circulation identified by Ibn al-Nafis. They also knew how to carry out small painless surgical operations using anaesthetics. The Canon of Medicine written by the Persian physician Ibn Sina, born in Afshana near Bukhara around 980, and known in the West as Avicenna, was the most widely used medical treatise, in both the East and the West, right up to the 17th century.

Benefiting from the translation of Greek and Hindu works to Arabic, outstanding progress was made in mathematics, particularly in algebra, trigonometry, and geometry. One of the first mathematicians in the Baghdad House of Wisdom was Al-Khwarizmi. From the Latinised form of his name, Algoritmi, came the word 'algorithm'. He wrote several important books, one of which, with the abbreviated name of Kitab al-Jabr wa-l-Muqabala, is the foundational text of modern algebra and provides an exhaustive account of how to solve polynomial equations up to the second degree. The word 'al-Jabr' in the title of the book, meaning 'completion', is the origin of the word 'algebra'.

Zero and the other nine integers, said to be Arabic, and the decimal system, came from India and were later adopted and widely used in Islam. The concept of zero as a number was first developed in India, where the Sanskrit word for zero or void is 'sunya', translated by the Muslims as 'sifr'. Leonardo de Pisa (1175–1240), the scholar who started the dissemination in Christian Europe of the Arabic integers and the algebraic calculus of the Muslim mathematicians, Latinised the word 'sifr' as 'zephirus', and this was later transformed into 'zero' (Seife, 2000).

Using the translation of Ptolemy's Almagest into Arabic, and once again aided by Persian and Indian texts, Muslim astronomers significantly developed astronomical observation. One of their most outstanding achievements was the elaboration of astronomical books and tables that allowed the calculation of the future positions of the Sun, Moon, planets, and stars. These tables, called zijes, became an essential tool for all astronomers for several centuries. From among the more than two hundred known zijes, some of the most remarkable were written by Abu Mashar al-Balkhi, Al-Khwarizmi, and Al-Battani, and these all had a considerable influence in Medieval Europe. The translation into Latin of Al-Battani's zij, known as *De Motu Stellarum*, greatly influenced Tycho Brahe and Johannes Kepler. Apart from satisfying a form of curiosity that we would characterise today as scientific, the observation of the movements of heavenly bodies and the elaboration of tables was fundamental to the practice of astrology. Remarkable astronomical observatories were constructed, housing instruments of large dimensions, such as those in Baghdad and Maragheh. The latter, built in 1259 near Tabriz in the northwest of what is now Iran, was in many ways the prototype of a large research institute, with a library and around ten resident astronomers, among them visiting scholars from various regions, as far afield as China.

There were significant contributions in physics as well, especially in the field of optics, with the explanation of the refraction of light, and the rainbow, which results from the refraction of sunlight in raindrops. In 984, the mathematician and physicist Ibn Sahl, who worked in the Abbasid court of Baghdad, wrote a remarkable book called *On Burning Mirrors and Lenses*, where he describes various types of mirrors and lenses that could cause ignition by concentrating the Sun's rays. There he derived for the first time the law for the refraction of light (Rashed, 1990), known to us as Snell's Law, after the Dutch mathematician and astronomer Willebrord Snellius (1580–1626), who rediscovered it in 1621.

The systematic search for knowledge through philosophy and science eventually created a reaction in the Islamic world. The suspicion grew that the new knowledge obtained through reasoning and analysis was overshadowing and undermining the central objective of reaching the union with God. This tendency is clear to see in the writings of Al-Ghazali (1058–1111), one of the most important Islamic theologians, who greatly influenced law, theology, and philosophy. In his book *The Incoherence of the Philosophers*, he underlined the inadequacy of reason outside a restrictive sphere, beyond which the answers should come from various forms of mysticism, and defended the overriding need to find the way to the revelation of God. According to his writings the revelatory truth is by far the most important objective but it cannot be obtained by reason. He was a pioneer as regards the methodology of scepticism, and he contributed to shifting Islamic philosophy away from Greek influence. Al-Ghazali practiced Sufism, a mystic and ascetic movement of reaction to the more legalistic and ritualistic ways of Islam. Sufists consider that the supreme truth is unreachable through knowledge based on reasoning. To reach it requires an intense personal experience of faith, which will show the way to the union with God. The emphasis on frenzies that may lead to new revelations, and on the limits of reason, contributed to weakening the leadership of philosophy and science that had prevailed in the Islamic golden age. Nevertheless, Sufism was a decisive factor in the worldwide spread of Islam, because the Sufis could easily make compromises with local customs and beliefs. By breaking down into a large number of orders at the regional level, it became easier to accept and incorporate the beliefs and traditions of the converted areas.

The caliphate that led to the great dynasties of the Umayyad and Abbasid, culminating in the Ottoman Empire, lasted until the 20th century. It was abolished in 1924 when Turkey became a secular state. Meanwhile, the European colonial expansion into Islamic nations and the following creation of states, often in a hurried and arbitrary way, was not well perceived by the Ummah, and injured its pride in the Islamic religion and culture. The frequent failure of Western development models and the strong impact and influence of Western culture in the new independent states tended to provoke a deep cultural, political, and psychological disorientation. All this has exacerbated the nostalgia for the past glories of Islam and tended to radicalize the way of thinking and the behaviour in some sectors of contemporary Muslim societies.

These tendencies may endanger or even make it impracticable to export the values of Western political modernity, in particular as regards the practice of secularism, democracy, and freedom of expression. In contemporary Muslim societies, there is an intense debate between the moderates, and the conservatives and fundamentalists. The first group accepts the coexistence of the Islamic faith with the liberal influences of Western culture, while the others repudiate them and seek to return to the more austere practices of the Islamic religion, and to forms of governance based on the Sharia. In spite of this potentially divergent scenario, there is a strong development of science and technology throughout the Muslim countries, which remains as an unchallenged link with Western culture. At the same time, Islam, which now has about 1 100 million followers worldwide, succeeds in keeping alive its ability to proselytise, and is clearly continuing to expand.

2.38 Hinduism

Hinduism is one of the oldest expressions of spirituality and religiosity, which has resisted innumerable pressures and challenges since its appearance around 3 500 years ago (Mahadevan, 1956). Its origins are probably related to the Indo-European nomads from the great plains north of the Caspian Sea and Central Asia, who invaded India in successive waves, through the mountain passes of the Himalayas, Iran, and the Indus river valley. They called themselves Aryans, meaning noblemen or landlords, and established a Vedic religion based on the Vedas, texts composed between 1500 and 800 BC, considered to be sacred, eternal, and revealed through a kind of intuition to the rsis, archetypal sages and visionary men with supernatural powers.

In contrast to other religions, Hinduism does not have a privileged founder. The first Vedic texts were followed by the epic poems of Ramayana and Mahabharata, which recount the conflicts between the Aryan tribes and the peoples of the Indus valley, and the victories of the Aryans that allowed them to impose and consolidate their power.

One of the more specific characteristics of Hinduism is the system of social stratification called varna, meaning colour in sanskrit, whose origins, although obscure, are related to the period of conquest and assimilation of the indigenous peoples. Initially, there were four social orders called castes or jati in Hindu, meaning birth. In the central place of power were the Brahmins, priests, spiritual and intellectual leaders of society. The Kshatriyas were rulers and warriors, keepers of sovereignty and order, responsible for the administration of the public goods. Third were the Vaishyas, or producers, who as farmers, merchants, and traders guaranteed the economic well-being of the society, and last of all came the Shudras or labourers and serfs. Later on, these social orders subdivided into a complex caste system, which, although abolished in present day India, is still noticeable in the mentality and behaviour of the people. The caste of a newborn is the same as that of his parents, and determined by his karma, the law of consequence with regard to action that is the driving force behind the cycle of reincarnation and rebirth. Under the law of karma, everything one does in life, whether in thought, words, or deeds, will determine one's destiny in future existences. Besides the karma, there is the atman, a true self, an innermost and unseen force, present and operative in every form of life, not just in humans, which persists in the transmigrations of life and is independent of external appearances.

While the atman is enrapt in earthly desires, it continues to samsara, flowing continuously or transmigrating in the form of humans, animals, or plants, in situations and environments as diverse as paradise and hell, but always under the relentless command of the law of karma. The reincarnation in human form is a rare opportunity to advance towards deliverance from the cycles of death and rebirth or, in other words, to attain moksha, the release from samsara and the ultimate goal of Hinduism. There are three ways to moksha: through actions and works that follow the dharma, which represents one's righteous duties and also the order and customs that make life and the universe possible; through knowledge and philosophy; or through devotion to the gods. Each action has a consequence that manifests itself in the form of future reincarnations: the morally good actions have positive consequences and the morally unacceptable actions have negative consequences. Suffering, pain, and misfortune in a current life are not the fault of others or of the gods, but the consequence of a bad karma accumulated from previous incarnations.

From among the thousands of Hindu deities, three especially important ones — Brahma, Vishnu, and Shiva — have millions of followers and represent the fundamental trilogy of the interactive forms of divine manifestation. Brahma is the god of creation, the omnipresent spirit, absolute and impersonal, the origin and support of the universe. The Hindu mythology represents Brahma with four heads and four arms, galloping on a swan, and his consort Saraswati is the goddess of knowledge, music, and the arts. Vishnu, the pervading essence of all beings or the one taking different forms, is the god of love and benevolence, who maintains the order of the cosmos. According to tradition, he has nine avatars, among them Rama, Krishna, and Gautama Buddha, and his tenth, called Kalki, will lead to the end of our current era, known as Kali Yuga and considered to be an age of darkness. In his representations, he often rests on Ananta, the immortal and infinite snake. His wife Lakshmi is the goddess of good fortune and wealth. Finally, Shiva, the most popular, is the god of generation and destruction, worshipped through the power of the lingam, a symbol of generative energy, usually represented as a phallus. He is a dynamic god who destroys order and opposes Vishnu. The representations show him in deep meditation or dancing the tandava, a cosmic dance that is the source of the creation and destruction cycles. Shiva's mount is a hump-backed bull and his wife is Parvati, daughter of Himavan, lord of the mountains and personification of the Himalayas, and mother of Ganesha, the god with an elephant's head.

2.39 Siddartha Gautama and Buddhism

The 5th and 6th centuries BC were times of doubt and dissension in the Vedic religion of northern India. The disenchantment with the caste system, the sacrifices, and the rituals drove many to give up their social position in order to follow an ascetic and wandering life, away from the crowds. One of these dissidents was Siddhartha

Gautama, later called Buddha, which is a title meaning the enlightened or awakened one.

All knowledge about his life and teachings was transmitted orally over four centuries, until it was finally written down in the first century BC in the *Pali Canon*. The Canon texts, often called *Tripitaka*, meaning three baskets, were kept in three receptacles: the basket of discourses, the basket of disciplines, and the basket of miscellaneous texts, mainly consisting of philosophical and psychological analyses (Armstrong, 2001).

According to these sources, Siddartha was born in Lumbini, in the foothills of the Himalayas, in today's Nepal, in the kingdom of the Shakyas. He was the son of Suddhodana, chieftain of the Shakyas, who belonged to the Hindu Kshatriya caste. Probably born at the end of the sixth century BC, he had a long life of about 80 years, dying probably in 483 BC. He was well educated, brought up in great comfort and luxury, and lived with his wife and young son in his father's home in Kapilavatthu. However, when he was 29 years old, he decided to leave home, to become one of the many ascetic travellers that wandered through the magnificent forests on the banks of the River Ganges. Along with five companions, he submitted himself to a very austere regime of extreme self-mortification that left him near death, but concluded that asceticism was not the effective path to absolute truth, freedom, and deliverance.

Finally, in Bodh Gaya, an ancient sacred place in Bihar, he withdrew in profound meditation under a bo tree (*Ficus religiosa*), reaching deeper and deeper states of consciousness, and thereby revisiting his former existences until he could finally comprehend the cause of the cycle of rebirths. He thus understood the origin of human suffering, attained enlightenment, and became the supreme Buddha.

In his first sermon in Sarnath, near Varanasi, he proclaimed the spiritual path of the Middle Way, equidistant from asceticism and passion. Buddha reduced the various Hindu concepts associated with reincarnation to a single one called dharma or cosmic law, summarized in the *Four Noble Truths* which constitute the essence of Buddhism. The first is that all composite things are in a state of dukkha, meaning transience, and all that arises from the personal experience of transience, such as dissatisfaction, frustration, unhappiness, suffering, pain, and misery. All events and phases of a lifetime, from birth to death, involve dukkha. In the Universe, everything is transitory and in a state of continuous change. Consequently, since a permanent reality cannot exist inside or outside us, Buddha denied the existence of the atman of Hinduism.

The second truth explains that the cause of dissatisfaction associated with dukkha is a consequence of tanha, meaning literally thirst and figuratively all forms of desire and craving, such as craving for material possessions, power, personal attachments and relationships, intellectual gratification, fame, and even the desire for life or for death. According to Buddhism, tanha works on the misunderstanding that desires, whatever their nature, can be fully and permanently satisfied. One's tanha leads not only to one's own dukkha, most likely in the form of suffering, but also to the suffering of others.

The third of the noble truths indicates that the end of suffering comes from the quenching of tanha, which leads to its destruction. Dissatisfaction ends when one reaches nirvana, a state of liberation, disengagement, and full peace in which the fires of desire, hate, and ignorance, interpreted as unfamiliarity with the Buddhist teachings, are fully extinguished.

The last of the noble truths reveals that the way to freedom from dukkha is the Middle Way, which requires eight categories of liberating procedures called Astangika Marga, meaning the Eightfold Path. In its essence, it is a pathway of mental and physical detachment, self-denial, and renunciation. Those who follow the Eightfold Path will eventually reach nirvana, becoming free from suffering and from the cycle of rebirth.

Buddha refused to confirm the existence of a supreme god, challenged the immortality of the Hindu gods and goddesses, and insisted that his spiritual way of life, including the experience of the liberating state of nirvana, is within reach of every man and woman. Furthermore, he rejected the Hindu caste system and criticised its theological foundation on the basis that it is incompatible with the non-existence of the atman or true self which, according to Hinduism, transmigrates from life to life.

For more than forty years, Buddha wandered through the vast Ganges plain, surrounded by his followers and preaching his doctrine. He died in Kusinara, a small village in the jungle, on his way to Lumbini where he was born. According to the *Pali Canon*, his last words to the monks that surrounded him were: "All composite things pass away. Strive for your own liberation with diligence."

Initially, Buddhism remained confined to its region of origin and did not spread significantly. Ashoka, one of India's greatest emperors of the Maurya Dynasty, who ruled from 269 to 232 BC, tormented by the violence and brutality of his warring activities, converted to Buddhism in 250 BC and promoted its diffusion from Afghanistan to Ceylon.

From the 12th century onwards, Buddhism suffered a decline in India but expanded outside its borders, in three distinct geographical directions and in a large multiplicity of schools: to the north in Tibet, to the east in China, Korea, and Japan, and to the southeast in Burma, Laos, and Thailand. In the 20th century, due to migration from Asia and local interest, it has spread throughout Europe and the United States. In India, a movement reappeared, mainly in the lowest castes, generated by its defence of equality.

There were many more religious movements in India. One of them Jainism, had its origin, like Buddhism, in the period of reaction to Hinduism in the 6th century BC. Mahavira, the founder, and a contemporary of Buddha, defended a form of radical asceticism that led him to roam the towns and plains of India naked proclaiming the principle of 'ahimsa', or non-violence, in thought or deed against any living creature, human, or animal. Jains are strict vegetarians. They should not molest, injure, or kill, no matter how insignificant a living creature may be, an extreme way of rejecting the Hindu sacrifices. Jainism, despite the relatively small number of current believers, has had a great influence on religious, social, political, and economic life in India.

Sikhism is another religion from the subcontinent, a contemporary of Protestantism, created in the 14th century by a guru called Nanak, who tried to synthesize Islamism and Hinduism, with a concept of god close to that of Islam, and polytheistic form of devotion like the Hindus. Nanak also rejected the caste system and other discriminatory Hindu practices.

2.40 The Great Religious and Philosophical Systems of China

The classical period in China (722 BC–221 AD) was characterized by the gradual decline of the feudal system, the impoverishment of the old noble families, the rise of a middle class of farmers and merchants, and violent civil disorders that culminated in the reunification of China in 221 by the Emperor Shih Huang Ti. This period of conflict and change stimulated the search for new political, social, and philosophical solutions. It was in this context that Confucianism and Taoism emerged.

Confucius, the Latinised version of K'ung Fu tzu or Master K'ung, was born in 551 BC and died at the age of 72. He was a remarkable thinker, teacher, and philosopher, whose teachings sought to develop the virtues needed to maintain order and progress in the nation and harmony in the family. Apparently, and according to tradition, he persistently sought a career in politics, but without success. His concern was mainly with ethics and codes of moral, social, and political conduct rather than theological questions. He is credited with saying that it is more productive to establish a plan for life on Earth than to speculate about life after death.

The main virtues defended by Confucianism are: Rén or benevolence, charity, and love of others as a manifestation of humanity; Yì or good conduct, honesty, morality, and duty to one's neighbour; Shù or reciprocity, altruism, and consideration for others; Xin or faithfulness and integrity; and Li or the virtue of correct behaviour, good manners, politeness, ceremony, and worship. This set of virtues supports an ideal lifestyle that had the purpose of defending the imperial power structure and the stability of family life, privileging civil servants above the common people.

Confucianism is primarily an ethical and philosophical system and not a religious system in the sense that it left aside the questions of divinity and deities. Although Confucius did not talk about the gods, Confucianism became a religion with temples in towns across the whole of China, where people worshipped and still worship Confucius as the patron of students and scholars. His ethical principles are outstanding and were revolutionary at the time they were written.

Confucius enunciated the ethical principle of reciprocity or golden rule probably at about the same time as it was first written in the Leviticus of the Torah. Ethical reciprocity is very clearly defined in the version that reached us: "What one does not wish for oneself, one ought not to do to anyone else; what one recognizes as desirable for oneself, one ought to be willing to grant to others" (Analects XV: 23). The golden rule represents a very important ethical step forward in civilization, and is present in the teachings of the main religions. We may only conjecture as to how different the world and the relations between people would be if we did not have that rule, or at the other extreme, if it was actually followed and applied by everyone.

Confucius did not leave any writings, but the canonical texts of Confucianism written by his disciples well after his death greatly influenced China for more than two millennia until the present time. Those who had studied and followed the canons were the elite of the state. The high functionaries that supported the functions of the Chinese state and its imperial government were recruited from the elite. One of the most celebrated texts is the *Lun Yu*, or Analects of Confucius, which record his ideas and discussions he had with his disciples. It has a simple and attractive style, avoids dogmatism, and transmits a message of profound humanity, which was highly relevant in the past, as it is today. It is very probably one of the most influential books in the history of mankind, studied and appreciated in China as well as in other civilizations of East Asia and throughout the world.

During the Cultural Revolution in the 1960s Confucianism came under attack. However, there is an striking revival today. Many people pay homage to the teachings of Confucius and seek to embrace its principles, such as ruling by morality, filial piety, and discipline. Since 2007, the Chinese Government has sponsored the worship of Confucius on his birthday and supports the reintroduction of Confucianism into public life. This new movement is probably a way of filling the present ideological vacuum in the country. It also serves to counteract the loss of cultural identity resulting from an increasing globalization and to promote social stability, as well as a more cohesive and harmonious society.

Taoism is a religious and philosophical system which, compared to Confucianism, has a stronger religious component. Very little is known about its origins, but the founding text, called *Tao-te Ching*, is attributed to Lao Tzu, a mystic philosopher and teacher who was a contemporary of Confucius. Tao, an expression that can be translated as 'the way', is at the same time a code of conduct and a supernatural cosmic force, present in all phenomena, whose powers Taoism tries to capture and channel to the benefit of human lives. It proposes a pathway of strength, virtue, and physical and moral well being that guarantees a long and harmonious life. According to Taoism, the universe and its dynamic diversity result from the fluctuations and interactions between two opposing energies: yin, which represents all that is feminine, receptive, and soft, and also the Moon, the south, water, clouds, and the even numbers; and the yang, which represents all that is masculine, active, and hard, and also the Sun, the north, redness, and the odd numbers. The interaction between them is believed to have generated the five elements or phases, called Wu-Hsing — wood, fire, earth, metal, and water — which should not be understood literally as physical substances, but rather as metaphysical forces associated with the nature of these substances. These, in turn, produce the countless myriads of mutable forms in the universe, and also their dynamism and history. The three main virtues proclaimed by Taoism, known as the three jewels, are compassion, moderation and humility. A central concept in Taoism called Wu Wey, literally meaning without action, reveals the possibility of an alignment with Tao by placing our will in harmony with the Universe.

Taoism assimilated a large part of previous religious practices in China, including the cult of the ancestors, festivals, the belief in the occult and in magical and supernatural forces, and the deification of the forces of nature. It was greatly transformed by the influence of Buddhism, especially as regards its social organization, and ended up by diversifying into a large number of schools with different cult rituals and sacred books. Taoism frequently imposes itself on Chinese culture as a force that counters the dominion of Confucianism, inspiring an attitude of opposition to the established order.

2.41 Shintoism and Japan

Shinto is a form of religiousness specific to Japan with its origins in prehistory. The word 'Shinto' corresponds to the Chinese expression 'shén-dào', meaning 'the way of the gods', and it only started being used in the 6th century, probably to distinguish the local religious practices from Buddhism and Confucianism that had arrived in Japan. The Japanese translation is 'kami-no-michi' also meaning 'the way of the gods'. 'Kami', however, has a far broader significance, designating all that has an innate supernatural force, which can be gods, heroes, emperors, and forces or spirits of nature — animals, plants, rocks, mountains, and seas. There is no reference to a supreme or omnipresent god.

Shinto brings a strong awareness of the supreme value of nature for life, and this attitude has important implications at the environmental, ethical, and philosophical levels. The natural environment is sacred and should be revered and preserved as such, since it is the home of the 'kami'. Today it is still frequent to find a site in a forest or on a mountain chosen as a sanctuary for prayers and celebrations, as in the primordial animist cults of Japan, when temples did not yet exist. One of the most important characteristic of Shinto is the principle of unification of the religious and political dimensions of life. This prescription has contributed to establishing in the national conscience a strong code of conduct of supreme duty and loyalty to the nation and the Emperor, who is the leader and priest that serves the 'kami'. A good example is the ritual suicide of seppuku, also known in the common language as 'hara-kiri', used when one's behaviour is dishonourable or betrays the commitment of loyalty.

Since the 3rd century AD, Japan has been tied to China through commercial expeditions and wars, and has benefited from the influence of a cultural and religious civilization more advanced than its own. The Buddhist religion and the ethical, social, and political codes of Confucianism were well received and spread widely after the 4th century. Shinto, Buddhism, and Confucianism influenced each other without serious conflict or persecution, although they never fused or lost their identity. This constitutes a rare example of religious tolerance. Christianity arrived with the Portuguese when they reached the island of Tanegashima in the south of Japan in 1543, but later it was not possible to establish the same type of coexistence with the local cults due in part to its missionary and centralizing nature.

2.42 The Perennial Religious Conflicts

Throughout history, the relationship between one religion and the others, especially when they compete aggressively for the same space, or the relationship between different branches of the same religion, results from a complex process essentially determined by the need to affirm, maintain, and consolidate its own identity. Historically and in recent times there have been many examples of tension and confrontation, benign or violent, between followers of different religions. All of them aspire to bring a growing number of faithful into their fold, but through time, proselytism has taken a variety of forms. Christianity and Islam frequently assumed violent missionary practices that led to armed combat, the conquest of territories, and the extermination of non-believers.

The rapid expansion of Islam, initiated in the 7th century against the Byzantine and Persian Empires, was an offensive and violent campaign for the propagation of the Islamic faith. The Wars of the Cross, or Crusades, whose objective was to free Jerusalem from Islam, gave rise to episodes of extreme aggression and barbarity. The inquisition, established to eradicate heresies, also resorted to terrible forms of violence, torture, and death. During a long period in European history, from approximately 500 to 1800, Christians tried to convert Jews, and frequently, when not able to do so, persecuted and massacred them.

The expression 'Holy War' appears in the book of *Deuteronomy*, and is sometimes used to characterize the 'jihad' and the 'crusade'. 'Jihad' signifies at the same time the defence of Islam and Islamic communities from aggression by infidels and also the struggle against those who do not profess to Islam. The world is divisible into three regions: dar al-Islam are the territories where Muslims are in the majority and dominate, dar as-Sulh is a territory outside the control of Islam but having treaty relations with the Islamic state, and dar al-harb is the land of war, dominated by infidels and where an offensive jihad can be waged.

After the initial conquests and expansion of Islam, it was established in the canons of the jurists that the Caliph had the duty to invade any territory in the dar al-harb at least once a year to keep the jihad active. 'Ghazw' is an Arabic word that means a battle associated with the expansion of the Islamic territory. In the context of the jihad, the ghazi or warriors that participate in the ghazw had the objective of destroying or weakening the defences of the infidels in order to conquer their territory and to subjugate them. A transliteration of 'ghazw' led to the English, French, and Spanish 'razzia', and the Portuguese 'razia', meaning a plundering raid.

The practice of proselytism by the great Asiatic religions is distinct from the forceful missionary practices of Christianity and Islam. In Jainism, Buddhism, and Hinduism, ahimsa or non-violence prevails, but to different degrees. Nevertheless, some have the dharma of waging war in justified circumstances. Although most of the widespread violence between followers of different creeds has become a thing of the past, we still find tension and confrontation with strong religious motivations, and there is no sign of abatement.

2.43 Religious Diversity and Innovation

The religions briefly reviewed here constitute only a small part of the vast realm of past and present cults, although probably the most significant part from the point of view of their influence on human history, and the most successful in terms of their ability to increase the number of believers. The dominant characteristic of this admirable multitude of creeds is the diversity that results from the remarkable human capacity to construct very diverse conceptual systems, all of them capable of attracting recognition and the religious fervour of millions of believers. The diverse systems fill different spaces in an immense matrix that represents our undifferentiated disposition for religiosity and spirituality. They arose and developed in unexpected, contingent, and precarious ways, generally through the impetus of men with exceptional qualities and capacities. It is curious to note that the history of religions simulates some of the mechanisms of the evolution of species. Events of a random nature create different evolutionary forms that become adapted to use in the best possible way the available resources in the competitive medium in which they develop. These solutions, conditioned by the specificity and complexity of the environment, are not foreseeable, although they have *a posteriori* intelligible evolutions.

It was a long way to the discovery and development of the early forms of monotheism in the context of a dynamic opposition with polytheistic expressions. An enduring preoccupation was to find an acceptable balance between simplicity and complexity in the proposed religious systems, so that they could attract both the common people and the elite. Some systems put great emphasis on self-restraint and control of earthly desires, and on personal responsibility for wrong-doing, while others rely more on the adoration and supplication of merciful divine beings that transcend our understanding to pardon our wrong-doings. Some show a tendency, more or less explicit and assumed, to unify the religious, social, and political spheres, while others clearly separate them. Some religions reveal a strong will to engage in missionary, proselytising, and expansionist activities, while others are less forceful. As regards the environment, some religions profess various forms of animism and identify sacred forces and spirits in animals, plants, rivers, streams, mountains, and nature in general, while others do not. The position of a given religious system in the matrix defined by these various aspects determines to a large extent the world view that it generates in the faithful, in particular as regards social, political, governmental, economic, cultural, and environmental matters. Furthermore, it contributes to determining the way people understand their past, their present situation, and the future.

It is important to bear in mind that religions have, and will most likely continue to have in the future, a profound influence on contemporary societies, because they are frequently the primary builders of identity, and determinant as regards the development of codes of conduct, moral standards, and the vision and interpretation of the world. They constitute the most intimate and safest refuge for believers, sheltering them in moments of uncertainty, anguish, or crisis.

The European voyages of discovery in the Renaissance revealed the existence of a great variety of different cultures and religions and initiated a process of interaction between them. This process reached global proportions in the 20th century. Communication and dialogue between people with the most diverse religions, cultures, and languages, belonging to communities and countries with different levels of social, economic, industrial, scientific, and technological development, has now become relatively easy. The current high mobility of people, the extraordinary development of communication and information exchange, and the growing dissemination of science and technology tend to upset or inhibit the expression of some of the moral values and practices associated with religions. On the other hand, cultural and scientific development and the exchange of ideas and experiences throughout the world has generated new spiritual and religious movements, some syncretistic, involving several religions and seeking to integrate different elements of doctrine and practice, especially in an effort to conjugate the East with the West. Others invoke science, as is the case of the Church of Christ Scientist, better known as Christian Science, which arose in New England at the end of the 19th century, or the Church of Scientology, founded in 1954.

The new spiritual and religious movements that appeared in various continents are very diverse in their doctrines, practices, internal organization, proselytism, geographic distribution, and degree of acceptance by the societies in which they have developed. Their main common goal appears to be that of finding new forms of spirituality and religiosity, better suited to contemporary values and way of life. In purely quantitative terms, the total number of followers is very small compared to those of the great world religions, which continue to be the strongest expression of human religiousness. Karl Jaspers (Jaspers, 1949) has pointed out that the foundations of the dominant contemporary spirituality in the world are still the religions that emerged independently in the West, China, and India in the period from 800 BC to 200 BC, which he calls the axial age. In other words, we still have essentially the same spiritual and religious framework that flourished in that remarkable period of the history of humanity.

One may wonder whether there is any chance of new religions coming into being that would be capable of mobilizing a significant part of humanity. Would it be possible or is this type of phenomenon a thing of the past? Such a process would have to occur in a global socio-cultural context, dominated by the increasingly powerful world media and the direct means of communication through email and social networks. It would surely use the new communication channels and the ubiquity of the social media. The great religions of the past arose generally as reforming movements in reaction to the religious and moral practices specific to a certain region. It is very unlikely that new large religious movements will repeat this pattern. The present concerns of humanity have a much more materialistic nature. Nevertheless, if it were a reactionary process, it would have to occur in the context of the globalised confrontation of ideas and expectations, and address the contemporary frustrations associated with the development paradigm based on continuous growth, poverty and hunger, the profound and ever-growing inequities and inequalities, the increasing prices of commodities resulting in part from a scarcity of natural resources, and the degradation of the environment. At present, however, this is an unlikely scenario because there is still a relatively strong and generalized confidence in the capacity of the contemporary paradigm to satisfy expectations for development and for improved prosperity and well-being.

It is more likely that spiritual or religious movements may spring from recurrent financial, economic, social, and environmental crises, probably in association with high levels of human deprivation and increasing conflict. Such extreme situations could create a fertile ground for the emergence of new forms of spirituality or religiousness at the local and regional level, and these would probably propagate slowly throughout humanity. In any case, from the conceptual point of view, we are witnessing the decline of the axial age.

2.44 The Middle Ages and the Assimilation of Classical and Islamic Cultures by Europe

The expression 'Middle Ages' was first used by historians and academics in the 15th century to designate the era between the decline of ancient civilizations and the renaissance of classical culture that they were experiencing at the time. For them, the old world was synonymous with a high level of civilization, whereas the Middle Ages represented a decline into barbarianism, parochialism, and religious fundamentalism. Despite this negative vision, it was a period of profound transformation in Europe that decisively influenced its future. It was a time for Christianity to reform itself under new forms of temporal power and slowly assimilate other cultures. This process started in the 5th century with the fall of the Western Roman Empire in 476, under the pressure of waves of Germanic invasions, followed a few centuries later by the coronation of Charlemagne in 800 AD, and ending with the emergence of the Renaissance and the beginning of the Reformation in the 16th century, traditionally associated with the date of 31 October 1517, when Martin Luther posted his Ninety-Five Theses on the door of Castle Church in Wittenberg.

Medieval society was mostly rural, centred on the feudal domains and on the relations between landlords and serfs. It was in this period that a large number of towns started to develop near seaports, bridges over important rivers, important religious sites, or the residences of the nobility. These emerging towns were generally protected from external insecurity by large walls. They fostered the organization and development of commerce, the diffusion of culture, and the appearance of a new urban society, made up largely of artisans and tradesmen who gradually freed themselves from the feudal regime that prevailed outside the walls. These were the early bourgeoisie, identified with the new freedom of the city dwellers and independence from hereditary privileges.

Various proxy records of climate allow us to conclude that Europe was relatively warmer in the period between the 10th and 14th centuries. Agriculture developed significantly and became more efficient, trade expanded, and urbanization started to intensify significantly. From 900 to 1350, the number of cities rose by a factor of 10 (Barlett, 1993). A surge of monastery and church building, from small rural chapels to great cathedrals in the city centres, spread across Europe. Cathedrals, besides being the centre of religious activity, also ran schools for the education of the clergy and for the study and interpretation of ancient texts.

In the summer of 1085, Alfonso VI of León and Castile, in collusion with the Emir of Seville, whose daughter was one of his lovers, captured the Islamic city of Toledo. This first step toward the Christian recovery of the Iberian Peninsula, which lasted nearly 400 years, had other noteworthy consequences. Toledo was the largest and most central city of about 25 taifa, or small emirates, into which the Caliphate of Cordoba fragmented after its collapse in 1031. Its library was magnificent and contained a large collection of philosophy, science, and mathematics books from Islamic writers and Arabic translations of books written in Greek, Syrian, Persian, and Sanskrit. Some of the books from Toledo's library and from similar libraries in

other cities in Iberia and Sicily were translated into Latin and Hebrew and disseminated throughout Europe, starting the process of knowledge transfer that led to the
Renaissance. It was by this meandering path that the most important achievements
of the Greek, Roman, and Islamic civilizations became known in most of Europe.
Gerard of Cremona, John of Seville, and Adelard of Bath are some of the famous
translators of hundreds of books from Arabic to Latin that enriched the libraries and
schools where theology, law, medicine, and natural sciences were taught, located in
various European cities such as Salerno, Montpelier, Bologna, Paris, and Chartres
(Virk, 2003).

The spread of culture in Europe through books benefited greatly from the invention of paper. There is a record of a paper factory in the Islamic part of the Iberian
Peninsula during the 11th century. One of the first references dates from 1056, and
mentions Abu Masafya who owned a paper mill near the old irrigation canal in the
city of Statiba, today called Jativa, south west of Valencia (Bloom, 2001). The excellent paper that was made there from flax was called shabti, and was well known
throughout the Peninsula. The technique of making paper from the bark of trees,
flax, and other vegetable fibres was invented in China around the year 100, by Cai
Lun, a eunuch who was secretary to the Emperor He of the Han Dynasty. It was one
of the most important inventions of all history, and accelerated the development of
civilization, first in China and later in the Middle East and Europe.

The secret of paper making was a well-guarded secret in China, and it was only
revealed under coercion. In the year 751, near Taraz in Kazakhstan, the Arabs of the
Abbasid Caliphate allied with the Tibetans to defeat the Chinese forces of the Tang
Dynasty at the battle of the river Talas. It was a decisive battle in the extreme west of
the Chinese Empire that opened the way to Islamic control over a large part of Central Asia. Surprisingly, it was also responsible for the transfer of paper technology
to the West. Two of the Chinese prisoners of war caught during the fighting knew
the secret of its fabrication and were forced to reveal it in Samarkand. A little later,
paper began to be produced in Baghdad and Damascus, and the technology spread
along the north of Africa until it finally arrived in the Iberian Peninsula. Before this,
in Europe, papyrus was used, and later parchments made from calfskin, sheepskin,
or goatskin. The reproduction of a book was time-consuming and very expensive,
accessible only to the great ecclesiastical centres and the very rich. Paper made from
flax, and other vegetable fibres was strong, flexible, and economical.

Midway through the 16th century, a new discovery lowered the cost of books,
making them accessible to an ever-growing number of people, and accelerating the
spread of knowledge. Johannes Gutenberg, in Mainz, recognized the potential economic value of a technique for printing with characters made from forged metal that
was developed in Korea at the end of the 13th century. With great skill, he developed
movable-type printing and managed to print a Bible with many hundreds of pages
in 1455. From then on, the printing press spread throughout Europe and, by the end
of the 15th century, about 5 million volumes had been printed.

Texts from Aristotle, enriched with comments by their translators, offered a rational explanation of the world obtained through the use of reason, and based on
fundamental entities and causal properties. The problem was to make this vision

compatible with religious dogma, and to address the recurrent challenge of reconciling reason with faith. It was the Dominican theologian and philosopher St Thomas Aquinas who skilfully solved the problem by arguing that reason is a will of God and should do service in the cause of faith, so that, properly interpreted, the two could not be contradictory. He distinguished truths known by reason from the higher truths know by revelation. With this new formulation, it became possible to promote the idea that nature operates in accordance with laws that can be deciphered by reason and to encourage the study of nature as a better way of knowing God. This conceptualisation was decisive in creating the environment necessary for the emergence of modern science in the heart of Christianity from the middle of the 16th century onward.

2.45 The Crusades and the Black Death

In the mid-Middle Ages there were more pressing concerns. A short time after the conquest of Toledo, on 27 November 1095, in the synod of Clermont, in Auvergne, Pope Urban II appealed to Christians to fight to free Jerusalem from Islam. This was a way of asserting the identity and power of Christianity, which would also help to diminish the endemic violence of feudal society. Popular response to the call to the crusade was enthusiastic, and for nearly 200 years, the nobility and the people of the Christian European nations wholeheartedly participated in it. The extremely violent wars of the cross created an enormous rift between Christianity and Islam, and the dominion over Jerusalem was short-lived.

Soon after, the relationship between Christians and God was put to a hard test. In the summer of 1346, the Genovese colony of Caffu in the Crimea was besieged by the Tartars. At the same time, a deadly epidemic called the 'Black Death' spread amongst them. The corpses of the victims were piled up at the city walls, and in a final desperate attempt to break the resistance of the Genovese, the bodies were catapulted into the city. The besieged very quickly surrendered and fled in their ships to the West, but many of them were already infected. In October 1347, one of the ships arrived in Messina, Sicily, with the majority of its crew dead. By January of 1348, the Black Death arrived in Genoa, and in the winter the pandemic reached Venice, Pisa, Florence, Paris, and London, spreading throughout the whole of Europe.

No one, including doctors, could understand the spreading mechanism. Panic ensued. Those that could, fled and took refuge in isolated places, others threw themselves into carnal pleasures and the most outrageous orgies in the expectation of death, while others tried to assist the sick and dying. The affected populations were left to their own fate, and a black banner was flown from the church tower. It was firmly believed that God was punishing humanity for its sins. Religious fervour and fanaticism were exacerbated, and violent persecutions were launched against some minorities, especially Jews and lepers. There were severe socio-economic consequences and the Church was greatly debilitated.

It is now well established that the Black Death was bubonic plague, whose causative agent is the bacterium *Yersinia pestis* hosted by rats and transmitted to people by fleas or in some cases directly by breathing (Besansky, 2010). The Black Death is the second of three large epidemics of plague that have occurred in historical times. A recent reconstruction of the bacterium family tree has shown that they all had their origin in China (Morelli, 2010). The Black Death reached Europe through the Silk Road and arrived in Crimea with the Tartars. About 30% of the European population, approximately 30 million people, died, and in China the mortality was even greater. Although it caused dramatic suffering the bacterium has no special interest in humans whom it kills by accident since its natural hosts are various species of rodents, such as rats, marmots, and moles, which are common throughout China.

2.46 The Ideals of the Renaissance

The transition from the Middle Ages to the Renaissance and the Reformation was slow and gradual, although it can be roughly situated around the year 1450. Initially, the interests that gave rise to the European Renaissance were cultivated only by minority elites in a few cities. The most significant aspect was the emergence of a tendency to confine the power of the Church to the religious sphere and to focus the influence and relevance of religion on the realm of individual conscience.

A certain freedom of thought was generated, slowly encouraged by the perception that it could lead man to knowledge and to better living conditions. It was admitted that the use of reason and the various human skills, all emanating from God, to satisfy curiosity about nature and the world should be exploited to discover the laws of the Universe and thereby improve our well-being on Earth. This new tendency opposed the widespread conviction in the Middle Ages that the destinies of men and women are guided by divine providence and dominated by the incomprehensible forces of nature and the Universe that surround them. Nevertheless, during the Renaissance, many of the occultist practices of the Middle Ages continued, such as divination, augury, necromancy, sorcery, witchcraft, and soothsaying. Magic and superstition continued to dominate the imagination of the greater part of the population, especially the destitute and less well educated.

In the middle of the 16th century, after the Reformation and during the counter-Reformation the scientific revolution took its first timid steps which would eventually give birth to modern science and technology. This was a period of increasing divergence between Islam and the cultivation of science. An eloquent example of this growing separation was the fate of a large astronomical observatory built in Istanbul. The observatory was proposed to the Sultan Murad III by the astronomer Taqui al-Din (1526–1585), an important figure in contemporary Islamic science, who had written several books on astronomy, optics, and clock-making. He wanted it to rival the famous observatory of Tycho Brahe, whose very accurate and complete data were essential for the establishment of Kepler's laws of planetary motion. The Sultan was eventually convinced and agreed to the construction, which was completed in 1577. A few months later, Taqi al-Din observed a comet and interpreted the phenomenon as a sign that the Ottoman army would conquer Persia. However, the prognostication did not happen and instead a devastating plague broke out in some parts of the empire. Confronted with this error, the Sultan ordained the destruction of the observatory in 1580 by a squadron of his janissaries (Sayili, 1960), only three years after it was built.

This episode illustrates that Islam did not conform to the emerging methodology of modern science characterized by the systematic observation and interpretation of phenomena, devoid of any connotation or meaning of human value. In the following centuries, Islam made extensive use of the military and medical applications of science, but did not participate actively in its revolutionary development.

Chapter 3
The Contemporary Situation

3.1 The Energy Challenge

Let us return to the present and to the great challenges we face nowadays. We live in a remarkable period of history in which we enjoy a magnificent quality of life, characterized by levels of well-being and comfort, never before attained. We have food security and relatively easy access to health care, housing, energy, good quality water, education, professional training, culture, information, very good means of communication, exceptional mobility, and an almost limitless variety of goods and services. But this more or less luxurious way of life is only accessible to about twenty per cent of the world population. However, it is the current world paradigm and serves as both a model and a goal for the rest of humanity. The long-term survival of the paradigm and its application to a growing number of people depends on the development models adopted at national, regional, and global levels, and on how they achieve a sustainable use of natural resources and avoid dangerous environmental degradation. In this context, energy has special relevance due to its central position as regards social and economic development. Finding sustainable energy sources and systems to support world development over the next 100 years and beyond is surely one of our most important and difficult challenges at the beginning of the 21st century.

Let us therefore start with energy. Before the industrial revolution, we used almost exclusively the mechanical or kinetic energy delivered by humans and a few domestic animals, usually called somatic energy. Besides this form of energy, we had access to the thermal energy obtained by burning wood and biomass, the kinetic energy of water in rivers and streams that power watermills, and the kinetic energy of wind that moves sailing boats and powers windmills. Later, with the discovery of engines that could convert the chemical energy stored in fossil fuels into mechanical energy, humankind entered a new energy era, where the energy consumption per capita increased amazingly, especially in the industrialised countries.

It is important to emphasize that energy exists in various forms. There is gravitational potential energy, which converts into kinetic energy when we drop an object, or when water flows over a waterfall. Heat or thermal energy is another form of energy. Radiant energy, like the solar radiation emitted by the Sun, is the energy carried by electromagnetic waves. Electrical energy is the potential energy stored in an electric field, corresponding to the potential energy of a charged particle in an electric field. Chemical energy is the energy contained in some molecules that make up living beings, such as glucose, produced in photosynthesis. Nuclear energy is the potential energy stored in the nuclei of atoms, which is released in some nuclear reactions, in particular nuclear fission and fusion.

These various forms of energy are convertible into one another by means of energy converters, but the conversion process is never entirely efficient, which implies that part of the energy is not recoverable. Photosynthesis converts solar radiant energy into chemical energy with a very low global average efficiency in terrestrial ecosystems of about 0.3%. In humans, the conversion of the chemical energy contained in food into mechanical somatic energy has an average efficiency of 18%, while in a horse, the efficiency is only 10%. The internal combustion engine makes the same type of conversion with an efficiency of 15–25%. However, electric engines, which convert electrical energy into mechanical energy, have a far greater efficiency, that can reach 90%. The photovoltaic cells that convert radiant energy into electrical energy have lower efficiencies, between 20 and 30%.

The main issue regarding energy supply is that it should mach energy demand, ensuring that users have the necessary form of energy in the necessary amounts whenever and wherever they require it. This objective implies the availability of sufficiently diversified primary energy sources and an efficient energy conversion system that delivers the mix required by the demand in an economically practicable way. The investments made in energy conversion and supply systems over many years have been gigantic, and this implies that any attempt to alter the primary sources would automatically involve very high costs.

Initially, before the discovery of agriculture, humans only had access to somatic energy. They used muscle power, derived from the conversion of the chemical energy stored in the plants and animals consumed in their diet. As agriculture became more efficient and hence able to produce a larger amount of food crops, the availability of energy rose considerably. The pyramids, the Great Wall of China, and other remarkable architectural works required enormous amounts of energy. Their construction was only possible due to the very large number of people involved. The mechanical energy delivered per second by the human body in a burst of effort is about 100 W. This value constrained human enterprises for many centuries, especially as regards urbanization, mobility, and warfare.

3.2 Watermills and Windmills

About 8 000 years ago, after the emergence of agriculture, humanity developed the ability to build machines that can use natural energy sources, thus beginning a long process of liberation from the dependency on somatic energy. The Greeks, the Romans, and the Chinese were probably the first builders of waterwheels with vertical or horizontal axles to use the kinetic energy of rivers and streams, about 2000 years ago. In China, the first references to watermills date from the Han dynasty, which lasted from 202 BC to 220 AD. They were used to grind cereals and to power trip hammers and the bellows in furnaces for iron smelting. There are indications that watermills were in use from the first century BC in Greece. In other parts of Europe, this technology was used on a large scale only in the Middle Ages, when the lack of available labour made the mills profitable. In England there are records of more than 5000 watermills in 1086, whereas one hundred years earlier there were only about a hundred (Burger, 2003). They were used mostly for grinding cereal grain, but also for producing olive oil, making paper, sawing wood, crushing minerals, and also operating the bellows in forges.

The first windmills we know of were built at the beginning of the 7th century in Seistan, a windy and dry region across Iran's eastern and Afghanistan's southwestern territories. They served to mill cereals and raise water from the streams to irrigate vegetable patches and gardens (Hill, 1992). The technology may be even older, and may also have existed in China, but there are no reliable historical records. The Seistan windmills with a vertical axle did not spread very far, and it was only in the 12th century that we find references to windmills in Europe (Sorenson, 1995). The axle became horizontal, mounted on a rotating structure to allow orientation by the wind. They were used to mill cereals and also to drain marshy areas, especially in Holland. Both windmills and watermills helped to develop the economic and financial activities in Europe due to the investment and specialised workers needed to build and maintain them in good working order.

3.3 The Discovery of the Steam Engine

The wood consumption to build ships and houses, to cover household demand, and to feed the furnaces in foundries and those used to produce bricks and glass, began to dramatically reduce the forest area in Europe. The solution was to increase the use of coal, abundant in England and in other regions of Central Europe. First, however, the mines had to be drained of flood waters before coal could be extracted. Thomas Newcomen sought to solve the difficulty, and in 1712 he built a steam driven machine that was successfully used to pump the water from the English coal mines. The model was based on an earlier prototype of 1675, built by Denis Papin, but modified in a way that allowed the intervention of atmospheric pressure in the movement of the piston (Rolt, 1977). Despite their enormous success, the new machines were very inefficient.

James Watt solved this problem. He had remarkable abilities as an experimenter and also benefited from working in a university environment. In collaboration with Matthew Boulton, he managed to build steam engines with an efficiency of around 5% and with a power output of 20 kW, equivalent to the maximum continuous and simultaneous power output of around 200 men. The crucial innovation introduced by Watt was to produce an adiabatic expansion in the interior of the cylinder, using the thermal energy of the water vapour.

The 1780s marked the beginning of an era of durable and widespread conversion of the chemical energy stored in coal and other fossil fuels into mechanical energy. Compared with windmills and watermills, steam engines had the enormous advantage of not being fixed in specific sites, and even being able to power ships and locomotives. This is one of the first examples of a positive feedback in the dynamics of the industrial revolution. When the steam engine was used to power locomotives, it facilitated the transport of coal, and this in turn led to an increase in the use of this primary energy source.

3.4 The Beginning of the Intensive Use of Petroleum

Coal provided the main energy supply for the industrialization process of the 19th century, during which production increased from about 10 to 1 000 tons per year. Petroleum or crude oil exploration started later, although it was in fact used in more remote times. In the interior regions of China, it was extracted from the subsoil at least as early as the 4th century, to isolate the salt that is sometimes associated with it. In the Middle East, the streets of the new city of Baghdad, built in the 18th century BC, were paved with asphalt obtained from abundant deposits of petroleum in the region.

The history of the intensive exploitation of petroleum begins in 1853 with the discovery of a process for distilling kerosene from petroleum by the Polish scientist Ignacy Lukasiewicz. Kerosene was very useful, because it replaced whale, fish, and vegetable oils with a great saving, especially as fuel for public lighting systems. In the second half of the 19th century, a notable series of discoveries led to a sudden increase in the use of petroleum as the primary source of energy. In 1856, the first petroleum refinery began to operate in Pleiesti, Romania, soon to be followed by many more. In 1859, Edwin Drake, originally a railway engine driver in New Haven, Connecticut, revolutionised the petroleum industry when he succeeded in extracting it from the subsoil by boring through the rocky layers, near Titusville, in the north west of Pennsylvania. Shortly afterwards, in 1876, Nikolaus Otto built and successfully used the first four-stroke internal combustion engine. It was the first competitive alternative to the steam engine. After a few decades, petroleum exploration became commonplace around the world and oil was used in an increasingly intensive way.

3.5 The Degradation of Energy in Heat Engines or the Increase of Entropy

Heat engines — steam engines, internal combustion engines, diesel engines, turbine generators, jet engine turbines, Stirling engines — all have the same basic function, which consists in taking the heat from a high temperature source, transforming part of that energy into work (mechanical energy), and transferring the remainder to the surroundings, which functions as a cold source. No matter how efficient the engine is, it is impossible to convert all the thermal energy from the combustion of a fossil fuel into work. This limitation results from the laws of physics and is insurmountable. Although the total amount of energy is the same before and after combustion, the energy in the final state has a distinct quality. This change corresponds to a variation in a remarkable physical property called entropy. The German mathematician and physicist Rudolf Clausius (1822–1888) introduced the concept in 1865 when he established the foundations of thermodynamic theory using the work of the French military engineer Nicolas Leonard Sadi Carnot (1796–1832), who studied the optimal functioning of heat engines.

To understand the concept of entropy it is important to bear in mind that the thermal energy of a piece of matter, gas, liquid, or solid, is the sum of the kinetic energies of the random motions of its atoms or molecules. Consider a certain amount of gas. If its temperature increases, this means that the average velocity of the constituent particles has increased. Let us compare two samples of the same gas with the same amount of thermal energy, but at different temperatures. At the higher temperature, the number of particles is smaller but the average kinetic energy is higher. At the colder temperature the total energy is shared by more particles, and that increases the number of possible states in which the system can be found, each state being characterized by the velocity of each particle. Entropy is a measure of the random motions of the constituents, and consequently a measure of the amount of thermal energy that is not convertible into work, that is, into ordered movement. The sample of gas at a lower temperature has more entropy than the gas at a higher temperature, even though they both have the same total energy. We can extract more work from the state with smaller entropy, i.e., the more ordered state, than from the one with greater entropy, i.e., the more disordered state. In the sample at the lower temperature, the energy is relatively degraded in terms of the capacity of the system to produce work.

Let us consider two bodies at different temperatures that are placed in contact with each other. Due to the propagation of heat, they will eventually reach a state of equilibrium in which they have the same temperature. The flux of thermal energy from the hotter to the colder body causes an increase in entropy, so that the final state has more entropy than the initial state, although the total thermal energy of the system remains constant. Before, the system was more ordered than afterwards, when it reached the equilibrium. The principle that thermal energy tends to flow to lower temperatures is the equivalent of saying that entropy always has a tendency to increase. In an isolated system, i.e., one that does not exchange energy or mass with the exterior, entropy cannot decrease. Consequently, an isolated system always ends up becoming incapable of producing work. The production of work is only sustainable in an open system, one which can exchange energy or mass, or both, with the surroundings. The rate of work production depends on the rate that it receives energy and transfers entropy to the surroundings. In a famous article published in 1865, Clausius (Clausius, 1865) formulated the first and second laws of thermodynamics by saying that: "The energy of the Universe is constant; the entropy of the Universe tends toward a maximum." This increase in total entropy is irreversible and determines the direction of time or the arrow of time.

3.6 How to Reduce the Entropy of a System?

The entropy of an open system can be reduced but the price is necessarily to increase the entropy of the system's surroundings. Consider a refrigerator. By transferring the thermal energy taken from the food products being cooled to the kitchen, the refrigerator is increasing the temperature of the heat source and consequently reducing the entropy in its cold chamber. However, this is only possible due to the increase in entropy associated with the process of electricity production that ensures that the refrigerator functions. If this electricity comes from a thermal power station, the entropy increase results from the burning of the fuel used in the power station.

Photosynthesis is a remarkable example of an organization process where entropy is reduced. By transforming solar radiation into chemical energy, a plant diminishes its entropy at the expense of the increase in the entropy of the system formed by the plant and its environment and the radiant energy generated by the nuclear reactions in the central region of the Sun. The preservation of life on Earth and the growth and development of the human population are organization processes in which entropy is reduced at the expense of an increase in the entropy of the whole environment, resulting from the use of solar energy and the exploitation of natural resources. Without some way of reducing entropy, the survival of living beings becomes impossible.

The concept of entropy is very general and applicable to several fields of knowledge. For example, in the theory of information, it is possible to define a concept of entropy that can be applied to a message and constitutes a measure of its degree of randomness. The theory of entropy can also be applied to the economy, where it establishes the limits of growth (Georgescu-Roegen, 1971).

The maximum efficiency of a thermal engine, first obtained by Carnot for his ideal engine, is determined by the second law of thermodynamics, and rises with the difference of temperature between the hot and the cold sources. In a real engine, it is not possible to achieve this maximum efficiency due to various energy dissipation processes in the engine mechanisms, such as friction. The general rule in thermal engines is that only a third of the initial available energy, contained for instance in a fossil fuel, is transformed into a useful form of energy, electricity for example. The remaining energy is lost, in the sense that it cannot be used. Note the huge waste of energy, part of which, as we have seen, is due to the laws of physics, which we cannot change.

3.7 Positive and Negative Aspects of Our Dependence on Fossil Fuels

It is estimated that around 1890 the global quantity of energy generated from fossil fuels — coal and petroleum — became larger than that generated by biomass (McNeill, 2000). At about the same time, large scale electrification began in cities and rural areas, spreading throughout the world until the present time and providing light, heat, coolness, refrigeration, and many other services. Global energy generation grew by a factor of about five in the 19th century, and accelerated to a factor of sixteen in the 20th century (McNeill, 2000). This was a remarkable feat of human ingenuity and endeavour. It significantly improved the quality of life, especially as regards mobility and domestic comfort, created new lifestyles and new patterns of behaviour and consumption, and provided the conditions that allowed for the very large increase in population and economic growth since the end of the 19th century.

However, not all the consequences of this very intensive energy use are positive. The greater part of the world energy consumption is based on fossil fuels, that is, petroleum, coal, and natural gas, the latter mostly since 1950. In 2003, the primary sources of energy at the global scale had the following breakdown: fossil fuels 80% (petroleum 34.4%, coal 24.4%, natural gas 21.2%), nuclear 6.5%, hydroelectricity 2.2%, combustible renewable and waste 10.8%, and modern renewable (geothermal, solar, wind, tides, waves) 0.5% (IEA, 2004a). The total value of this supply of energy was 4.4×10^{20} J, which corresponds to an annual average power of 1.4×10^{13} W. To obtain less abstract information, let us use the human scale. The average energy used throughout the world in 2003 was the equivalent of the continuous, maximum, and simultaneous working power of 1.4×10^{11} men during one year, or in other words, 22 men per capita. There is an enormous difference between this gigantic energy consumption and that of 300 years ago, before the industrial revolution.

We depend heavily on fossil fuels, and this has the disadvantage of generating a serious global environmental problem with regard to the climate system. Carbon dioxide (CO_2) is formed in the combustion of coal, petroleum, and natural gas. Part of it dissolves in the ocean or is absorbed by plants through photosynthesis. The other part accumulates in the atmosphere and changes its composition. Being a greenhouse gas, it intensifies the natural greenhouse effect, increasing the global average temperature of the troposphere, the lower layer of the atmosphere. Furthermore, this type of climate change leads to more intense extreme weather and climate events and a rise in the average sea level. There are already unequivocal signs of climate change caused by the anthropogenic emission of greenhouse gases, among which CO_2 plays a preponderant role. If this form of pollution is not controlled and slowed down, climate change will intensify dangerously throughout the 21st century, leading progressively to more severe impacts. One aspect that is extremely important to bear in mind is that, in terms of CO_2 emissions, coal is the most polluting fossil fuel, followed by petroleum, then natural gas. To produce the same amount of energy in combustion, coal produces almost twice the emissions from natural gas.

Another perverse consequence is that the inequalities in social and economic development and in the quality of life between developed and underdeveloped countries have increased with the new energy paradigm. Access to an intense and growing consumption of energy began in Europe and North America, and only afterwards, from around the middle of the 20th century, did it start to spread widely throughout the world. Nevertheless, about one fourth of the world population still does not have access to commercialised energy, and depends on biomass for heating and on somatic energy to provide mechanical energy. In 2008 approximately 1 500 million people did not have access to electricity (IEA, 2009). The very strong growth in the use of fossil fuels was a decisive factor for the improvement in the quality of life of the wealthier nations, but it was also one of the main causes for widening the social, economic, and cultural gap between developed and developing countries during the 20th century.

Currently, fossil fuels sustain the greater part of food production, the functioning of industry, cities, transportation and, generally speaking, all the main infrastructures of modern society. Crude oil is used to make fertilizers and pesticides that maintain a high level of world agricultural production. It is also used to produce a wide range of extremely useful chemicals such as plastics, building materials, solvents, and a variety of products for industrial, pharmacological, and medicinal purposes. It has the enormous advantage of being the raw material for the production of liquid fuels with a high energy density that are easy to store and transport and are relatively safe to handle. About 90% of the energy consumed by the transport sector worldwide comes from petroleum. In spite of the unique and varied properties of crude oil, most of it is used to produce gasoline, diesel, jet fuel, and fuel to be used in thermal power plants, with the aim of generating energy at a relatively low cost, although this cost has been rising over the last few decades.

3.8 When Will Fossil Fuels Be Exhausted?

The period since the beginning of the industrial revolution is insignificant when compared to the length of time it took to form fossil fuels, on the order of tens of millions of years. The exploration of existing deposits is a very long way from being compensated by the formation of new ones, which means that, if the present rate of consumption is maintained, these deposits will inevitably run out sooner or later. Estimates indicate that, with the present rate of consumption, coal reserves will be the longest lasting, followed by natural gas and petroleum.

Up to half way through the 20th century, geologists and geophysicists specializing in petroleum exploration maintained an optimistic attitude, based on reports that the rate of growth of reserves was superior to consumption. However, in 1956 Marion King Hubert, a geophysicist who worked the greater part of his professional life with the Shell Oil Company, forecasted that the amount of petroleum extracted annually from the subsoil of the USA, with the exception of Alaska, would peak around the 1970s and then fall off rapidly (Deffeyes, 2001). That forecast came true, and today the United States imports about 60% of the petroleum it consumes.

Generally speaking, the variation in the annual rate of consumption of a non-renewable natural resource as a function of time is approximately described by a Gaussian curve, which has only one maximum, is symmetrical, and has the shape of a bell. The annual rate of world petroleum consumption should follow a similar pattern. According to Hubert, when the maximum of the curve is reached, the price will spiral, because the rate of demand is not satisfied by a comparable rate of growth of new deposits. There are many other factors that influence the price of petroleum, including geopolitical conflicts, wars, terrorism, and the increased consumption of countries with emerging economies, such as China and India. Nevertheless, Hubert's main message was that the risk of crisis will not arise when the reserves run out, but much earlier, when the annual rate of world consumption reaches its peak, known as Hubert's peak or peak oil.

There are various ways of forecasting when peak oil will occur, based on the past behaviour of demand, the rate of discovery of new deposits, and estimates of total reserves. Note that the concept of global reserves is dynamic, given that technological innovation can lead to the discovery and exploitation of new deposits. Moreover, current global reserves are uncertain due to the manipulation of national reserve figures by some governments. The International Energy Agency has estimated in 2008 that the world's mature conventional oil fields are declining by 6.7% per year, a rate that is expected to grow to 8.6% in 2030 (IEA, 2008). To offset this decline requires the discovery of new conventional oil fields at an annual rate equivalent to the total annual oil production of Kuwait. In any case, the localization in time of peak oil also depends on the future evolution of consumption, and this depends on the emergence of new, economically competitive primary energy sources and new processes to produce petroleum and other liquid fuels with similar properties from other fossil fuels in an economically viable way.

Given the great advantages of petroleum at the present time, especially the fact that it is a liquid fuel with a good cost/benefit ratio, the growth in consumption over the next few decades will be chiefly determined by the growth in world population and the rate of development in countries with emerging economies. The rise in crude oil prices increases investment in renewable energies and in the implementation of more efficient energy systems, and it also tends to lead to energy savings. In an environmentally optimistic vision, if a worldwide effort is really made to reduce global CO_2 emissions into the atmosphere and hence avoid dangerous anthropogenic interference in the Earth's climate, there will be added pressure to reduce the consumption of fossil fuels.

Taking into account the various factors, it seems likely that Hubert's peak for conventional oil will occur within the next 20 years, for natural gas probably later, and for coal much later still, after the 2040s. There is great uncertainty in these estimates, but there is also a growing consensus that we have entered the twilight zone for the fossil fuel era. With current rates of consumption, the known reserves of conventional crude oil and natural gas will not last much beyond 2100. It is striking that the peak in the per capita world oil consumption was already reached in the

1970s. It is therefore highly improbable that petroleum and natural gas will constitute significant primary energy sources in the 22nd century and beyond. As for coal,
the reserves are far greater, and will probably support relatively high consumption
after the end of the century.

3.9 How Will We Replace Fossil Fuels?

What energy forms can be used to replace fossil fuels? What are the primary energy
sources capable of satisfying the growing demand, caused by the growth of the
world's population and by the expectation of global socio-economic development,
particularly in the emerging economies, with a comparable cost/benefit ratio to fossil
fuels? In a recent report, the International Energy Agency (IEA, 2008) forecasted an
increase of 40–50% in the worldwide demand for primary energy up to 2030, and
a doubling up to 2050 to achieve annual global values on the order of 8.4×10^{20} J.
The demand for electricity, due to the versatility of the applications of this energy
form, is likely to increase faster, probably leading to a doubling of consumption
over the next 30 years. What options do we have to fulfil these projections? Before
analysing them, it is important to emphasise that the projections of the IEA are
based on the continuation of the current paradigm of continuous economic growth
supported by an intensive and increasing energy use worldwide, especially in the
developed countries and the emerging economies.

To support the large increase in demand that is expected up to 2050, it will be
necessary to intensify the use of non-fossil fuels, especially renewable energies.
Nevertheless, fossil fuels will continue to support a large part of the demand during the twenty-first century. As the conventional oil reserves start to decline, petroleum will be increasingly extracted from non-conventional crude oil sources such
as extra-heavy oils, oil shale, and oil sands, also called tar sands. Petroleum can
also be extracted from coal and natural gas through synthetic fuel processes such as
the Fischer–Tropsch process. The main problem is the increased cost of production
compared to conventional oil. The evolution in the price of petroleum will be a major determining factor in decision-making as regards the large investments needed to
produce non-conventional oil. If the increase in the price of oil becomes very large,
this will have profound consequences on the economy, especially in agriculture, the
automobile industry, and aviation. A preview of the expected impacts has already
been witnessed in the strong surge in the oil price that occurred in the summer of
2008. Meanwhile, as the price rises, the risks of geopolitical conflicts, centred on
access to and control of petroleum supplies by countries with larger populations and
larger consumption per capita, are likely to increase.

Large reserves of petroleum are not necessarily a blessing for the countries that
have them. If a country depends on petroleum to assure the larger part of its GDP,
this will not have necessarily positive effects, either economically or politically.
The liquidity created in oil-rich countries by petroleum exportation may tend to
diminish taxes and with them government responsibility in maintaining strong state

institutions and functions, especially in authoritarian and corrupt regimes. On the other hand, the immediate access to that immense wealth diminishes the pressure to encourage a diversified social and economic development based on effective efforts in education, professional training, research, and institutional capacity, all essential factors in strengthening a democracy. In some countries that are highly dependent on the production of petroleum and do not have a strong and stable democracy, the rise in price has often contributed to a worsening situation as regards human rights and civil liberties (Auty, 1993).

3.10 Extraction of Petroleum from Oil Sands and Oil Shale. Shale Gas and Extreme Energy

The new age of non-conventional oil production has already started. The extraction of petroleum on a large scale from oil sands or tar sands, a mixture of rock, sand, and bitumen, began in February 2006, near Fort McMurray in the region of the Athabasca River, situated in the north of the Alberta province in Canada. It marked the beginning of a profitable type of exploration that is very likely to be practised frequently in the future in other regions of the world with rich deposits (McCullum, 2006). The oil sands are being explored in open-cast mines in regions that had previously been covered with forests. The bitumen is extracted from the sand using facilities similar to gigantic washing machines that consume huge quantities of energy and water. The water becomes heavily polluted and forms large artificial lakes. The final transformation into petroleum is expensive, consumes large amounts of energy, and produces much more CO_2 than conventional petroleum extraction from oil wells. About half of Canada's present greenhouse gas emissions come from the lucrative petroleum exploration from oil sands in Alberta (Le Monde, 2007). The volume of water consumed in the process used to separate the bitumen from the sandy soil is 2 to 4.5 times larger the volume of petroleum produced. Furthermore the extraction of oil from tar sands consumes an estimated 20 to 45 m^3 of water per megawatt hour, which is nearly 10 times that for conventional oil extraction.

In spite of the elevated costs of production and the extremely negative environmental impacts, it is very likely that this type of extraction will develop further, particularly if the oil price continues to increase. If we calculate the amount of petroleum held in tar sands together with conventional petroleum, the three countries with the largest reserves are, in decreasing order, Venezuela, Saudi Arabia, and Canada. According to some estimates, non-conventional oil is expected to make up 35% of the world's supply by 2015.

It is also possible to extract petroleum from oil shale, but the costs are even higher. Large scale extraction has already been tried by Exxon in the 1970s in the state of Colorado, USA, but it was abandoned in 1982 because it became economically unprofitable following the fall in the price of crude oil. With the recent price escalation, there is renewed interest in this process, and Shell has already developed a new technology for extracting petroleum from oil shale. The amount of energy used will be enormous. The energy efficiency of the process can be assessed by the energy return on investment (EROI), defined as the ratio of the amount of energy produced to the amount of energy used in the production process. The EROI of oil extraction from oil shale is about 3.5, while current oil extraction from traditional oil wells has an average value of around 20. In the 1930s, it reached values above 100. In spite of these shortcomings, it is important to bear in mind that the reserves of petroleum in oil shale in the USA, in the states of Colorado, Utah, and Wyoming are very large and would be able to satisfy the petroleum requirements of the country for many decades.

There are also unconventional sources of natural gas. The extraction of natural gas from shale, or shale gas, is rapidly becoming a very lucrative business due to a new drilling method of hydraulic fracturing, or fracking, developed in the USA. A well is drilled deep underground and turned horizontally into the shale layer. Charges are detonated in the horizontal pipe causing fractures in the shale and liberating the gas that is extracted by pumping into the fissures a high pressure fluid of water, sand, and chemicals. The new technology has been able to tap enormous quantities of gas from shale at very competitive prices but the negative environmental impacts are enormous. The greenhouse gas footprint of shale gas is greater than that for coal, conventional oil, and natural gas on a 20 year time horizon of exploration of the well, because a significant amount of methane escapes into the atmosphere at the time the well is hydraulically fractured, during the drill after fracturing, and through vents and leaks over the lifetime of the well (Howarth, 2011). Furthermore the shale gas extraction produces large amounts of contaminated water that constitutes an environmental and health risk.

This new energy source is very abundant worldwide and could satisfy a large part of the world natural gas demand but at a very high environmental cost. In 2011, nearly a quarter of the natural gas supply in the USA comes from shale gas, driving down prices for consumers. In Europe the technology of fracking to extract shale gas has met strong opposition in France due to environmental concerns and was banned by the government in May 2011. However, it is likely to be adopted by other EU countries and eventually also in France. The possibility of using other technologies with smaller environmental impacts is being considered. Shale gas offers a very good example of the dilemma between access to relatively cheap energy and protection of the environment.

3.11 The Question of Coal

Finally, let us turn to coal. Although it is a fossil fuel whose combustion produces the greatest amount of CO_2 compared to petroleum and natural gas, it seems likely that it will be intensively exploited over the next 150–200 years. The reasons are varied: the cost is relatively low, there are sufficiently abundant reserves to satisfy global demand for most of this century and it is clearly a preferable alternative to the large scale use of nuclear energy from fission. It also has the advantage that it can be transformed into diesel or other synthetic petroleum products, although these processes are currently very expensive.

The intensive use of coal as a primary source of energy will create an enormous increase in CO_2 emissions, with very negative consequences as regard climate change. This problem can only be solved by capturing and storing the CO_2 produced in the combustion of coal to generate electricity or liquid fuels, thus preventing its release into the atmosphere. After capture, the CO_2 can be stored in underground geological formations such as depleted oil and gas fields, non-mineable coal beds, and deep salt deposits. An important concern with carbon capture and storage (CCS) is to prevent leakage of CO_2 into the atmosphere. However, the main problem is that CCS will considerably increase the price of electricity produced from coal. There is already some valuable experience with successful industrial-scale CCS projects, such as Sleipner in the North Sea, operated by Norway's Statoil, Weyburn–Midale in the USA, and the Schwarze Pumpe power station in Germany. Further projects are being planned both in the USA and in the EU.

3.12 What Are the Primary Sources of Energy on Earth?

What alternatives are there to fossil fuels on Earth? Planet Earth is an open system that receives energy almost exclusively from the Sun. There are only two other sources of energy, but their flux is much lower than that of solar radiation. One source, located in the central region of the Earth, where the temperature is on the order of 5 600 K (Alfe, 2002), produces a very small heat flux at the surface. The other source of energy is the ocean tides, which result from the differential gravitational forces exerted on Earth by the Moon, and to a lesser degree by the Sun.

Solar energy was responsible for the formation of fossil fuels through photosynthesis over hundreds of millions of years. The same form of energy powers the water cycle responsible for the flow of rivers and streams. The Sun is also the main source of energy for the general circulation of the atmosphere, which determines the winds and the ocean currents. Finally, solar radiation makes the growth of plants and trees possible through photosynthesis. Solar thermal and photovoltaic, hydroelectricity, wind, biomass, and energy from ocean waves and currents are all renewable energies that derive either directly or indirectly from solar radiation. The only renewable energies with a different origin are the ocean tides and the geothermal energy that results from the thermal energy stored in the Earth interior, which is generated by some radioactive elements, mainly potassium 40 (^{40}K), uranium 238 (^{238}U), uranium 235 (^{235}U), and thorium 232 (^{232}Th).

The other source of energy that we can now access is nuclear energy, a form of potential energy that exists in atomic nuclei and which can be converted through nuclear reactions into kinetic energy. At the moment, we only have nuclear fission reactors for which the fuel is normally uranium enriched in the isotope ^{235}U. By absorbing a neutron, this nucleus splits into two fragments and emits a small and variable number of neutrons (whose average is 2.43) with different kinetic energies.

Some of these neutrons then collide with other nuclei of ^{235}U and initiate further fission reactions. In the nuclear fission reaction, a very small part of the mass of the ^{235}U nucleus is converted into the kinetic energy of the nuclei and neutrons produced in the reaction, in accordance with Einstein's celebrated formula $E = mc^2$, where E is energy, m is mass, and c is the speed of light in vacuum. Due to the very large factor c^2, the amount of energy produced by the complete fission of a given mass is enormous relative to the amount of chemical energy stored in the same mass of fossil fuels, for example. The kinetic energy resulting from the complete energy conversion by fission of 1 gram of ^{235}U is equal to the chemical energy stored in one ton of petroleum.

3.13 Nuclear Fission Reactors: Advantages and Risks

The nuclear fission energy industry entails three types of problems involving significant risks that are difficult to reduce. The first derives from the long-lasting radioactivity of some of the elements produced in the nuclear fission reactions and contained in the waste resulting from the operation of nuclear reactors. The fission of ^{235}U produces nuclei with a wide range of radioactive and chemical properties. Some, such as americium (^{241}Am), are produced in small quantities, but remain radioactive for millions of years. Others like caesium (^{135}Cs) and strontium (^{90}Sr) have a higher radioactivity and a long half-life, on the order of tens or hundreds of years. The most problematic is plutonium (^{239}Pu). Generated from uranium ^{238}U, it has chemical properties that make it extremely toxic and, by being fissionable after purification, can be used to make atomic bombs or as the fissionable element in second generation nuclear reactors.

About 27 tonnes of spent nuclear fuel are taken each year from the core of a nuclear reactor that generates 1000 MW of electricity. This used fuel can be treated entirely as waste or it can be reprocessed through mechanical and chemical treatments that separate it into plutonium, uranium, and high-level radioactive waste solutions. The ultimate disposal of these wastes requires their isolation from the environment for long periods of time. The most favoured method is burial in dry underground geological formations that are considered stable for periods of the order of 100 000 years. In the USA, a mountain in Yucca, Nevada, was chosen, but the construction of the infrastructure for storing the waste in the interior of the mountain has been hotly contested. Meanwhile, the waste has been accumulating since the 1940s, in the nuclear reactor installations, on military bases, and in research laboratories. In Europe, Japan, China, and other countries with a significant production of high level radioactive waste, the situation is similar. There are studies for storage in stable geological structures, but no decisions have yet been taken. In February 2009, President Barack Obama of the USA stopped funding the Yucca project and initiated the development of a new plan for the long term management of radioactive waste from nuclear reactors. The main idea is to form a community of countries that would share their nuclear power technologies and store all their high-level radioactive waste after significantly reducing its volume by reprocessing.

The other problem is the question of security. A major effort has been made to increase the safety of nuclear reactors. From 1990 to 2000 the number of incidents inside nuclear power plants involving some risk, even small, fell by a factor of twenty (Reeves, 2003). Although this is good news, we should be aware that it is impossible to completely exclude the possibility of human error, and that accidents in nuclear reactors can be very serious in terms of the loss of human lives and severe and prolonged health impacts, as was the case with the Chernobyl accident on 26 April 1986. Furthermore, there are the security risks generated in nuclear power plants by some kinds of natural disasters. A dramatic recent example was the nuclear accident caused by the 9.0 Tohoku earthquake and subsequent tsunami of 11 March 2011, on the Fukushima Dai-Ichi (Dai-Ichi meaning first) multi-reactor power site in Japan. On 12 April, the IAEA rated Fukushima I as a major accident, corresponding to level 7, the highest, on the International Nuclear Event Scale. This attribution implies that the event produced a major release of radioactive material with widespread and dangerous health and environmental effects.

A recent review by the physicist Thomas B. Cochran (Cochran, 2011), who works in the nuclear program of the US Natural Resources Defence Council (NRDC) indicates that there have been twelve nuclear power accidents with fuel-damage or partial core meltdown. These include the earliest meltdown, which occurred in 1959 at an experimental sodium reactor located at the Santa Susana Field Laboratory, California, USA, the Enrico Fermi Unit I reactor, Michigan, USA in 1966, The Saint Laurent A-1 power plant, Loir-et-Cher, France, in 1969, the Three Mile Island Unit 2 power plant, Pennsylvania, USA, in 1979, Chernobyl Unit 4 power plant, Ukraine, in 1986, and Fukushima I, Japan in 2011. Since the beginning of nuclear commercial energy, 582 reactors have operated worldwide for a total of about 14 400 reactor-years. Thus, at present, the historical frequency of core-damage accidents is approximately one in 1300 reactor years, a probability that is considered unacceptable by the NRDC (Cochran, 2011). Some of the accidents occurred in reactor types that are now out of the market. However, there are many operating nuclear reactors with obsolete designs that should be phased out rather than having their licenses extended.

Another disquieting aspect, related to security, is the fact that nuclear reactors can be a target for terrorist attacks. The proliferation of nuclear reactors tends to increase the risk that nuclear weapons may be used by groups or governments with terrorist or extremist tendencies. The International Atomic Energy Agency has reported frequent incidents of illicit trafficking of nuclear materials, and has developed many schemes in collaboration with its member states to combat such activities and to protect against nuclear terrorism (IAEA, 2007). Here again it is impossible to reduce the probability of a devastating terrorist attack on a nuclear reactor to zero.

In part as a result of these security risks, there is a lack of confidence or even antagonism toward nuclear energy in public opinion in many countries, especially in the more industrialized nations. Several European countries have abandoned their nuclear programmes or are involved in an intense and sometimes passionate debate over their future. In others there are signs of a renewed interest in nuclear energy. Finland has four nuclear power reactors and is presently constructing a fifth. The more robust and ambitious nuclear energy programmes are found in developing countries, particularly in the emerging economies. At present China and India are leading world investments in nuclear energy. China is planning to build 32 nuclear reactors up to 2020 (IAEA, 2006) and is presently constructing 27, while India is building 5 (ENS, 2011). As of 19 January 2011, there were 442 nuclear power plants operating in 30 countries with an installed capacity to generate 375 000 MW of electricity (ENS, 2011). The USA leads with 104, followed by France with 58, and Japan with 54. On the same date, 65 plants were under construction in 16 countries, with an installed capacity of 63 000 MW (ENS, 2011). A significant number of the 442 nuclear reactors are getting old. Due to the radioactivity inside a reactor, its life is usually limited to about 60 years, but dismantling operations are long and costly.

In 2008, 13.5% of world electricity was generated by nuclear power plants, corresponding to 2 731 TWh/year, while 67.5% had its origin in fossil fuels (41.0% coal, 5.5% oil, and 21.0% natural gas), and 19.0% in renewable energy (IEA, 2009). To replace all the fossil fuels used to generate electricity by nuclear energy would have required about 2 195 additional nuclear reactors in addition to the 439 that were operating worldwide in 2008. This substitution would create tremendous security problems and is clearly inconceivable. Most energy scenarios project nuclear energy to continue generating only about 13% of world electricity up to 2035.

Apart from security concerns, there is the additional problem that the presently known terrestrial reserves of uranium would only be sufficient to sustain global energy consumption for a period of 5 to 25 years (Hofferth, 2002). The oceans contain large amounts of uranium, about 3 grams per cubic metre of sea water, but although methods are known for extracting it, the cost of its use on a large scale would be enormous.

Another solution would be to use reactors that generate fissile material at a greater rate than they consume the nuclear fuel. These are the so-called breeder reactors. One possibility is to use ^{239}Pt and ^{238}U, the latter being the most abundant in natural uranium. Plutonium is the fissionable element and ^{238}U plays the role of a fertile element by transmuting into plutonium 239 through the absorption of rapid neutrons. With this technology, the energy that can be potentially extracted from uranium increases by a factor of about 100, but it also radically increases the security problems and the risk of proliferation of nuclear weapons due to the processing of large amounts of plutonium.

The answer to these problems may be found in the fourth generation nuclear reactors that are now being planned, but are not expected to be available for commercial use before 2030. Their primary goals are to use nuclear fuels in a sustainable way, to improve nuclear safety, and to minimize production of long-lived and high-level radioactivity wastes. There are essentially two competing alternatives. One is the fast breeder reactor that uses fast neutrons to produce the fission reactions, transmutes most of the uranium and produces only a small volume of radioactive waste with shorter half-lives, of the order of decades. The other alternative is to use thorium 232 (^{232}Th) as the nuclear fuel, transforming it into uranium 233 (^{233}U), which is a fissionable element. The advantage is that thorium is about three times more abundant in nature than uranium. Furthermore, thorium can be used in nuclear reactors in ways that practically eliminate the production of long-lived radioactive waste.

Despite its complex problems, nuclear fission energy is a very important component in the primary energy sources on a global scale, especially in countries with emerging economies, where growth in energy demand is very high. It has a great advantage relative to fossil fuels, in that CO_2 emissions into the atmosphere are relatively negligible, since they are almost exclusively associated with the life cycle of the nuclear fuel. As regards the future, it will be important to stimulate scientific research to improve the overall security of nuclear energy use, particularly in the design of new nuclear power plants and in the search for solutions to the problems of radioactive waste.

Before the Fukushima I accident there was an upsurge in the interest in nuclear fission energy by the industrialized countries (The Economist, 2007). For the industry to consolidate, consumers would have to be persuaded that it is a safe technology and more competitive in terms of cost, so that it can become less dependent on governmental support. Today, after Fukushima, the development of nuclear power has become much less certain. Fukushima has led to a re-evaluation of nuclear power programs and a review of nuclear safety in many countries. A significant number of old nuclear plants or ones located in seismic zones are likely to shut down. Anti-nuclear feelings have increased all over the world, particularly in some industrialized countries, such as Germany, France, Spain, Taiwan, and the USA. The Chinese government has initiated a review of nuclear power plant safety and hinted that the goals of their nuclear program will be lowered.

Some have gone so far as to suggest that, in the light of the nuclear disaster in Japan, nuclear power should be completely abandoned. The main argument is that humans are imperfect and fallible creatures, unfit to exploit the strongest fundamental force present in the universe in a safe and controlled way. This is a particularly relevant discussion in a globalizing society that currently requires the use of ever more complex and potentially dangerous technologies to sustain the paradigm of continuous economic growth.

3.14 The Promise of Nuclear Fusion

A more promising nuclear solution in the long run could be controlled thermonuclear fusion, although its success as a workable commercial energy source is still uncertain. The fuel is practically inexhaustible and, just like the other nuclear power sources, it has the advantage of making a much smaller contribution to CO_2 emissions into the atmosphere than fossil fuels. The basic idea is to use energy produced in nuclear fusion reactions similar to those occurring inside stars, and in particular in the Sun.

To achieve this goal, the most favourable reaction is the fusion of the nuclei of two hydrogen isotopes, namely deuterium (^{2}H) and tritium (^{3}H), which are made of a neutron and a proton and two neutrons and a proton, respectively. When the two nuclei approach each other very closely, the nuclear force comes into play, producing a helium nucleus (^{4}He), made up of two neutrons and two protons, and a free neutron, which is released with very high kinetic energy. It is this energy, first converted into heat, that can later be converted into electrical energy in the fusion reactor. The problem lies in getting the two nuclei close enough, since they both have positive electrical charges and thus repel one another. The solution that is currently being investigated in experimental fusion reactors involves confinement of the gas at a very high temperature, in such a way that the random thermal motions are fast enough to create collisions in which the probability of their coming close enough together would be significant. When a gas reaches a very high temperature, the atoms and molecules ionise. The resulting medium of electrically charged particles, electrons and atomic nuclei, is called a plasma.

In the Sun, the very high temperature plasma where fusion reactions occur is contained by gravitational forces, that is, by the weight of the spherical layers that surround the central region of the star. On Earth, we cannot use this mechanism to contain the plasma. The solution involves confinement in a vacuum chamber by means of an intense magnetic field which, by acting on the electrically charged particles, stops them from escaping, at least for a while.

The objective is to produce a deuterium and tritium plasma with a high enough density and temperature, and to be able to contain it for long enough to allow nuclear fusion reactions to occur, and thereby obtain useful energy. Note that, to produce the magnetic fields and very high temperature plasma, huge amounts of energy are needed. The process is only viable if the plasma is contained long enough for the energy produced by nuclear fusion to be greater than the energy spent on heating and containing the plasma.

It has not yet been possible to build a prototype and create the conditions that would produce energy from nuclear fusion in a controlled and sustainable way. In 1997, the Joint European Torus project achieved a peak of fusion power of 12.9 MW, corresponding to 60% of the input power during 0.5 seconds. The main ongoing fusion research programme, called ITER (International Thermonuclear Experimental Reactor), involves the collaboration of the European Union, the Russian Federation, China, India, Japan, South Korea, and the USA. It is budgeted at 10 billion Euros and will last for 30 years. Its objective is to build a reactor that will reach a power output of 5×10^8 W for about 500 seconds. That would already be much more than was obtained from previous experimental fusion projects.

Only with the knowledge and experience resulting from the construction and operation of the ITER reactor will it eventually be possible to plan a nuclear fusion reactor for the commercial production of electricity (Hiwatari, 2005). Due to the time normally involved in planning, licensing, and constructing commercial nuclear reactors, the most optimistic estimates consider that it will not be possible to start using energy produced by fusion for 50 years. So we cannot expect nuclear fusion to contribute in a significant way to replacing fossil fuels and controlling CO_2 emissions into the atmosphere until at least the 2060s.

Fusion reactors produce much less radioactive waste than fission reactors, given the very different nature of their fuels, and those that are produced do not have long half-lives. The risks involved do not include the possibility of uncontrolled nuclear chain reactions, as may happen with fission reactors. There are, however, various problems regarding the use of energy from fusion that have not yet been entirely solved. The neutrons produced when helium is formed by nuclear fusion are very energetic and frequently produce nuclear reactions by colliding with the walls and protecting materials surrounding the fusion chamber and the reactor. These reactions lead to the formation of other elements, some of which are radioactive, and to a gradual alteration of the structure and composition of surrounding materials. There is still a lot to learn about the consequences of these processes before a fusion reactor can be safely maintained in full operation.

Regarding the fuel, deuterium is a stable hydrogen isotope which combines with oxygen molecules to make heavy water. It exists naturally in significant amounts in the oceans, which means that the supply is practically inexhaustible. Curiously, this deuterium is very old, given that it formed in the primordial nucleosynthesis in the first few minutes after the Big Bang. The case of tritium is different, being a radioactive isotope with a half-life of 12.33 years that turns into helium 3 (^{3}He) through the emission of an electron and an antineutrino. A continuous supply of tritium must be produced, and this can be done by nuclear reactions, the most convenient being the collision of neutrons with lithium 6 (^{6}Li) nuclei. The supply problem depends therefore on the availability of lithium. It is also important to emphasize that tritium, from the chemical point of view, behaves just like hydrogen and constitutes a hazardous radioactive gas, difficult to contain inside the reactor. Several problems regarding the use of tritium as a fuel have not yet been fully solved.

Nuclear fusion energy is clearly one of the possible future solutions to the growing demand for energy, in the mid- to long-term future, particularly as fossil fuel reserves become exhausted. However, it is still only a hope that may be turned into a reality, something that is very unlikely to happen for at least half a century. If we are able to construct economically viable, safe, and sustainable nuclear fusion power plants, it will be necessary to adapt the energy infrastructure at the national and regional levels to the new reactors, which generate an extremely larger power output compared with fossil fuel power plants.

3.15 The Promise of Modern Renewable Energies

Besides fossil fuels and the various forms of nuclear energy, we also have renewable energies. Currently, the most important source of renewable energy at the global level is biomass, in the form of wood and organic residues that are used mostly in developing countries. These energy sources, classified by the International Energy Agency as combustible renewable and waste, corresponded in 2003 to 10.8% of world energy supply, that is, 4.7×10^{19} J (IEA, 2004a). Only 14.5% of this energy was produced in the OECD countries. The total of all other renewable energies represents only 2.7% of the total primary global energy sources. The most important was hydroelectricity, with 2.2%, whereas the others, sometimes referred to as modern renewable — geothermal, solar, wind, tides, and waves — accounted for 0.5%, of which the greater part, 65.5%, was generated in the developed countries, in contrast with the situation regarding biomass. These numbers reveal clearly the long road that must be followed before modern renewable energies can sustain a significant share of the global energy supply. The annual rate of growth of these energies is vastly superior to the total of all other renewable energies and to the totality of the primary energy sources. In 2003, these rates were 8.2%, 2.3%, and 2.1%, respectively. Among the modern renewable energies, wind is growing fastest worldwide, followed by solar and geothermal.

It is revealing to compare the mix of global primary energy sources in 2007, four years later. The share of combustible renewable and waste decreased from 10.6% to 9.8%, while hydroelectricity remained at 2.2%, and the modern renewable energies increased from 0.5% to 0.7% (IEA, 2009). This is a very large increase, but the global share of modern renewable energies remains extremely small compared with fossil fuels.

3.16 The Limits of Hydroelectricity Exploration and the Promise of Hydrogen

Hydroelectricity is one of the primary energy sources that produces the least pollution and is relatively safe. The main impact of reservoirs on the environment is the reduction of agricultural land and the destruction of local ecosystems rich in biodiversity. There is also the problem of retention of sediments carried by rivers, which contributes to coastal erosion. In the case of some very large dams, the impact can be serious, as in the Aswan dam, which has changed the flooding cycles and contributed to soil degradation in the Nile valley and delta.

In 2004, about 16% of world electricity generation came from hydroelectric power plants. In some countries, the figure is much higher, as in Norway, with 99.5%, Brazil with 87%, Canada with 59%, and Sweden with 54%. However, in developed countries, the capacity for generating hydroelectricity has come close to its limits. In Europe and North America, electricity generation began to stagnate in the 1980s. The regions where an expansion is still possible are Central Asia, sub-Saharan Africa, and Central and South America. It is obviously not possible to increase hydroelectricity generation indefinitely. Estimates of the global capacity for hydroelectricity are of the order of 2.5×10^{19} J (Nielsen, 2006), a value that is less than half the annual global generation of electricity, and a third of the projected figure for the year 2020. In accordance with IEA scenarios, it is unlikely that the generation of hydroelectricity will be more than 1.5×10^{19} J up to 2030. These future projections indicate that hydroelectricity is a primary energy source with a limited capacity for expansion worldwide in comparison with other renewable energies. Hydroelectric dams have the very important property that they can be operated to store energy in the form of water pumped from a lower elevation to the higher reservoir using low-cost off-peak electric power. The stored water can then be used to generate electricity during periods of high demand. This is the largest-capacity form of grid energy storage presently available.

Hydrogen is an energy carrier, not a primary energy source. It is attractive because its combustion produces water and therefore no pollution. However, free hydrogen does not occur naturally so it has to be produced by the electrolysis of water, from fossil fuels, or through biological processes. All these mechanisms require the use of energy. Nowadays, the majority of the hydrogen comes from fossil fuels and therefore involves the emission of CO_2 into the atmosphere. These emissions can be avoided by using renewable energies or nuclear energy to bring about the electrolysis of water. The use of wind and solar energies to produce hydrogen through the electrolysis of water has the important advantage of avoiding the intermittency problem associated with those energies.

Hydrogen has the disadvantage that it is a gas at most temperatures and highly flammable. It will burn at a wide range of concentrations from 4% to 75% (percentage volume of hydrogen in air), which makes it dangerous to handle. In spite of this difficulty, hydrogen is currently widely used in fuel cells that generate electricity from the combination of hydrogen and oxygen to produce water, and also some heat. This is an example of a chemical energy conversion into electricity. Hydrogen fuel cells are a promising technology as a power source for electrically driven vehicles. They can also be used to supply the energy needs of buildings and portable electronic devices. The possibility of an intensive use of hydrogen as an energy carrier has led to a vision of a hydrogen economy. However, there are significant hurdles in the path to this goal, mainly regarding security in transport applications.

3.17 The Future of Solar Energy Opportunities

Solar energy presents significant advantages as far as environmental impact and safety are concerned. Total solar irradiation, that is, the total solar electromagnetic energy that reaches the top of the Earth's atmosphere, per second and per unit area perpendicular to the direction of the Sun's rays at the mean Sun–Earth distance, is called the solar constant, and has an approximate value of $1\,367$ W/m^2. The total solar power received by the Earth, equal to the product of the solar constant by the area of the disc corresponding to the illuminated hemisphere, is 1.74×10^{17} W. In one year, the total solar energy received is therefore about 5.5×10^{24} J. However, about 30% of the radiant energy received from the Sun is reflected to outer space by the Earth's surface, the clouds, and the atmosphere. The Earth only absorbs about 3.85×10^{24} J annually. In order to maintain the Earth's energy balance, this amount of energy is returned to space, but in the form of infrared radiation, not visible radiation.

The Earth is in a state of equilibrium, in which the total amount of radiated solar energy received from the Sun is equal to the total energy emitted by Earth. There is no loss of energy, but there is a degradation of energy: in other words the entropy of the energy emitted by the Earth is much larger than the entropy of the solar radiation it receives, because the lower energy photons of the outgoing radiation are much more numerous than the higher energy photons of the incoming radiation. Essentially, it is this enormous difference of entropy that has allowed for the organization, complexity, and sustainability of all living beings on Earth for thousands of millions of years.

To have a less abstract notion of these numbers, note that the solar energy received by the Earth during one year is about 7 700 times greater than the energy supplied by all the world's primary energy sources in 2007. Thus at present we use much less energy per year than the total amount of radiant energy received from the Sun. The main problem with solar energy resides in the fact that it is diffuse, intermittent, and dependent on local climate conditions. Solar energy can be directly converted into electricity with the help of photovoltaic cells. It can also be converted into thermal energy using solar panels that operate at low temperatures. Another way is to build concentrating solar power systems that use lenses or mirrors and tracking systems to focus the sunlight. The concentrated beam is used as a heat source in a conventional power plant that converts thermal energy into electricity, or in photovoltaic surfaces. The contribution of solar energy to world energy production is still very small, with a value of 0.039% in 2003. This corresponds to a value of 1.7×10^{17} J. However, the growth over the last three decades has been remarkable, with an annual rate of 29%. The greater part of the power generated from solar energy is thermal solar energy at low temperatures, around 88.4 GW, compared with 5.4 GW from photovoltaic cells, in 2005 (REN21, 2006).

The relatively low efficiency of photovoltaic converters, on the order of 10–30%, combined with their high cost makes this energy source uncompetitive when compared with fossil fuels, at least for the time being. Solar photovoltaic cells are likely to become competitive between 2020 and 2030, depending on the investment made in research and technological development and on the price evolution of fossil fuels. Maximum efficiencies of the order of 40% have been reported recently. However, a significant increase in the use of photovoltaic solar energy would require huge investments worldwide. To be able to supply half the current global electricity demand, the production of solar cells would have to be increased by factors of about 3 000.

The large scale production of electricity by means of photovoltaic cells also requires large areas, especially at higher latitudes. To get some idea of these areas and the technological, social, and environmental challenges they involve, let us assume that the total energy requirement of a country is satisfied by photovoltaic cells. Assuming that these cells have a conversion efficiency of 15% and an efficiency of 50% at the local level where the energy is used, it would be necessary to cover with cells an area equivalent to 24% of Belgium, 15.4% of Japan, 10.3% of Germany, but only 0.3% of Australia (Weisz, 2004).

Notwithstanding these limitations, photovoltaic solar energy will become more competitive in the future through investment in research and development to lower the cost of the panels and increase their efficiency. An optimised use of solar

energy could also be achieved with systems of photovoltaic solar panels installed
in satellites in geostationary orbits, transmitting the energy to Earth by means of
electromagnetic radiation in the microwave frequency range. This space technology
is likely to be developed only in a more distant future.

Very ambitious projects for the large scale use of solar energy are being developed right now. The Mediterranean Solar Plan, launched in 2008, aims to promote
various forms of sustainable development in the countries bordering the Mediterranean. One of the objectives is to generate 20 GW of renewable energies by 2020,
3 to 4 GW by photovoltaic systems, 5 to 6 GW using the wind, and 10 to 12 GW
with thermal solar concentration systems. The DESERTEC project pursues the objective of a sustainable supply of electricity for Europe, the Middle East, and North
Africa up to 2050. As regards the European Union, the objective is to cover about
15% of the energy demand with solar energy at an estimated cost of about 400 billion Euros over 40 years. These and similar projects in other regions of the world
require very large investments in new energy infrastructures, particularly in smart
electricity grids and high-voltage direct current grids to minimize energy losses in
the transmission of electricity over long distances.

3.18 The Future Potential of Wind Power

A very promising primary energy source that has grown very rapidly recently in
some countries is wind power. In 2003, it corresponded to 0.051% of the total primary energy sources, but it grew at an annual rate of 49% between 1971 and 2003.
The greater part of this type of energy, 72%, is produced in just four countries:
Germany, Denmark, Spain, and the USA. However, other countries are investing
decisively in wind power.

Just like any other primary source of renewable energy there are limits to the
production capacity. Estimates point to about 7.2×10^{19} J, which is equal to about
three times the maximum projected total output of hydroelectricity and also equal
to the total production of electricity projected for 2020. Clearly wind power has
the potential to supply a significant part of the global electricity generation demand
over the next 20–30 years, if the necessary investments are forthcoming. Although
priority can be given to remote areas, which implies a higher cost associated with
energy transmission, the strong wind power growth will populate the countryside
abundantly with wind turbines. It will therefore be necessary to accept this intrusion
into the environment and the associated impact on the aesthetics of the landscape as
a lesser evil, in view of the advantages of this energy form.

3.19 Geothermal Energy and the Promise of Oceanic Energies

In 2003 geothermal energy represented 0.416% of the world energy supply, which is 4.6 times more than the sum of solar and wind power. However, between 1971 and 2003 the average annual growth of solar and wind power was much higher than for geothermal energy, which grew at an annual average of 7.6%. Used as a source of heat for millennia, geothermal energy has recently been converted into electricity. The countries that make most use of this energy form are the USA, Indonesia, and Japan. Theoretically, geothermal energy can be exploited anywhere in the world, but there are only a few areas, particularly those that show signs of volcanic activity, where exploitation is economically competitive. The annual global potential for electricity production has been estimated at 4.3×10^{19} J (Bjornsson, 1998), a little more than half the potential of wind power. Note, however, that, as with the other renewable energy sources, this type of estimate involves a large uncertainty because it does not include future technological innovations that could make exploitation more competitive. Further into the future, science and technology may find ways to convert the gigantic amounts of thermal energy stored in the interior of Earth into usable energy forms.

The oceans also constitute an enormous reserve from which we can extract useful forms of energy. The kinetic energy from the cyclic movements of the tides can be transformed into electricity by installing turbines or other devices in river mouths. Naturally, the efficiency of the energy conversion increases with the amplitude of the tide. Amplitudes in the range of 5 metres or higher are considered competitive, but only about 40 such locations on Earth fit the bill. Systems of turbines can be located on the river beds in estuaries so as not to affect marine transportation. In France, in the estuary of the River Rance in Brittany, the power output is 240 MW. The amount of energy generated from tides is very small, contributing only 0.000 5% of the total primary energy sources (IEA, 2004b).

The ocean waves that are generated and moved by the winds transport huge amounts of kinetic energy. The power with which they break upon the coastal regions of the world is estimated to be between 2 and 3×10^{12} W (McCluney, 2005). However, given the use of coastal regions for various purposes, and also for environmental reasons, it would only be possible to exploit wave energy economically in a few places. Energy converters can remain on the surface or on the bottom of the sea near the coast. There are many projects for the construction of wave farms, some at an advanced stage, but none are being exploited commercially yet. The global potential of energy from tides and waves is estimated to be 7.9×10^{19} J and 6.5×10^{19} J, respectively (Nielson, 2006).

The kinetic energy from oceanic surface currents can also be converted into useful energy, but this potential source has not yet been fully developed. One of the most advanced projects is the Dalupiri project, which plans to install a system of turbines between the islands of Samar and Dalupiri in the Philippines, to produce 2 200 MW of electricity from the strong currents running between the two islands.

We could also exploit the thermal gradient between the surface and the deep layers of the ocean, using a fluid converter placed in a partial vacuum. Ammonia

(NH_3), for instance, can be evaporated by the temperature at the surface, and its expansion used to move a turbine. After passing through the turbine, the gas condenses by being cooled in the deeper waters, and this completes the cycle. Although the efficiency is low, due to the small temperature difference, the process could become economically competitive over long periods of time. The potential for this type of energy conversion, taking into account the huge area of the oceans, is enormous, and has been estimated at 7.2×10^{21} J (Nielsson, 2006).

3.20 Advantages and Disadvantages of Biofuels

Biomass in the form of firewood and other vegetable and animal products is the main source of energy in the least developed countries: in sub-Saharan Africa and South Asia it dominates household domestic consumption. On a global level, fossil fuels only exceeded the use of biomass as a primary source of energy at the end of the 19th century. In fact, the very first automotive vehicles with an internal combustion engine, constructed by Nicolaus Otto, Rudolf Diesel, and Henry Ford, were designed to work with liquid biofuels, especially bioethanol. Now we are witnessing the reversal of the process. There is a clear tendency for a greater usage of liquid biofuels in transport, as a means of increasing the energy security of countries that depend heavily on petroleum imports, and also to limit CO_2 emissions.

A biofuel is a solid, liquid, or gaseous fuel derived from biomass, or in other words, raw materials obtained from recently living vegetable or animal organisms. The most important in terms of world production are liquid biofuels, especially bioethanol and biodiesel for use in motor cars. Currently they represent about 1.8% of the liquid fuels used globally in transportation. Bioethanol is ethanol produced from fermenting sugars in biomass materials such as sugar cane and corn, and to a lesser extent from millet, wheat, beet, and cellulosic materials. The largest producers are the USA and Brazil, with almost identical world production percentages, at around 45%. Biodiesel is produced principally from vegetable oils extracted from the seeds of various plants, such as rape, soybean, genus jatropha plants, or used oils. Germany is currently the largest producer of biodiesel in the world, with about 50% of the total production.

It is important to evaluate the energy efficiency of biofuel production, or in other words, the relationship between the energy generated from the combustion of one unit of mass, and the energy consumed in the production of that unit, including transport and conversion into the final product, and also fertilizers, pesticides, and herbicides used to produce the crop. Generally speaking the energy efficiency is greater in the biofuels produced from crops grown in tropical and sub-tropical regions than from those grown in the more temperate areas, although there are other important factors, such as the type of plant, soil, and cultivation practices. This difference results essentially from the greater amount of radiant solar energy received per unit of area in the lower latitudes. The efficiency of Brazil's bioethanol production from sugar cane is much higher than that of the USA's production from corn.

The efficiency calculation is essential when making a quantitative assessment of the advantages in replacing fossil fuels by biofuels as a means of reducing the emissions of greenhouse gases, given that the greater part of the energy used in the cultivation originates from those fuels. The reduction of emissions by the use of bioethanol is on the order of 30–40%, when produced from corn and other cereals in the USA and Europe, and about 90% when produced from sugar cane in Brazil (IEA, 2004c). Another important aspect is that the cultivation of plants with a high ratio of nitrogen/carbon composition, such as rape, requires the use of fertilizers that release large amounts of nitrous oxide (N_2O) into the atmosphere, a gas with a strong greenhouse effect. Recent calculations indicate that in such cases the emissions of greenhouse gases can be greater than those that would result from using petroleum instead of biodiesel (Crutzen, 2006).

The production of biofuels is being actively stimulated by the governments of various countries, including subsidies to farmers and other financial incentives. In the USA, the subsidies remain high, although the rise in oil prices has made ethanol production competitive with petroleum. An increasing percentage of corn production is used to produce ethanol, and this is likely to go above 50% in the next ten years. Corn production for the food industry has decreased, causing a serious rise in its price and in the price of other cereals on a worldwide scale, partly because the USA ensures 40% of the world production of maize. This new trend was one of the main causes for the world food price crisis of 2007–2008 and also for the increasing trend in prices in 2010 and 2011.

The growing use of biofuels has established a strong link between the prices of energy and food, with profound implications for producers in the two sectors, for consumers, and for world food security, especially in developing countries. If the price of petroleum remains high, or increases, the production of liquid biofuels will become more attractive than the production of cereals and other food crops. These will become more expensive, aggravating the problems of hunger and malnutrition for hundreds of millions of people. To get an idea of the competing values here, note that to fill the tank of a sports utility vehicle with 100 litres of pure ethanol requires about 200 kg of corn. This amount of corn has enough calories to feed one person for one year (Runge, 2007). The developing countries will be the most affected, and particularly those which are large importers of petroleum and have food supply problems. Recent estimates (Runge, 2007) indicate that in 2020 the price of maize may have risen by 41%, soy, rape, and sunflower by 76%, and cassava by 135%, due to an increasing demand for biofuels. These increases could be avoided if production became more efficient, or if it were possible to produce ethanol from other raw materials, such as trees and grasses.

First generation biofuels are made from food parts of plants, such as seeds and grains, using conventional technology. Second generation biofuels use biomass consisting of the residual non-food parts of plants and trees and involve new technologies to extract useful sugars from the woody and fibrous biomass. Another possibility is to use non-food crops such as switch grass and jatropha.

The species of the genus jatropha are flowering succulent plants, shrubs and trees, often with red flowers, which are very resistant and well suited to semi-arid regions, and whose seeds can be used for the production of biodiesel. It has the great advantage that it does not need soils or climates that would be useful for the cultivation of food crops. In the future, jatropha will probably be an important raw material for the production of biodiesel in emerging countries. Both India and Brazil are making large investments to stimulate its cultivation.

From the environmental point of view, biofuels can also have negative impacts. In some regions, especially the tropics, the search for land to produce biofuels, combined with the search for agricultural land, is intensifying deforestation and increasing the loss of biodiversity. Examples are Malaysia and Indonesia with the production of palm oil, and Brazil with the production of sugar cane.

A critical product selection and management approach that takes into account competition with food production and potentially negative environmental effects can reduce the more serious impacts of biofuel production. Despite the various negative aspects, biofuels constitute an important energy form that should be further developed because it reduces our dependency on fossil fuels and emissions of CO_2 into the atmosphere. Although they have relatively higher carbon intensity compared with solar, wind, and geothermal energy, liquid biofuels are very important for carbon savings in the transport sector for which there are few other options in the short and medium term.

3.21 Energy: The First Conclusions

In the above we have presented a very brief history of energy use and a rather succinct description of the current situationaround the world. The intensive energy use afforded by the availability of relatively cheap fossil fuels sustained the industrial revolution and allowed us to achieve today's admirable civilization, flourishing mainly in the industrialized countries. However, fossil fuels have the problem of producing CO_2 emissions, and at the same time they are a non-renewable resource. Modern renewable energies should thus be actively developed, although their small share of the global primary energy sources means that they will not be able to replace fossil fuels in the short to mid-term. It is essential to understand these difficulties and to promote a more sustainable use of energy, implementing more efficient energy systems and undertaking savings in consumption.

The end of the fossil fuel era or the hegemony of coal, petroleum, and natural gas is inevitable. We are entering a difficult transition phase that will require large investments in scientific research, technological development, and new infrastructure in the energy sector. The interim period between the fossil fuel era and a new era, still largely undefined, of renewable energy has been called the era of extreme energy (Klare, 2009). The name is particularly adequate to describe the current situation with regard to oil and natural gas. At the present time, most oil and natural gas comes from conventional sources, typically large underground reservoirs on land or beneath shallow coastal waters.

In the future, non-conventional oil and natural gas sources, such as oil sands, oil shale, and shale gas will be increasingly explored, with the serious disadvantage that their greenhouse gas footprints are larger than coal or conventional oil and natural gas. Furthermore, conventional sources of oil and natural gas are becoming increasingly difficult and problematic to exploit. New and extensive oil fields are found almost exclusively offshore, at great depths, or in remote regions where exploration is difficult, such as the climate-altered Arctic.

Deepwater Horizon, an ultra-deepwater offshore drilling rig, was supposed to reach a depth of 5 596 m while drilling a well for BP in the Macondo Prospect oil field in the Gulf of Mexico. On 20 April 2010, an explosion on the rig caused by a blowout killed 11 men and injured 16 others, releasing about 4.9 million barrels of crude oil into the ocean. It was not until 19 September that the well was declared effectively dead. The oil spill, one of the worst in the history of petroleum exploration, caused extensive damage to marine and coastal ecosystems as well as to the Gulf's fishing and tourism industries.

The era of extreme energy is very likely to create extreme environmental risks. The main challenge is to accelerate the transition to the renewable energy era. Ensuring an abundant and relatively inexpensive energy supply is an essential condition for continuity of the current paradigm of economic growth. The extent to which this paradigm will be affected by the new transition is an open question that will be addressed in Chap. 5.

But let us now consider another natural resource that is fundamental for development — water.

3.22 The Cycle, Reserves, and Use of Water on the Blue Planet

The most remarkable and significant feature of Earth, compared with the other planets in the Solar System, is that it has water in abundance in the liquid state. We live on a blue planet, whose oceans were essential for the appearance of life. The cells which are the basic structural unit of all living organisms are almost entirely made up of water.

Water is a renewable resource, made continually available by way of the hydrological cycle, driven by the solar energy. One of the fundamental properties of this cycle is that, in the oceans, the amount of rainfall is smaller than the amount of water that is evaporated. So water is transferred from the oceans to the continents in the form of water vapour that originates precipitation, feeds the streams and rivers, and replenishes the water in soils, lakes, aquifers, glaciers, and ice fields. The annual outflow from the continents into the oceans, variable from year to year, has an average value of approximately 40 000 km^3 (Shiklomanov, 1993). In the period from 1920–1995, the annual average was 42 000 km^3. About two thirds of this volume runs off in floods, so that only approximately 14 000 km^3 can be regularly used. In fact, this value corresponds globally to an average annual water availability per capita greater than $2\,000\ m^3$, which would be sufficient to satisfy the water requirements of every inhabitant on Earth.

The reality, however, is very different. The problem lies in the considerable asymmetries in the temporal distribution of precipitation through the year, and especially in the spatial distribution of precipitation around the globe, which is frequently completely unrelated to the population distribution. The spatial asymmetries result from the general circulation of the atmosphere and the geographic distribution of oceans, continents, and mountains. In certain regions in north east India, the annual precipitation reaches more than 10 metres, while in some deserts, like the Atacama in Chile, there is frequently no rainfall during the entire year. A long strip of semi-arid, arid, or desert regions stretches from the north east coast of Africa, crossing great swathes of the Middle East and Central Asia up to northern China. In this vast area, which supports a significant part of the human population, rain is scarce, falling only during a period of a few months, and with a large interannual variability.

Of the total reserves of water on Earth, with a volume of 1.4×10^9 km^3, 97.5% is salt water in the oceans and salt lakes. Only the remaining 2.5%, corresponding to 3.5×10^7 km^3, is freshwater, and of that, the greater part, 68.7%, is stored in glaciers and ice fields, while 30.1% in aquifers. Most freshwater resources are obtained from rivers, lakes, soil humidity, and low-lying aquifers. These correspond to about $200\,000$ km^3, which is much less than 1% of the total amount of freshwater. It would be possible to increase these resources by exploiting deeper aquifers, but with costs that are not currently economically viable.

Since the emergence of agriculture, about $10\,000$ years ago, most of the water has been used for crop irrigation. The great civilizations of Egypt, Mesopotamia, and China were heavily dependent on the use of water from their large rivers to develop agriculture and trade through fluvial transport. Currently, the greater part of the global use of water is for agricultural irrigation, estimated at about 70%, but it should be noted that 15 to 35% of withdrawals are unsustainable (WBCSD, 2009). The gigantic increase in world water consumption caused by better living conditions, new lifestyles, and the increase in the global population represents one of the most serious challenges currently facing humanity. Estimates indicate that the use of water rose by a factor of 40 from 1700 to 1990 (McNeill, 2000), and that it tripled in the last half of the 20th century (Brown, 2006). During the same period of fifty years, the per capita use of water increased by more than 60%. Currently, the annual global extraction of water has reached about $4\,000$ km^3, and the average world value per capita in 2001 was 650 m^3, with huge disparities, going from $1\,900$ m^3 in the USA to about 250 m^3 in Africa (Gleick, 2004). It is estimated that about 1 100 million people do not have access to safe drinking water, while $2\,400$ million do not have basic sanitary installations (UNEP, 2008). It is also estimated that 450 million people in 29 countries suffer from water shortages at the present time (UNEP, 2008).

It is important to distinguish between the volume of water extracted from rivers, lakes, and aquifers and the volume of water actually used. Only about 55% of the water extracted is used, due to evaporation and leaks in irrigation systems and losses in transport and supply systems. In the richer countries, the consumption per capita started to level off from the 1970s, due to increased efficiency in distribution systems and anti-pollution legislation. In the poorest countries, as consumption increases, the problem of pollution worsens. For thousands of years, the rivers served as sewers for human activity and it was possible to dilute the residues, making pollution almost innocuous due to the large volumes of water involved. But this solution is no longer possible.

It is calculated that, in the developing countries, around 90% of human waste and 70% of industrial waste is thrown into rivers, streams, and lakes without any previous treatment. More than 3.6 million people die each year from diseases caused by unsafe drinking water, lack of sanitation, and insufficient water hygiene (WHO, 2008), which is about seven times more than the recent average annual figure for the victims of war and armed conflicts. This is without doubt the most serious problem of pollution in terms of human lives. And the problem of water scarcity is likely to get much worse. Over the next few decades, water shortages will be more frequent in poorer countries where resources are already overstretched and the population is growing rapidly. Large urban and peri-urban areas will require new infrastructures to provide safe water and adequate sanitation.

3.23 Competition for Water by Various Sectors of Human Activity, Inequality of Access, and Pollution

Over the last few decades the competition between the agricultural and domestic sectors and industry for the consumption of water has been aggravated. This question illustrates the risks involved in our present development paradigm coupled with a continued increase in the world population. At the moment, the distribution of global consumption is roughly 70% for agriculture, 20% for industry, and 10% for the domestic sector. To be able to feed the world population, it will be necessary to considerably increase the extraction of water for irrigation, but in strictly financial terms the productivity of water in industry is much greater.

Let us consider an example. One thousand tons of water are needed to produce 1 ton of wheat, whose value is about US $ 270 at the end of 2010, while the same quantity of water could be used to produce about 70 tons of steel with a value on the order of US $ 40 000. In the context of a policy that strongly favours economic expansion and the creation of jobs, the scarcity of water tends to place agriculture in a secondary position. On the other hand, rising urbanization in large cities is another powerful source of competition for water usage, particularly in regions with water scarcity, like Mexico City, Cairo, and Beijing.

Many countries are exploiting their water resources in an unsustainable way in order to face growing water needs. That is the case in China, India, Pakistan, the USA, Mexico, and most countries in the Middle East and North Africa. A recent report from the World Bank (Financial Times, 2007) warns that, according to its estimates, the water availability per capita in the Middle East and North Africa will be reduced to 50% in 2050, with very serious social and economic impacts. This region currently houses 5% of the world population, but benefits from only 1% of the water resources. The volume of water used is superior to precipitation, which means that water resources are being exploited above their regeneration capacities. It will be necessary to develop sustainable water management policies, integrating both the agriculture and energy sectors.

Desalinization is a possible solution, but it is an expensive answer since it requires a lot of energy. Reverse-osmosis plants require about 3.7 kWh of electricity to produce 1 000 litres of drinking water. Recent technological developments have shown that solar energy can be used directly to desalinize water, thereby considerably reducing the use of electricity and hence also the cost of desalinization. Although the lowering of the water tables in aquifers due to their overexploitation may sound relatively abstract, since it is not directly visible, the decrease in the average runoff of some of the world's great rivers, like the Colorado, the Huang He, the Indus, and the Ganges is clearly evident to many millions of people. Furthermore, some lakes are decreasing in volume or even disappearing.

A large part of the water in the rivers Syr Darya and Amu Darya that feed the Aral Sea was diverted to irrigate extensive cotton plantations in the 1950s, decreasing the river flow into the lake by 90%. Consequently, the area of the lake was dramatically reduced and the salt concentration increased. The fish died, destroying the fishing industry that had an annual production of 50 000 tons. In some locations, the water line of the lake receded 250 km, exposing a sandy bottom, now covered with salt. Every day, tons of sand and salt transported by the wind are polluting the surrounding areas and decreasing agricultural productivity. Although the development of agriculture was indeed the initial motivation, it ended up becoming a victim of the environmental degradation that resulted from unsustainable water exploitation. Currently, there are encouraging signs that the recently launched recovery plans for the Aral Sea are improving the environmental situation. The lake will never be recovered, but the situation in the river deltas has been improving, mainly thanks to more diversified and sustainable agriculture.

Nowadays the competition for water resources reaches extremely basic situations, as in the case of drinking water versus the industrial production of soft drinks. There is clear evidence that the construction of a Coca-Cola plant in the village of Plachimada in Kerala, India, in 2000, has led to a scarcity of drinking water in the region due to over-extraction of ground water (The Times of India, 2011). Coca-Cola indicates that it now requires only 2.1 litres of water to produce one litre of soft drink, which is a considerable technological improvement, but still a huge problem in regions with scarce water resources and large populations. Furthermore, the Kerala plant caused health problems through environmental and soil degradation due to the disposal of sludge containing cadmium and chromium, and a decline in agricultural production. There are similar situations in other parts of India where aquifers are overexploited, especially in the summer, when the demand for water becomes critical and the production of soft drinks increases to satisfy the higher demand.

3.24 Overexploitation of Marine Resources, Ocean Pollution, and Sustainability of the Coastal Regions

The oceans cover about 71% of the Earth's surface, contain about 97.5% of the water reserves of the planet, and play a fundamental role in the climate system. They support the marine ecosystems that provide fish and other food resources in abundance. Furthermore they serve as a sink for a large part of the wastes resulting from many human activities.

For many thousands of years, fishing activities were relatively inefficient, but recently, in the last half of the 20th century, the situation has changed radically due to improvements in fishing technologies and fish conservation. The annual world capture production of fish, crustaceans, and molluscs increased by a factor of 4 from 1950 to 1990 and later stabilized between 80–90 million tons (FAO, 2009). Values around 85 million tons probably constitute the maximum sustainable limit for fishing resources. The greater part of the catch is made up of forage fish found in the lower levels of the marine food chain, such as herring, sardine, and anchovy. In recent decades, the capture of these three species amounted to about 25% of the world total. This selective exploitation seriously disturbs the ecology of coastal marine ecosystems (Jackson, 2001).

There are also signs that the decline in stocks extends to the great ocean expanses. Recent estimates (Myers, 2003) indicate that the global ocean has lost about 90% of the great predators — tuna, swordfish, marlin, cod, skate, and flounder — since the beginning of the industrialization of fishing activities in the 1950s, with potentially serious consequences for the preservation of the marine ecosystem. It is probable that some of these predators at the top of the food chain will disappear by the middle of this century (Jenkins, 2003).

Unsustainable fishing, destined partly for the transformation of fish into sub-products such as fish flour and fish oils, for agriculture and other ends, has been at the origin of various serious critical situations of overexploitation of stocks, like cod since the 1960s, the northeast Atlantic herring since 1968, and anchovies on the Peruvian coast since 1972. These crises led to the establishment of exclusive economic zones, stretching 200 nautical miles from the coast, and to the implementation of policies for the sustainable management of fish resources, particularly within the scope of the United Nations. But many regional stocks continue to be over-fished and show clear signs of biological degradation. There are many reasons to be concerned, especially since we do not know much about the structure and functioning of marine ecosystems, nor about the effects of over-fishing.

A particularly dramatic case is the relentless hunt for tuna. As tuna production continues to increase, breeding stocks plunge, particularly those of the southern and Atlantic bluefin. If the over-fishing is not stopped, the Atlantic bluefin that spawns in the Mediterranean could disappear from its waters before 2015. Although the fishing companies know that the only way to save the business is to save the species, the recommended temporary bans on fishing have been rejected.

Another sign that we are reaching the limits of sustainability in the oceans is the proliferation of jellyfish. Their numbers are increasing, particularly in Southeast Asia, the Mediterranean, the Gulf of Mexico and the North Sea due to over-fishing, eutrophication, and increasing surface seawater temperature caused by global warming. This phenomenon tends to decrease fish stocks since jellyfish feed on fish eggs and larvae. Furthermore, larger nitrogen and phosphorus run-off into the oceans causes red phytoplankton blooms that create low oxygen zones where jellyfish survive but fish do not.

The oceans are affected, for more or less long periods of time, by innumerable sources of pollution: pathogenic elements from urban and livestock industry sewers; excess nitrogen, phosphorus, nutrients, and sediments from agriculture and from coastal and mining activities; and persistent toxic substances like PCBs, DDT, pesticides, detergents, and heavy metals from cities, agriculture, land and sea transportation, and petroleum industries. There is also the pollution associated with oil spills from tankers and from petroleum exploration in the sea, plastics, and radioactive materials. Given that the larger part of these polluting agents originate on land, their effects tend to accumulate in coastal areas, especially adjacent to regions with a high-density population, and semi-enclosed seas where the circulation and renewal of water is limited. All these facts become more worrying given the general lack of detailed knowledge about the way the oceans and their ecosystems are affected and react to the various kinds of pollution on different time scales.

The coastal areas are simultaneously one of the regions of Earth with the greatest biological productivity and with the highest population densities. About 40% of the world's population live in a 100 km wide strip along the coasts, and this constitutes only 20% of the total land surface. This enormous and expanding population inevitably exerts a great pressure on coastal resources. A large number of coastal habitats in various regions of the world, especially coastal wetlands, mangroves, and coral reefs, are in danger of being destroyed due to overexploitation of resources, pollution, sedimentation, and erosion, in addition to urban and industrial development, and tourism activities. The sustainable management of these coastal zones is thus one of the main challenges that humanity faces at the beginning of the 21st century.

In the middle and long term, the most serious risk for coastal areas is the increase in the average sea level due to climate change. According to current estimates, the ice sheets of Greenland and West Antarctica are likely to initiate an irreversible process of thawing if the global average temperature of the lower atmosphere increases between 1.5 to 3°C above its pre-industrial value (Gregory, 2004). An increase of 3°C corresponds to greenhouse gas emissions that were not sufficiently reduced to keep their atmospheric concentration below the range 535 to 590 ppmv in equivalent CO_2 (concentration of CO_2 that would cause the same level of radiative forcing as the mixture of greenhouse gases present in the atmosphere). The complete melting of the Greenland ice sheets would take many centuries and raise the average sea level by about seven meters. In Antarctica, the western region is more vulnerable and likely to melt before the eastern region. The complete melting of the Antarctica ice sheets would take much longer and raise the average sea level by more than 60 meters. Although this is a long term scenario, it is important to realize that it is a likely scenario if greenhouse gas emissions are not curbed fast enough. If it does happen, most of the present coastal regions would disappear, causing profound transformations in human societies throughout the world.

3.25 Deforestation Continues to Worsen

Forests were one of the first ecosystems to be intensely exploited and devastated by man. It is known that, in interglacial periods, the area naturally covered by forests is much greater than during glacial periods. When the last glacial period ended about 12 000 years ago, forests initiated a process of conquest of the vast spaces previously covered by steppes and tundra. However, this expansion did not last long. The history of deforestation begins with the emergence of agriculture and is responsible for a 20–25% reduction of the global forest area (Mather, 1990; FAO, 2001). Is very difficult to make a precise assessment of the historical evolution of forested areas on a regional and global scale because different definitions and concepts of forestry have been used and the quality of the data has a strong spatial and temporal variability. Currently, the situation has greatly improved with access to remote-sensing data, which provides a much more reliable form of monitoring.

There are essentially two main reasons for cutting down forests: to obtain wood for housing construction and industrial purposes, in particular to make wood products, and to generate energy and land use change, mainly for agriculture, to create new settlements, for urban expansion and open-cast mines. Throughout history, the destruction of forests was particularly intense in the Mediterranean, the Middle East, India, and East Asia following the emergence of the great civilizations, in Europe during the Middle Ages, and in Africa and North America after colonization. At the present time, there are only three large forest areas with a significant spatial continuity: the river basins of the Amazon and the Orinoco in South America; the northern part of Eurasia, from Scandinavia to the Far East; and the Congo River basin in Africa. The other vast forest regions have disappeared or have been reduced to scattered groups of relatively small areas, as is the case in Central and Southern Europe, the Anatolian Peninsula, India, Indonesia, China, Madagascar, and the Atlantic coast of Brazil.

Currently, the forests in the temperate zones of North America and Eurasia have stabilized or are increasing, while the tropical forests continue to decline strongly. The end of the long period in which the temperate forest area decreased is a consequence of a much lower population growth, less demand for agricultural land due to increased productivity and food imports, and high imports of tropical varieties of wood. To a certain extent, there has been a transfer of deforestation from the developed countries to the developing ones. However, within the temperate zone, there are still some regions where deforestation is a problem, such as some parts of the Mediterranean and Australia, mainly due to forest fires. Globally, the FAO estimates that the average annual reduction of the forest area was 8.9 million hectares in the 1990s and 7.3 million from 2000 to 2005 (FAO, 2005).

In tropical regions, there are two distinct types of forest: the rainforest and the seasonal tropical forest or monsoon forest. These two forest ecosystems are mainly found in Central and South America, West and Central Africa, and South and Southeast Asia, covering an area of approximately 1 500 million hectares. By halfway through the 1990s, intensive deforestation had reduced that area by about one third (McNeill, 2000). Of the three areas with tropical rainforests, the highest rate of deforestation is in South America, followed by Southeast Asia and Africa. The main driving forces for deforestation are subsistence and intensive agriculture, ranching and pasture, and logging. In South America the main driving force is ranching and pasture, while in Southeast Asia and Africa it is agriculture, both subsistence and intensive. One of the key pressures underlying deforestation is the production of livestock feed, which is needed to satisfy the global increase in meat consumption, driven largely by the emerging economies.

Using high-resolution satellite images, it has been possible to estimate that the average annual rate of tropical forest destruction between 2000 and 2005 was 5.4 million hectares and that about half of the deforested land is located in Brazil. If this rate of deforestation continues, the tropical rainforests will disappear in the next 100 years. The situation in Amazonia is deteriorating. Some estimates indicate that in the next 20 years, with the present rate of deforestation, 40% of the Amazonian forest will be destroyed and a further 20% degraded.

What are the reasons for the concern about the rate of destruction of the tropical rainforests? The first comes from the fact that deforestation increases erosion and fluvial transport of sediments, degrades soils, and upsets the water and carbon cycles. In the Amazon area for example, about half the rainwater is returned to the atmosphere through evapotranspiration. Cutting down the forest significantly reduces this circulation, and according to models used to simulate the climate system, it tends to reduce local precipitation, with negative consequences for the equilibrium of the remaining forest and for agricultural productivity. It is also likely to have an impact on the climate on a global scale (McGuffie, 1998).

Secondly, forests are one of the main sinks of atmospheric CO_2 at the global level. Large scale deforestation interferes with the carbon cycle by increasing emissions of CO_2 into the atmosphere through the decomposition of vegetable matter and forest fires. Currently, land use change is the second largest global source of CO_2 emissions, with about 20% of the total, the first source being emissions from fossil fuel combustion. By cutting down trees, we are contributing to the problem of anthropogenic climate change. By foresting and reforesting, we are contributing to solving this problem.

Thirdly, the tropical rainforest is the land ecosystem with the greatest biodiversity. Although it covers only about 6% of the Earth's surface, it is estimated to be home to between 50% and 90% of all the species of flora and fauna on our planet. The rate of destruction of the tropical rainforest is nowadays one of the main threats to land biodiversity. There are many illustrative examples of the great diversity of plants and animals that inhabit the tropical rainforests. In a single hectare of Amazonian rainforest in Ecuador scientists found 473 species, 187 genera, and 54 families of trees, more than half the total number of species of trees in the whole of North

America (Valencia, 1994). Many tropical tree species with localized distributions depend on specific ecological relationships with other species, particularly pollinators, so that the destruction of small areas can be fatal to their survival. Fragmentation of tropical rainforests increases the mortality among the older trees, whose age could be many hundreds of years, and their population may not recover. It is very probable that, halfway through this century, the great forest blocks of the Amazon and Congo Basins will be strongly reduced and fragmented with as yet unknown consequences for biodiversity and for the regional and world climate.

The upper aerial layer of the tropical rainforest or forest canopy, sometimes called the last biotic frontier, is a habitat that has supported the evolution of thousands or even millions of species of plants, microorganisms, insects, birds, and mammals rarely or never found on the forest floor. It is now clear that this ecosystem plays an essential role in the maintenance of biodiversity and the resilience of the forests it inhabits, although there is still much to learn about the way it functions.

Besides logging, forests all over the world are also being more frequently destroyed by large uncontrolled wildfires. Forests fires are a natural occurrence which contributes to the renewal of the ecosystem, but today a large majority are of human origin. Forest fires of great intensity and frequency, like those that have happened in the Amazon and in Southeast Asia in the last few years are a significant cause of biodiversity loss and bring into question the sustainability of the rainforest. In the last few decades, extensive and relatively more frequent wildfires have also been reported in Australia, the south west of the USA, and southern Europe. In July and August 2010, several hundred wildfires broke out across Russia during a drought, following the hottest heat wave ever recorded in Russian history. The increasing number of wildfires is partly caused by more frequent droughts and heat waves, a tendency that is very likely to have its origin in anthropogenic climate change.

3.26 Why Is Biodiversity so Important?

Biodiversity is the abbreviated form of 'biological diversity', a term first used in 1985 by Walter G. Rosen, in preparatory meetings of the National Forum on Biodiversity, which took place in Washington in 1986 (Wilson, 1988). Nowadays it is a word frequently used by environmentalists, economists, and politicians, having acquired a high degree of notoriety, both at the national and international levels. Biodiversity is the diversity of life forms at all levels of biological organization. Today's biodiversity is the result of 3.8 billion years of life's evolution.

It is important to realize that biological diversity manifests itself in three fundamental hierarchical levels of biological organization: genes, species, and ecosystems. The genetic diversity results from the genetic variation between individuals of the same population and between different populations of the same species. The diversity of species represents the different types of animals, plants, and other organisms that constitute a certain ecosystem, and the diversity of ecosystems represents the diversity of habitats that are found in a given area. Recently, various authors

(Szaro, 1990; Noss, 1992; Wilson, 1988) have defined a fourth type of biodiversity identified as landscape diversity, based on the concept that the landscape is a heterogeneous area characterized by a pattern of geological formations, soil types and uses, and interactive ecosystems.

The importance of biodiversity results from the implications it has in practically all areas of life and human activity. There are various definitions for the concept. One of them, included in the Convention on Biological Diversity (UNEP, 1992), defines biodiversity as the variability among living organisms from all sources, including among other things terrestrial, marine, and other aquatic ecosystems, and the ecological complexes of which they are part; this includes diversity within species, between species, and of ecosystems. One other definition, adopted by the Global Diversity Strategy (WRI, 1992), proposes that the biodiversity of a region is the totality of its genes, species, and ecosystems.

Natural ecosystems provide humanity with a vast and diverse set of services essential for life. On a global scale, the most important is that the biosphere is one of the components of the Earth's climatic system, playing a key role in assuring the stability of the chemical composition of the atmosphere and regulating the climate. Photosynthesis in green plants has an essential function in the carbon cycle. Observations and studies made to quantify it indicate that, in the period from 1990 to 1999, plants and other organisms with photosynthesizing abilities were responsible for sequestering from the atmosphere on average 5.1×10^{12} kg of CO_2 per year (IPCC, 2001). It is difficult to imagine the devastating consequences that would result from the elimination or serious curtailment of that function of the biosphere. One should also remember that the oxygen we breathe in the atmosphere, corresponding to about one fifth of its volume, and the oxygen dissolved in the oceans, which is essential for marine life, was accumulated there over thousands of millions of years through the process of photosynthesis.

A particularly important service provided by natural ecosystems is the generation, maintenance, and fertility of soils. In one gram of agricultural soil, there are billions of bacteria and tens of thousands of fungi, algae, and protozoa. These microorganisms assure the conversion of nutrients, nitrogen, phosphorus, and sulphur into chemical substances that the superior plants, and in particular agricultural crops, can assimilate. The soil ecosystems are also responsible for decomposing the dead organic matter that accumulates in the forests and for recycling their nutrients.

One other crucial function of natural ecosystems is to prevent the dissemination of plagues and diseases that attack agricultural crops, livestock, and domestic animals. Herbivorous insects that may seriously endanger the agricultural productivity have their populations naturally controlled by predator insects that feed on them. The misuse of pesticides that attack the predator as well as the insects harming the plants has seriously disrupted this type of ecosystem service.

In the past, before the discovery of agriculture, the natural ecosystems completely satisfied the food requirements of human communities. Today, this service is restricted almost exclusively to the oceans and coastal zones that provide fish and other types of edible marine organisms. However, it is important to emphasize that all types of food plants that are nowadays cultivated came from wild species with special characteristics favourable to agriculture, which were developed in order to become more productive. Similarly, today's domestic animals were species with special characteristics discovered by man in the large realm of biodiversity of the Earth ecosystems.

This immense genetic library, made up of millions of different species and even more numerous genetically distinct individuals from various populations, continues to be indispensable for humanity. One of the greatest benefits is the potential development of new medicines. More than half the remedies used in the developing countries originated from plants. More than one hundred important drugs in medicine are extracted exclusively from plants, and new discoveries are made every year by exploring and investigating an enormous variety of organisms, from fungi to trees. Two well known and relatively recent examples are the Madagascar or rose periwinkle (*Catharanthus roseus*), and the Pacific or western yew (*Taxus brevifolia*), from which it is possible to extract powerful drugs to combat cancer. By systematically destroying ecosystems which are very rich in biodiversity, such as the tropical rainforests, we are causing or increasing the risk of extinction of many species, some still unknown, which may be of considerable value to humans for medicinal or other reasons.

Finally, biodiversity finds values also in the realms of education, aesthetics, ethics, and culture, more important in future terms due to their fundamental, structuring, and formative character. The beauty of living organisms, the variety of their forms, colours, structures, and functions, and the enchantment of wild regions and natural landscapes are of an inestimable and irreplaceable value for humankind. These values were deeply appreciated in the most diverse civilizations and they are profoundly bound up with our human nature. To deny or disregard them repudiates our cultures and our origins, and destroys the hope of living in harmony with our environment. It means that we will live in an ever more artificial, contentious, uncertain, conflictual, and insecure world. According to Paul Ehrlich, loss of biodiversity is the most serious threat of an environmental nature for our civilization (Ehrlich, 1991). It should not be forgotten that, in the devotional sphere, there are religions like Jainism that confer a sacred value to all living creatures, making their destruction completely unacceptable.

There are ever more people who appreciate direct contact with nature, who know how to identify and recognise a wide variety of species of fauna and flora in the regions of the world where they live, and who contribute to varying degrees in the conservation of nature. However, we are still a minority on a worldwide scale. To learn about, observe, and appreciate nature and its biodiversity when we are young is a precious experience whose benefits and joys will accompany us for the rest of our lives.

My own enthusiasm was in part born by observing wild orchids on the limestone heaths and in the pinewoods that surround Lisbon, where I was amazed by the wide variety of species such as, *Ophrys fusca*, *Ophrys tenthredinifera*, *Ophrys speculum*, *Orchis papilionacea*, *Himantoglossum longibracteatum*, and *Spiranthes spiralis* among many others. I admired the charm and exquisiteness of the forms and colours, the subtlety of the fragrances, the diversity of the species, and the complex mechanisms of pollination and germination of seeds. The number of seeds that result from a single flower is enormous but they are extremely small and have almost no food reserves, requiring the help of a mycorrhiza, a symbiotic association with a fungus, to be able to germinate.

3.27 Are We Witnessing the Sixth Great Extinction of Species?

Ecosystems are always changing and some species become extinct naturally. The history of life's evolution is a continuing process of adaptation by communities of plants and animals to environmental changes, random catastrophes, and permanent competition with other species. A species that is unable to overcome these challenges will soon become extinct. The global rate of extinction has been relatively stable for long periods of time, but there are clear records that there have been events of mass extinctions of species occurring over relatively short periods of time, on the order of hundreds to thousands of years. Over the last 570 million years, there have been five such major extinction events. The most recent, about 65 million years ago, eliminated the dinosaurs along with many other families of animals and plants, and was most probably caused by the collision of an asteroid with Earth. But the most severe mass extinction happened about 250 million years ago and the causes are still not well understood.

It is probable that the sixth episode of mass extinction is happening now, this time caused by certain activities of the human population. Around 1.75 million species have been scientifically identified and described, but the total number is likely to be vastly greater, on the order of 14 million (UNEP, 1995), or greater still according to other estimates. It is impossible to obtain precise values for the actual rate of extinction of species due to the difficulty in assuring a reliable monitoring system worldwide, and the fact that many species are still unknown. There is, however, a general consensus that the rate of extinction is greater, by a factor of 100 to 10 000, than the natural rate of extinction free from anthropogenic interference (Lawton, 2005). The value of this natural rate is in the order of one species per million and per year. Regarding birds, to mention just one concrete example, estimates indicate the extinction of 350 species up to 2050, which is about 3.5% of the estimated total number of species of birds (BI, 2000).

The fundamental reason for this phenomenon with irreversible consequences is the unsustainable exploitation of natural resources to satisfy the growing demand for services and consumption of goods by the human population. Most of the factors that increase the risk of species extinction have their origins in deliberate human activities, such as hunting for various purposes, or in unintended activities such as the destruction or modification of habitats, resulting from land use changes, and the spreading of invasive species. Each species has a certain vulnerability that depends on the characteristics of its habitat, the size and growth capacity of the population, its reproductive strategies, and its ability to adapt to environmental changes.

Over the last 30 years, biodiversity has decreased in a generalised way, but the freshwater ecosystems have been the most seriously affected, due to pollution and overexploitation of water resources. The decrease has been less pronounced in the marine and forest ecosystems, although it is still very significant there. Biodiversity loss is particularly high in developing countries in the tropical regions (Jenkins, 2003). It is important to bear in mind that most of the so-called 'natural' systems have been disturbed more or less significantly by direct or indirect human actions for the last 50 000 years.

Apparently, there are no signs of a profound functional crisis resulting from the replacement of natural systems, rich in biodiversity, by systems that include species introduced, modified, and managed by man, as practised on a large scale in agriculture and forestry (Jenkins, 2003). However, this form of interference has its limits. If the degradation and destruction of natural ecosystems continues indefinitely and in an uncontrolled way, there is a risk of creating critical irreversible situations, where the services provided by the ecosystems decline seriously. It is still not possible to characterise those limits with precision, but it is undeniable that there are thresholds beyond which the consequences will be extremely dangerous for humankind.

The main strategy for species conservation at the present time is the creation and maintenance of protected areas in biodiversity hotspots with an exceptionally high density of endemic species. However, investment is small compared to the scale of the problem. The current annual cost of the worldwide network of nature reserves is on the order of US $ 6 500 million, and about half that amount is spent in the USA alone. Recent estimates indicate that an investment of US $ 45 000 million would be necessary in order to prevent the dramatic loss of biodiversity that has been forecasted for the next few decades (Balmford, 2002).

3.28 Agricultural Development and World Food Security

Historical studies reveal a strong link between the increase in population and the development and flourishing of agriculture, ever since it appeared about 10 000 years ago. The human population at that time, which probably numbered between 5 and 20 million people (Cohen, 1995), grew until it reached about 700 million in 1730, essentially because of the greater ability to produce and distribute food products. Halfway through the 18th century, at the beginning of the industrial revolution, the increase in population became explosive. The main factors that contributed to this evolution were irrigation and the mechanization of agriculture, the intensive use of fertilizers and pesticides, and the increasingly widespread availability of energy.

It was also necessary to extend the area dedicated to agricultural and livestock production. The area expansion accelerated from the beginning of the 18th century through the colonization of North, Central, and South America, southern Africa and the growing use of soils for agricultural production in Russia, China, and India. The main part of these profound changes occurred at the expense of the great prairies in North America and Eurasia. From the beginning of the 1980s, the global area dedicated to agriculture and ranching began to diminish. Much of the land abandoned in the USA, Soviet Union, China, India, and sub-Saharan Africa is unsuitable for agriculture. This process amounted to the destruction of the original natural vegetative cover, and the deterioration and erosion of soils. In other cases, the land became unproductive due to unsustainable agricultural practices that did not ensure soil conservation, and in some regions led to desertification.

Currently, agriculture and livestock production cover about one third of the world area that is capable of supporting vegetation. Approximately 40% of the net primary terrestrial productivity, the useful chemical energy generated through photosynthesis by plants, algae, and bacteria on land, is fully used in various forms by the human population (Ehrlich, 1991). In the oceans, humanity appropriates about 2% of the net primary productivity, which implies a global value of around 25%. Despite the uncertainties associated with these values, they clearly show the enormous effort involved in feeding humankind, and its potential limits.

During the 20th century, the world population quadrupled, while the area dedicated to agriculture and ranching rose by a factor of two, which was only possible due to a huge increase in the efficiency of production. One of the main reasons for the greater productivity was the mechanization of agriculture on a large scale, with tractors and other machines powered by diesel or petrol engines, especially in the USA, Soviet Union, Europe, and Australia. This mechanization led to the relocation of part of the workforce to industry, which contributed significantly to the modern phenomenon of urbanization.

The other determining factor, dating from the beginning of the 1940s, was that geneticists managed to select crop varieties with a greater productivity, resistant to diseases, better suited to the technologies of irrigation and fertilization, and with advantageous characteristics for harvesting and commercialisation. The introduction of selected varieties of wheat, corn, rice, and other cereal crops and of modern agricultural technology in developing countries, known as the green revolution, remarkably improved the agricultural productivity of these countries in the 1960s and 1970s.

As in all human undertakings, there were also negative aspects. The green revolution raised the dependency on irrigation and the use of fertilizers and pesticides and promoted monocultures. It had a greater success in countries like Mexico, India, China, and South Korea than in countries with smaller water resources for irrigation and more difficult access to credit to acquire seeds, fertilizers, and pesticides, as in sub-Saharan Africa. Presently, the capacity of agricultural production, and consequently the ability to feed the growing world population, continues to be heavily dependent on advances in science and technology, as well as access to investment, energy, and water for irrigation.

In Africa, the spread of the AIDS epidemic among agricultural workers and their families is one of the most serious factors contributing to food shortages and hunger, and this situation is likely to persist in the future. According to FAO estimates, the number of hungry people in the world decreased from about 875 million over the period 1969–1971 to about 825 million in 1995–1997. However, from then onwards it has been increasing at an accelerating rate. In the years 2000 to 2002, about 856 million people were starving, 819 million in the developing countries, 28 million in countries with transition economies, and 9 million in the developed countries (FAO, 2004). The number of hungry people rose to 1.02 billion in 2009, up from 925 million in 2008.

This was a dramatic reversal of previous expectations, but at the end of 2010 the number of hungry people has again started to decrease. Still according to the FAO, the annual number of food crises has increased in the last two decades. At the beginning of the 1980s, there were around 15 per year, but after the end of the century it rose to 30. The main reasons for these crises are droughts and other extreme climate events, conflicts, large-scale refugee movements, and dysfunctional governments. Climate change tends to decrease the food security of the world because it induces an increased climate variability that reduces agricultural productivity.

The food security of the world has reached a critical point. The most recent sign of this situation was the crisis of 2007–2008, when food prices experienced their sharpest rise for 30 years. Very probably the crisis was not the result of temporary fluctuations related to the weather, markets, or governance. The crisis revealed fundamental imbalances in the world food system related to: the global rise in food demand; the increasing global meat demand driven by population growth, rising incomes, urbanization, changing diets, and the opening up of markets; and the competition for land availability between biofuels and food crops, and declining yield growth in cereals.

If these imbalances are not addressed in an adequate way, future crises are inevitable. In 1979, about 18% of the total world official development aid was directed to agriculture, but that percentage fell to 3.5% in 2004. World food security was not considered a major issue. Now, after the food price crisis of 2007–2008, national governments in developing countries are implementing policies to invest in agriculture, in particular to subsidize seeds and fertilizers and to intervene in the operation of markets.

The challenges lying ahead are enormous. According to the FAO, world food production needs to increase by 50% by 2030, and to double by 2050 in order to respond to the expected demand. This will require an annual investment of 30 trillion US $. Moreover, this will necessitate a new green revolution based on investment in science and technology to increase food productivity and adapt to a rapidly changing world. Wasted food is also an increasing problem that requires consideration. Recent estimates indicate that 30% to 40% of the food produced in developed countries and 10% to 60% of the food produced in developing countries is wasted without being used.

It is important to realize that climate change will affect food security in a negative way, particularly in the tropical and sub-tropical regions. On the other hand, some areas in northern Eurasia and northern North America will very probably become more productive. Climate change tends to reduce agricultural productivity due to increased climate variability, which includes more intense and frequent extreme weather and climate events, such as droughts and heavy rains that cause flooding, and increasing irregularity in rainy season patterns. Some of these trends are already having an impact on food production. Adaptation to climate change will require research and dissemination of crop varieties better adapted to a changing climate, improvement of genetic resources, improved soil management practices, and improved infrastructure for the efficient use of small-scale water resources.

One of the main objectives of the UN Millennium Development Goals is to reduce the number of undernourished people in the world by half in 2015, as compared with 850 million in 1990. This is an achievable target, but it will require large investments and considerable political determination on the part of the international community. Recent experience has been disappointing. In the last decade, investments in irrigation, rural infrastructures, and agricultural research have decreased, while at the same time, the scarcity of water resources in various regions has worsened. The FAO estimates an annual investment of US $ 44 billion to eradicate hunger by 2015. However, at a recent FAO summit in Rome in November 2009 the commitments from developed countries were a very long way from reaching that target.

There are also increasing challenges as regards the impacts of agriculture on natural resources and the environment. Population growth, rising incomes, and globalization are strongly enhancing the demand for meat and other animal products in many developing countries. For instance, the average annual meat consumption in China was 13.7 kg in 1980, but is currently about 59.5 kg (FAO, 2010). The average in developed countries is higher and increasing, although at a slower rate: it was 76.3 kg in 1980 and has reached 82.1 kg today.

These trends combined with the underlying population growth are unsustainable in the long term. Livestock production is raising the pressure on soils, water resources, and biodiversity. Furthermore, it is also contributing significantly to climate change. More than 18% of the total anthropogenic greenhouse gas emissions as measured in CO_2 equivalent come from the livestock sector, which is more than from the transport sector (FAO, 2006). The loss of world food security on a broad geographical scale is one of the most dramatic risks awaiting us in the near future (Brown, 2011).

3.29 The Great Extension of Drylands and the Risk of Desertification

About 41% of the total land area of Earth consists of drylands, regions in which agricultural production, forestry, and other ecosystem services are limited by water scarcity. According to the definition adopted by UNEP, drylands are areas where the ratio of the mean annual precipitation to the mean annual potential evapotranspiration is in the range of 0.05 to 0.65. This gigantic expanse of land includes the arid, semi-arid, and dry sub-humid areas, excluding the polar and sub-polar regions. Areas where the ratio is below 0.05 are hyper-arid zones, or deserts. The most surprising fact is that about 35% of the world's population lives in the drylands, which corresponds to about 2 400 million people. Many are living on the edge of survival and depend on access to land and water resources to guarantee their subsistence.

The situation has a tendency to worsen, given that a significant part of the drylands suffers from desertification.

The use of the word 'desertification' is relatively recent. André Aubreville, a French forester, introduced it in the middle of last century, after observing the development of desert conditions in some parts of West Africa, near the Sahara, where forests had recently been destroyed (Aubreville, 1949). Meanwhile, various definitions arose, along with some controversy over the nature, cause, and extent of desertification. Finally, in 1966, the United Nations Convention to Combat Desertification came into being and adopted the following definition:

> Desertification is land degradation in arid, semi-arid, and dry sub-humid areas resulting from various factors, including climatic variations and human activities.

Note the complexity of the concept, involving natural and anthropogenic processes in regions with quite diverse climates.

Various types of paleoclimatic and geological records show that deserts have their own dynamics determined by natural climate change, which increase or decrease their range and make them appear or disappear over time. The region of the Sahara provides an example of these processes: 10 000 years ago, large parts of the Sahara were covered with savannah type vegetation, but 5 000 years ago, due to changes in the general circulation patterns of the atmosphere, it became the hyper-arid region that prevails today. This process of desertification was obviously natural, bearing no relation to human activity.

One well-known example of desertification caused by human activity is what happened in the 1930s in some regions of the Great Plains in the central part of the USA. The combination of droughts and unsustainable agricultural practices pursued by European emigrants caused the desertification of these regions, which became known as the Dust Bowl. In the 1970s, the recurrent famine crisis in the Sahel region of Africa called the attention of the international community to the problems of desertification and to both its natural and man-made causes. Rainfall in the Sahel has a strong variability, but since the end of the 1960s, it suffered a reduction of about 20–30% (Hulme, 2001). At the same time, the region underwent profound political, economic, and social transformations. Intensive agriculture, based on monocultures, putting great pressure on the scarce water resources, replaced pastoralism and traditional agricultural practices adapted over many centuries to endure periods of drought.

The main cause of human-induced desertification is the overexploitation of natural ecosystems — soils, pastures, forests — for the production of food and energy. It results from an inadequate response to the pressures of a growing population that expect improvements in their social and economic situation in the context of the economic globalisation process. The unsustainable management of resources causes soil degradation and the emergence or aggravation of poverty and hunger. Excess accumulation of salts, or soil salinization, resulting from the rate of water evaporation being superior to rainfall degrades the soils in arid and semi-arid regions. The build-up of salts occurs mainly in intensively irrigated agricultural areas, but can be avoided by providing adequate drainage water to leach the added salts from the soil.

Recent estimates (Adeel, 2005) indicate that desertification affects about 10–20% of all the world's arid, semi-arid, and dry sub-humid regions, that is, between 6 and 12 million km^2. Nowadays it is one of the foremost global environmental problems, especially regarding the relationship with food production. However, desertification has other grave consequences. By reducing natural vegetation cover, there is an increased risk of losing biodiversity, as well as introducing into the atmosphere large quantities of dust and aerosols produced by dust storms. These particles in turn affect cloud formation processes and regional rainfall patterns. Desertification also interferes with the climate system because it increases the albedo (the percentage of solar radiation reflected by the Earth's surface) and perturbs the carbon cycle, since the reduced vegetation cover decreases the rate of photosynthesis.

Let us now leave the Earth's surface to consider one of the terrestrial systems most important to life — the atmosphere.

3.30 A Brief History of Atmospheric Pollution

The atmosphere is a gaseous layer that covers the Earth's surface, very thin compared to its radius, and retained by the gravitational attraction exerted by the planet. It is made up of a mixture of gases composed essentially of 78% molecular nitrogen (N_2) and 21% molecular oxygen (O_2). The other gases such as argon (Ar), water vapour (H_2O), carbon dioxide (CO_2), neon (Ne), helium (He), and methane (CH_4) are minority components, often referred to as trace gases, but some of them are very important for life on Earth.

The density and pressure of the atmosphere decrease exponentially with altitude, but temperature has a more complex behaviour: it decreases with altitude from the Earth's surface, but at a height of around 10 km it reaches its minimum and begins to rise again. This transition zone divides the atmosphere into two layers. The lower layer, called the troposphere, from the Greek *tropos*, which means to mix, due to the strong rising and turbulent air movements, is where the meteorological phenomena that characterise the weather happen. The upper layer, containing the ozone (O_3), which protects us from ultraviolet solar radiation, is called the stratosphere, and is about 50 km thick. Most of the mass of the atmosphere, about 75%, corresponding to 5.15×10^{18} kg, is contained in the troposphere. The atmosphere is essential to life and co-evolved with it throughout the history of the Earth.

The first instances of air pollution caused by man probably occurred inside the caves where they lived, after the discovery of fire. Many of these Paleolithic caves retain particles of smoke on their walls, from the fires lit by their inhabitants. It is therefore probable that the pollution caused by the burning of biomass, was the first form of atmospheric pollution that affected human health. Still today, millions of people burn biomass to prepare food and to achieve other ends inside the huts where they live in the less developed countries.

Large-scale metallurgy to extract metals from ores, after their discovery about 6 000 years ago, was probably the first form of atmospheric pollution caused by man on a regional scale. The analysis of the air from small bubbles in the ice layers laid down over the years, centuries, and millennia in Greenland, show abnormal concentrations of lead and copper attributed to the smelting of their respective mineral compounds. This methodology reveals relatively high concentrations of lead in the atmosphere during the period of greater development of the Greek and Roman civilizations (Hong 1994). As for copper, the periods of greater pollution coincide with the expansion of minted coinage at the peak of the Roman Empire, about 2 000 years ago, and with the intensive metallurgy associated with the strong economic development during the Sung Dynasty in China between 960 and 1279 AD (Hong, 1996).

Although there are various references to urban air pollution during the Middle Ages, these episodes were relatively benign compared to the pollution that resulted from the intensive use of coal as the main fuel that supported the industrial revolution after the end of the 18th century. The Midlands region of England is also known as the Black Country due to the pollution caused by burning coal for domestic use and by tens of thousands of coal-fired internal combustion engines, used in mines, transport, and factories, particularly of the wool industry.

3.31 Atmospheric Pollution after the Industrial Revolution

The combustion of coal produces CO_2, sulphur and nitrogen oxides, and a range of particulate matter, including carbon black, which constitutes the smoke. The amount of sulphur dioxide (SO_2) emitted depends on the amount of sulphur in the coal. This gas in turn reacts with oxygen and water vapour to produce sulphuric acid (H_2SO_4), one of the compounds responsible for the acid rains that seriously damage forests and vegetation, soils, surface waters, and aquatic ecosystems, buildings, and historical monuments.

In the 19th century, the high degree of atmospheric pollution in some cities in Great Britain began to affect public health, attacking the respiratory system and sometimes causing fatal lung diseases (Clapp, 1994). From the 19th century onwards, industrialization based on the use of coal spread from Europe to Russia, the USA, especially Pennsylvania and Ohio, and Japan, especially Osaka, accompanied by the problem of atmospheric pollution. London, Pittsburgh, and Osaka are examples of cities where the inhabitants suffered serious negative health impacts from very inefficient coal combustion for industrial and domestic purposes. One of the most damaging urban episodes of atmospheric pollution resulted from 'smog', a combination of dense fog and smoke and SO_2 coming from burning coal, that formed in London in 1952 and killed about 4 000 people.

From the beginning of the 20th century, the use of petroleum began to contribute in an ever-increasing way to higher levels of pollution. By the end of the century, emissions from motor vehicles were the main source of atmospheric pollution

(Walsh, 1990). Nevertheless, the prevailing perception and opinion was to consider that atmospheric pollution caused by fossil fuels was a secondary problem. The population grew, purchasing power increased, food quality and home comforts improved, and so did health care, and all of this was the result of industrialization and increased mobility achieved by an industrial revolution powered by coal and petroleum. For the captains of industry and local authorities, the belching smoke of factory chimneys was a sign of prosperity, progress, and power. In 1892, Colonel W.P. Rend, the millionaire businessman from Chicago said that (Rosen, 1995): "Smoke is the incense burning on the altars of industry. It is beautiful to me. It shows that men are changing the merely potential forces of nature into articles of comfort for humanity."

3.32 Current Challenges in the Fight Against Atmospheric Pollution, and Its Impacts on Health

The situation began to change in the 1940s, in part due to the protests of citizens and the consequent political reactions. The air quality in the cities of the industrialized countries improved significantly due to the replacement of coal by petroleum and natural gas, and to a lesser degree by hydroelectricity and nuclear energy. Furthermore, the large industrial installations that are such intensive energy consumers tended to leave the city limits and relocate in industrial parks. Nowadays, the developed countries have a vast array of legislation that makes the fight against atmospheric pollution compulsory. One example of success was the reduction in lead emissions from the use of petrol.

In 1921, an American chemist, Thomas Midgley, invented the technology for adding a lead chemical compound to petrol to make the combustion more efficient. The first company to commercialise the new fuel in Ohio considered it to be a 'gift from god', avoiding all mention of lead in the advertising. However, in the 1970s medical analyses revealed that a significant number of Americans had high levels of lead in their blood, and that this lead came from petrol. Thomas Midgley himself secretly suffered from lead poisoning caused by his own invention, but kept this fact hidden from the public. From that time on, leaded petrol was gradually removed from the market, first in the USA from 1975 onward, then Japan and Europe in the 1980s, and in many more countries in the 1990s. However, the problem still exists in some developing countries.

Meanwhile, most of the problems of atmospheric pollution have migrated to the countries whose economies are in transition, a consequence of urbanization and industrialization carried out with little concern for the environment. In the year 1000 the most populous city in the world was Cordoba, with 450 000 inhabitants. In 1800, at the beginning of the population explosion, driven to a large extent by the industrial revolution, the most populous city was Beijing, with 1.1 million inhabitants, followed by London, Canton, Edo (Tokyo), Constantinople, and Paris. In 1900 London was first with 6.5 million people, followed by New York, Paris, and Berlin. After

1950 the number of megacities with more than 10 million inhabitants increased rapidly, reaching 25 in 2006 (Brinkhoff, 2006). Now the vast majority are located in the developing countries. It is in the large urban areas of these countries that atmospheric pollution is most severe and difficult to eradicate, due to weak political support to implement effective legislation and practical measures and the relatively high cost of these measures.

The consequences of atmospheric pollution on human health have been devastating, although it is difficult to evaluate the number of victims with any precision. According to the World Health Organisation, urban pollution caused mainly by motor vehicles, but also by industry and coal- and petroleum-fired power stations, was responsible at the beginning of this century for the deaths of 800 000 people annually (Kenworthy, 2002). The estimates of the number of victims of atmospheric pollution during the 20th century point to numbers between 25 and 40 million people (McNeill, 2000), and to a clear worsening of the situation in the second half of the last century.

3.33 Local, Regional, and Global Atmospheric Pollution

Atmospheric pollutants have their origin in the sources that produce them and, according to their characteristics, have different residence times in the atmosphere before removal through specific sinks. Due to the general circulation of the atmosphere, it takes about one week to mix a polluting agent to an almost uniform worldwide concentration in the troposphere. If the residence time of the pollutant is less than the time it takes to mix it worldwide, then the resulting pollution only occurs at a local or regional level near the source. However, if the residence time is longer than one week, then the pollution becomes global.

Photochemical smog is an example of local pollution that results from chemical reactions involving sunlight, volatile organic compounds, and nitrogen oxides produced mainly by the internal combustion engines of motor vehicles. This kind of atmospheric pollution occurs mostly in large cities with heavy traffic and a sunny, warm, and dry climate. One of the constituents of photochemical smog is ozone, a powerful oxidizing agent that attacks the respiratory system and is extremely harmful to human health. The groups most at risk are people with cardiovascular problems and children.

Acid rain is a form of regional pollution that generally crosses the borders of various countries and thus becomes an international problem. It originates from SO_2 produced in the combustion of coal, and from nitrogen oxides produced by motor vehicles. Initially, it mainly affected the north of Europe, the west of the USA, and the west of Asia. The problem was recognized and researched from around the middle of the 20th century in Scandinavia, where acid rains result mainly from the burning of coal in Great Britain. It also occurs in the west of Canada and the northwest of the USA, due to pollution originating in the industrial region of the Great

Lakes and the Ohio River valley, and in Japan with origins in the Korean peninsula and China.

From the 1970s, protocols were established to reduce the emissions of SO_2, and these had a significant degree of success, especially in the USA and Europe. However, the problem persists in Asia, due to concerns that the solutions are expensive and would slowdown economic growth, since the main sectors responsible for this type of pollution are energy and transport. Furthermore, the matter is delicate because it involves international policy issues.

Let us now consider a global atmospheric pollutant.

3.34 The Amazing Story of the CFCs

From the end of the 19th century until the end of the 1920s, the substances used as refrigerants in refrigerators were toxic and inflammable, and constituted a major cause of accidents. The hope was therefore to synthesize chemical compounds with similar capabilities but that would not lead to such hazards. The research financed by three large American companies — Frigidaire, General Motors, and Dupont — led Thomas Midgley, the inventor of the lead additive to petrol, to discover, in collaboration with Charles Franklin Kettering, a miraculous new compound, called freon, the first of the chlorofluorocarbons (CFC). Apparently, the new synthetic compound had only advantages. It had the desired properties to serve as a refrigerant, it was not toxic, and it was practically inert from the chemical point of view. Furthermore, CFCs have other uses, as solvents and propellants for sprays. Only after the depression in the USA and World War II did the production of CFCs increase significantly, reaching about 750 000 tons in 1970. A great technological and commercial success had been achieved.

However, in 1974, Mario J. Molina and Frank S. Rowland warned that there was a risk because the use of CFCs could damage the stratosphere through chemical reactions that release atomic chlorine capable of destroying the stratospheric ozone (Molina, 1974). The CFCs were not completely inert from the chemical point of view after all, and constituted a polluting agent on a global scale due to their long residence time in the atmosphere. The prediction of Molina and Rowland, not yet confirmed by observation, passed almost unnoticed.

However, a few years later, in the period between 1980 and 1984, measurements of the ozone concentration in the stratosphere made by the British scientist Joe C. Farman of the British Antarctic Survey in Hadley Bay, Antarctica, revealed surprising results. He observed that, in the month of September, during the spring in the southern hemisphere, the concentration of ozone fell sharply, apparently without explanation, only to rise again some months later. The results were so extraordinary that the scientists initially doubted that the measuring instrument, a Dobson ozone spectrophotometer, was in good working order. A new instrument was sent from Great Britain, but the results obtained the following year were similar. There remained no doubts that, in the southern hemisphere, the stratospheric ozone concentration over Antarctica fell significantly, by more than 30% in the spring.

This phenomenon, known as the ozone hole, allows more ultraviolet radiation to reach the Earth's surface and has hazardous implications for life and in particular for human health, destroying plankton at the base of the oceanic food chain and affecting the photosynthesis of plants. In man it causes cataracts, tends to suppress the immune system, and increases the chances of skin cancer. Note that the stratospheric ozone is in a permanent dynamic process of formation and destruction, due to the continuous interaction of ultraviolet solar radiation with molecular oxygen. Its concentration varies with latitude and throughout the year. However, the Farman's results were surprising because the variation detected was extremely sharp and pronounced. The results of these extraordinary observations were finally published in Nature in 1985 (Farman, 1985).

Curiously, the values reached were so low that they were automatically rejected in observations made by the TOMS spectrophotometer (Total Ozone Mapping Spectrometer) on board a North American satellite, an eloquent example that, in science, one should always be prepared to observe and register the unexpected. Subsequent measurements revealed that an ozone hole was forming in the Arctic as well, although less intense and with a smaller area than the one in the Antarctic, and that the average concentrations of stratospheric ozone in the middle and high latitudes were decreasing slowly.

Meanwhile, research led to an understanding of these phenomena. The molecules that constitute the CFCs are destroyed by chemical reactions in the stratospheric polar clouds. These reactions are favoured by very low temperatures and produce molecular chlorine (Cl_2). At the beginning of spring in the southern hemisphere, when Antarctica begins to receive solar illumination, the Cl_2 is dissociated by solar radiation and forms atomic chlorine that reacts with the ozone and destroys it.

The seriousness of this problem demanded a response that was necessarily international, given that it was a form of global atmospheric pollution. In 1990 the United Nations adopted the Montreal Protocol on substances that deplete the ozone layer in the stratosphere. In 1998 the Protocol was changed in order to accelerate the reduction of emissions and to terminate the production of CFCs and other substances that harm the ozone layer by 2010. Due to the long residence time of the CFCs in the atmosphere, the reconstitution of the stratospheric ozone layer will take several decades. It is probable that the Earth's atmosphere will only see its natural protection against ultraviolet radiation completely restored in the 2070s, about one hundred years after the problem was created.

It is not possible in any reliable way to quantify the negative impact of increased ultraviolet radiation at the Earth's surface on human health. There are indications of an increase in cataracts and skin cancer, but it is difficult to evaluate the impact on the human immune system or on the environment (Gruijl, 1995; Turco, 1997; Norval, 2007).

3.35 Increasing Anthropogenic Emissions of Greenhouse Gases into the Atmosphere

There is another type of global atmospheric change, more serious and more difficult to mitigate than the changes caused by CFCs. The increasing use of fossil fuels initiated by the industrial revolution has caused gigantic emissions of CO_2 into the atmosphere, and part of this has accumulated there. The global atmospheric concentration of CO_2 had increased from the pre-industrial value of around 280 parts per million by volume (ppmv) to an annual mean of 389 ppmv in 2010 (NOAA, 2011). This is very likely the highest concentration in the last 650 000 years (Siegenthaler, 2005) and has probably not been exceeded since the Cretaceous era, the geological period when the average temperature of the atmosphere was much higher than today and the polar ice caps did not exist in either hemisphere.

The increase in the concentration of atmospheric CO_2 results predominately from the burning of fossil fuels — coal, petroleum, and natural gas — and to a lesser degree from land use changes, especially deforestation. Annual CO_2 emissions from fossil fuels and cement production increased from 6.4×10^9 tons of carbon (6.4 gigatons of carbon, GtC) per year in the 1990s to 7.2 GtC in the period 2000 to 2005 (IPCC, 2007). Annual CO_2 emissions from land use changes are estimated at 1.6 GtC in the 1990s, which corresponds to about 20% of the total (IPCC, 2007). However, there is still a significant uncertainty about this percentage. Recent estimates consider that its present value is lower, of the order of 12% (Le Quéré, 2009).

The main greenhouse gases present in the atmosphere are water vapour, whose concentration is very variable, depending on the region and the type of weather, carbon dioxide (CO_2), methane (CH_4), nitrous oxide (N_2O), ozone (O_3), and some other gases produced by chemical synthesis, such as the chlorofluorocarbons (CFC) and the hydrochlorofluorocarbons (HCFC). In the absence of greenhouse gases, the Earth's surface would radiate directly into outer space and the atmosphere would not absorb the infrared radiation. In this hypothetical situation, the average global temperature of the troposphere would be about $-18°C$, instead of the current $15°C$. It is this temperature difference, now with the value of $33°C$, arising from a natural greenhouse effect, that allowed for the appearance, evolution, and permanence of life on Earth. By significantly increasing the concentration of greenhouse gases the radiative balance in the atmosphere is perturbed and causes climate change.

Among anthropogenic emissions, those of CO_2 contribute most to the increase in the greenhouse effect. The other most important greenhouse gases emitted in human activities are methane (CH_4), coming from agriculture, especially livestock and paddy fields, nitrous oxide (N_2O) coming from agriculture and industry, and the CFCs and HCFCs used in various industries. At the present time, CO_2 emissions produce an average radiative forcing of 1.7 W m^2, which corresponds to about 63% of the global radiative forcing by the anthropogenic emissions of greenhouse gases (IPCC, 2007).

Svante August Arrhenius (1859–1927), a Swedish chemist, was the first scientist to state clearly that the rise in the atmospheric level of CO_2 produced by fossil fuel burning would raise the average surface temperature of the atmosphere (Arrhenius, 1896). However, he was not particularly concerned with the consequences of this trend. In fact, he wrote (Arrhenius, 1908):

> By the influence of the increasing percentage of carbonic acid in the atmosphere, we may hope to enjoy ages with more equable and better climates, especially as regards the colder regions of Earth, ages when the Earth will bring forth much more abundant crops than at present for the benefit of rapidly propagating humanity.

His colleague Walter Nernst (1864–1941) went a step further, proposing to start underground fires in disused coal mines to release CO_2 into the atmosphere (Coffey, 2008).

3.36 The Climate Has Varied Naturally Throughout Earth History

The Earth's climate has been changing since the atmosphere first formed about 4 000 million years ago. It is important to bear in mind that the climate of a given location or region is defined as the statistical description in terms of the mean and variability of the meteorological variables in that location or region for periods of months up to thousands or even millions of years. The standard period of time to define a climate, adopted by the World Meteorological Organization, is 30 years. There have been glacial epochs, with extensive ice fields formed in the polar regions, which alternate with relatively warm epochs when there is no polar ice. The current ice caps only started to form about 34 million years ago, first in the Antarctic and then much later, about 5 million years ago, in the Arctic.

Besides these long-period climate oscillations, driven by plate tectonics, by the formation and erosion of mountains, and by changes in the geographic position and flow of the ocean currents that transport heat from the equatorial regions to the high latitudes, there are other much shorter glacial and interglacial periods, during the glacial epochs. We can identify and study the cycles of the current glacial epoch in the ice sheets of Greenland and the Antarctic. Over the last million years, the Earth's climate has oscillated between cold glacial periods, with a duration of between 80 000 and 100 000 years, and relatively warm interglacial periods, with a duration of 10 000 to 20 000 years.

Milutin Milankovitch identified the main causes of these oscillations in the first half of the 20th century (Milankovitch, 1930). They result from astronomical forcing related to small variations in the eccentricity of the Earth's orbit around the Sun, the obliquity of the Earth's axis relative to the plane of its orbit, and the precession of the Earth's axis. The last glacial period started about 120 000 years ago, and reached its maximum about 20 000 years ago. At that time the average global temperature of the atmosphere was around 5–7°C lower, and the average sea level was about 100 m to 120 m lower than it is today.

We are now in a particularly stable interglacial period in which the temperature has had only small variations, less than 1°C per century, during the last 8 000 years.

The stability of the climate and the relatively high temperature has created especially favourable conditions for the development of diverse civilizations over the last 6 000 years. *Homo sapiens* has witnessed two glacial and two interglacial periods. He appeared in Africa in the penultimate glacial period and was also in Europe for the last.

3.37 Recent Signs of Anthropogenic Climate Change

At present, there is a strong consensus in the scientific community that anthropogenic emissions of greenhouse gases are intensifying the natural greenhouse effect in the atmosphere. These emissions are causing a climate change that will very likely intensify during the 21st century. The signs that this climate change is happening are becoming ever more obvious and unequivocal. Global average surface temperature (land and ocean) has increased by 0.8°C since pre-industrial times and by 1.0°C over land alone (IPCC, 2007). In the European land area, it risen by 1.2°C. In the Arctic, the average surface temperature increase has been higher, about twice the global value. Every single one of the 13 years from 1997 to 2009 was one of the 14 years with the highest global average surface temperature since 1850, the date when temperature first began to be regularly measured with thermometers.

One of the most notorious signs of recent climate change is the retreat of the vast majority of mountain glaciers and the reduction of the ice field mass at high altitudes (Oerlemans, 2005). Climate change is also becoming very clear in the polar ice caps, especially in the Arctic. In this polar region the area of summer sea ice has fallen by 16–20% over the last 30 years. Note that the melting of the floating sea ice in the polar regions does not contribute to a change in mean sea level, except for small second order effects related to temperature and salinity (Jenkins, 2007). As regards ice above sea level the area of the Greenland ice fields annually subjected to summer thaws increased by 16% from 1979 to 2002 (ACIA, 2004). The increasing amount of melted ice above sea level that is not replaced by new winter ice contributes to the rise in the average sea level.

In Greenland the annual rate of this meltdown doubled from 1996 to 2005 and has currently reached a value of 239 km^3 (Zwally, 2006). To get an idea of what this means, note that it corresponds to more than 10% of the water volume of all the rivers on the planet. The loss of mass of the Greenland ice sheet is estimated to contribute currently to a global mean sea level rise of 0.5 mm per year (Wouters, 2008). In the west of Antarctica and in Greenland, the melting of the surface ice creates small melt ponds that apparently accelerate the thawing process. This liquid melt water descends into moulins, vertical shafts that carry the water to the base of the ice sheet, diminishing the stability of the ice and accelerating the melting process (Kerr, 2006). Furthermore, glaciers whose base is in contact with seawater are melting more quickly due to the increased temperature of the ocean surface layers.

The reduction in sea ice area in the polar regions generates a positive feedback for global warming because the ice reflects 50 to 80% of solar radiation, whereas the oceans reflect only about 5%. There are clear indications that the deposition of black soot, which includes carbon black and organic carbon, is also accelerating the melting of ice fields and ice caps. In Antarctica, the warming has been more pronounced in the Antarctic Peninsula, but there was a significant warming trend over the whole continent in the period 1957 to 2006 by $0.1°C$ per decade (Steig, 2009). Antarctic ice shelves that slow the movement of continental ice sheets toward the ocean are exhibiting increased calving and melting, especially in the Antarctic Peninsula.

The rate of global mean sea level rise during the 20th century was in the range of 1.2 to 2.2 mm, with a best estimate of 1.7 mm, about 10 times more than the average rate in the last 3 000 years (IPCC, 2007). This increase is mainly due to the thermal expansion of the surface layer of the ocean, whose temperature rose due to the increase in the average global atmospheric surface temperature, and to a lesser degree, to the thawing of the mountain glaciers and ice fields. There is also a contribution from the melting of the ice fields in Greenland and Antarctica located above sea level, whose evolution is more difficult to predict. Nevertheless, this contribution is very likely to increase and become dominant if anthropogenic emissions of greenhouse gases are not controlled.

The mean global surface temperature of the ocean has increased by about $0.5°C$ since 1970 and this heat propagates very slowly downwards, amplifying the thermal expansion of the oceanic mass. Consequently, the oceans have a very slow response to the anthropogenic emissions of greenhouse gases into the atmosphere. Recent estimates indicate that the ocean has absorbed about 80% of the thermal energy resulting from anthropogenic intensification of the greenhouse effect (IPCC, 2007).

In the last few decades, extreme weather and climate events have become more intense. These include heat waves, heavy rainfall events, and droughts. There are indications that the number of very intense tropical cyclones ranging from categories 4 to 5 has been increasing recently due to increases in sea surface temperatures (Mann, 2006). According to insurance company records, the annual cost of damages caused by catastrophic weather and climate events rose from about US $ 4 000 million in the 1950s to US $ 40 000 million in the 1990s, after adjusting for inflation (IPCC, 2001). This tendency is due in part to socio-economic factors, such as demographic growth, increased urbanization, and infrastructure investment in vulnerable areas, such as low coastal zones, but also to the greater intensity of severe weather and climate events.

Observed changes in precipitation, with increases in the high latitudes and decreases in the subtropics, have been at the upper limit of model projections. In particular, increased precipitation has been observed since the beginning of the 20th century in some eastern parts of North and South America, northern and central Europe, and northern and central regions of Asia. On the other hand, climate observations for the same period show decreased rainfall in southern Europe, the Mediterranean, the Sahel, the southern part of Africa, and some parts of Southern Asia. Moreover, the area affected by droughts has increased in most regions since the 1970s.

3.38 Future Climate Scenarios

To understand the current climate change and to obtain projections of future climate change, it is necessary to use general circulation models that simulate the behaviour of the Earth's climate system. This is a complex system formed by various interacting subsystems: the atmosphere, the hydrosphere, including the oceans, lakes, and rivers, the cryosphere, formed by the snow and ice masses, the biosphere, formed by all ecosystems, and the lithosphere. All these subsystems are linked together by fluxes of mass, energy, and momentum. If we are interested in knowing the future evolution of the weather over intervals of time from one hour to a few days, it is a good approximation to consider only the atmosphere subsystem of the full climate system. However, if we are interested in knowing the future behaviour of the climate, which implies time intervals of the order of at least a few decades, it is necessary to consider also the hydrosphere, the cryosphere, and the biosphere. For still longer intervals of time, on the order of ten to a hundred million years or more, we must take into account the lithosphere, due to the impact on climate of the relative motion of the continents, mainly through its effects on the ocean circulation and orogenesis.

General circulation models are the best means we have to obtain scenarios of the future climate. They are the fundamental tools for the assessment of the potential future impacts that climate change is likely to have on the various socio-economic sectors and biophysical systems, such as water resources, agriculture, coastal zones, marine resources, forests, biodiversity, oceans, fisheries, human health, urban areas, energy, tourism, migrations, insurance, and financial services. The future climate projections, usually covering the period until 2100, depend on the scenarios adopted for the emission of greenhouse gases into the atmosphere until the end of the century. If global emissions continue to rise significantly in an uncontrolled way, the increase in the global average surface temperature will be larger and the impacts more serious. On the other hand in a scenario where it is possible to control and therefore reduce global emissions, the increase in the global average surface temperature will be smaller and the impacts less serious.

The construction of emission scenarios is based on socio-economic scenarios that describe a variety of future worlds with particular social, demographic, economic, technological, and environmental characteristics. These scenarios do not claim to represent a likely reality, but rather a range of possible coherent outcomes. They were first developed by the Intergovernmental Panel on Climate Change (IPCC), created in 1988 by the United Nations Environment Programme (UNEP) and the World Meteorological Organization (WMO). This international institution differs from other scientific and technological panels because it includes representatives from the governments of UN member states as well as scientists, engineers, economists, and sociologists. In no other type of scientific evaluation process do so many experts from such a wide range of scientific backgrounds and countries meet to analyse a given problem in a coordinated and coherent way. It is important to emphasize that the main objective of the IPCC is to provide a critical up-to-date review of current scientific knowledge regarding climate change issues, based on peer review publications, rather than to produce new science.

How can we validate future climate scenarios and how reliable are they? The main proof of the robustness of the general circulation models is their capacity to reproduce satisfactorily the behaviour of the climate system over the last 150 years. Other forms of validation consist in accurately reproducing the response of the atmosphere to transient forcing. One example is the response to aerosol, dust, and ash emissions in a major volcanic eruption like the Mount Pinatubo eruption in the Philippines in 1991. In this case a negative anomaly was observed in the global average surface temperature, lasting about 5 years after the eruption, with a value of around $-0.25°C$, and this was well reproduced by climate models (Hansen, 1992).

However, climate scenarios obtained with these models still involve uncertainties, mostly related to our present inability to simulate meteorological phenomena on a scale of kilometres or hours, especially cloud formation processes. One weak point of current general circulation models is that their spatial resolution is about 300 km, and this makes them difficult or inappropriate to use for impact assessments at the regional or local scales. To construct scenarios with a higher spatial resolution, from 50 km down to 10 km, one must use dynamical or statistical methods for downscaling the global scenarios. Increasing the spatial and temporal resolution of global circulation models would require a supercomputing performance that is not yet available (Palmer, 2005). This may become possible in the near future. At the time of writing, in 2010, computers perform in the teraflop (2×10^{13} floating point operations per second) to petaflop (10^{15}) range, but given the current rate of technological development they may reach zetaflops (10^{21}) within 10 to 20 years. Currently, the fastest supercomputer is the Tianhe-1A machine at the National Supercomputing Center in Tianjin, China, working at the rate of 2.5 petaflops.

The last report from the IPCC, using various climate models and emission scenarios, projects an increase in the global average surface temperature of 1.8 to $4.0°C$ for the period of 2090–2099, relative to 1980–1999 (IPCC, 2007). The temperature increase is expected to be greater in continental regions than in coastal areas and oceans, and greater at higher latitudes, especially in winter. In the Arctic, the increase in temperature will probably reach values between 4 and $7°C$ (ACIA, 2004). With higher mean temperatures, the atmosphere is able to retain more moisture, which leads to an intensification of the water cycle and an increase in global rainfall. There will also be significant changes in the spatial distribution patterns of rainfall, with an increase in the higher latitudes, some equatorial regions, and Southwest Asia. On the other hand, reduced rainfall is expected in the middle latitudes, such as the Mediterranean and southern Europe, and in some equatorial regions, such as Amazonia.

Extreme climate phenomena are very likely to become more intense, including heat waves, severe droughts, heavy precipitation events leading to floods, and tropical cyclones. According to the latest IPCC report, the average sea level is expected to rise from 0.18 to 0.59 m by the end of the century (IPCC, 2007). However these estimates are very probably too low. More recent studies project increases of 0.75 to 1.90 m in the period of 1990 to 2100 (Rahmstorf, 2007; Vermeer, 2009). One of the main reasons for the uncertainties in the rate of sea level rise is the difficulty in eva-luating the contribution from the melting of the Greenland and Antarctic ice sheets. In the medium and long term, sea level rise is one of the most damaging impacts of climate change. Recent climate models indicate that an increase of 3°C in the global average surface temperature will lead to the irreversible thawing of the Greenland ice sheet over a period of time of several hundred to more than one thousand years (Gregory, 2004), raising the sea level by about 7 m. In the Antarctic, the western ice sheet is less stable and more vulnerable to thawing than the eastern one. Complete melting of the west Antarctic ice sheet would raise the sea level by about 5 to 6 m, while complete melting of the Antarctic ice sheet would raise it by about 60 m.

Abrupt climate change, resulting from relatively rapid forcing that compels the system to change suddenly to a new state of equilibrium, is a possibility. This is much more difficult to simulate and predict than gradual changes, since it results from non-linear processes in unstable situations. Paleoclimatic studies clearly reveal examples of abrupt climate change in the last 120 000 years, especially in the tran-sition to the last glacial period and in the transition from that period to the current interglacial. The analysis of climate proxies indicates that the mean global tempe-rature varied by about 5°C over short periods of time of the order of decades. One of the best known examples, called the Younger Dryas, happened about 12 800 to 11 500 years ago and probably resulted from the interruption of the thermohaline circulation in the North Atlantic, where deep water currents are formed following a strong heat exchange between the ocean and the atmosphere. It is possible that the current global warming will also cause a weakening or, much less likely, an inter-ruption in that circulation, due to the increased inflow of freshwater resulting from the thawing of ice fields and the increase in precipitation in higher latitudes. This would trigger a localized cooling in the North Atlantic region.

3.39 Impacts and Vulnerability to Climate Change

The degree to which a given human or natural system is affected, either adversely or beneficially, by climate related stimuli, characterizes its sensitivity to climate change. The adaptive capacity of these systems is the ability to adjust to climate change, including climate variability and extremes, to cope with the impacts, to moderate potential damage, and to take advantage of potential opportunities (IPCC, 2007). Vulnerability describes the degree to which the human or natural system is able or unable to cope with the adverse effects of climate change, taking into account its level of exposure, its sensitivity, and its adaptive capacity. In general, developing countries are more vulnerable to climate change. They have less adaptive capacity because of their limited access to economic and financial resources and to knowhow. Countries with high population densities situated in dryland areas with water scarcity problems or in low coastal zones prone to flooding are among those that will be most affected by climate change.

The adaptive capacity in different regions of the globe where the impacts of sea level rise are potentially very adverse provides a clear example. There are about 100 million people living in coastal zones with a maximum elevation of one meter above mean sea level. In Bangladesh, there are about 6 million people. Assuming that by the end of the century the sea level rise is around one meter, it is very clear that developing countries with low-lying coastal areas, such as Bangladesh, Mozambique, and Egypt, especially the Nile Delta, will have much less capacity to adapt than developed countries like the Netherlands and the USA, particularly concerning Florida.

People from the most vulnerable areas will migrate, increasing the number of climate refugees. There is no agreed definition of environmental and climate refugees within the United Nations system, but there is a consensus that their number is very likely to increase dramatically in the future (OCHOA, 2009). According to the International Organization for Migration, future forecasts of environmental migrants vary from 25 million to one billion by 2050, with 200 million being the most widely cited estimate.

There is still a long way to go before we can have a detailed and reliable evaluation of the impacts of climate change and the associated damage and costs, especially at the local level, in the various countries of the world. What we can safely say is that, the faster and the more intense our interference with the climate system, the more probable will be its unexpected and abrupt responses, and the more serious the impacts. In the short to medium term, some of the climate change impacts are likely to be positive in some regions of the world. For example, the global average temperature increase will favour the expansion of forests and agriculture towards higher latitudes in northern Eurasia and North America. However, in the long term, after 50 years or more, if global warming continues unchecked, the vast majority of the impacts will be negative.

The impacts of climate change in a given country or region should be evaluated in an integrated way, taking into account as many of the socio-economic sectors and biophysical systems as possible. Moreover, it should be based on a coherent system of future climate and socio-economic scenarios. Only in this way will it be possible to make credible estimates of the impacts in the various sectors and of the synergies between them. It is a necessary first step to evaluate the costs of the potential damage and to devise and implement the most appropriate adaptation measures that minimize the adverse effects of climate change. This type of study is an essential tool to inform and raise the awareness of stakeholders. Adaptation to climate change is a long term process that is implemented mainly at the regional and local levels and should involve municipalities and public and private stakeholders.

One of the main negative impacts of climate change worldwide is on water resources. Projections based on current scenarios indicate that, by 2050, the annual river runoff and water availability will increase by 10 to 40% in the higher latitudes and in some regions of tropical rainforest, and decrease by 10 to 30% in the drylands of the mid-latitudes and in the dry tropical regions. These changes tend to aggravate existing asymmetries and will probably have disastrous consequences in many already water-stressed countries.

Estimates indicate that, by 2020, 75 to 250 million more people in Africa will be exposed to serious water shortages due to climate change (UNFCCC, 2007). On the one hand, the areas affected by droughts will increase, and on the other, the more intense heavy precipitation events will increase the risk of flooding. In the high mountainous regions, the increasing rate of glacier thawing will cause landslides, flash floods, and glacial lake overflow. Furthermore, it will affect the water resources by changing the river runoff regimes.

The main rivers with sources in the Himalayas and the Tibetan plateau — the Indus, Ganges, Brahmaputra, Yangtze, and Huang He — are fed by rains in the monsoon season, but in the dry season, they depend heavily on melt water from glaciers. These rivers provide water to more than a billion people. In a few decades, the retreat of glaciers will significantly decrease the runoff, particularly in the dry season when there is a greater need for water to irrigate rice and wheat, threatening food security and causing huge environmental problems. It is estimated that by 2050, more than 1 000 million people will be affected by this problem in central, south, east, and southeast Asia. The same kind of negative impacts are also expected on the western slopes of the Andes, affecting the livelihood of tens of millions of people.

Initially agricultural productivity will have a tendency to increase in the middle and high latitudes, but to decrease in the low latitudes, raising the risk of hunger. However, if the increase in the mean global surface temperature is higher than 3°C (IPCC, 2007), food productivity is very likely to decrease on a worldwide scale.

The combined effects of climate change and other pressures exerted by human activities, such as habitat change and destruction, pollution, and overexploitation of ecosystem services will significantly damage the majority of the world's ecosystems. The capacity of ecosystems to act as carbon sinks is likely to increase up to 2050 and then to decrease, potentiating global warming (IPCC, 2007). There will be a strong tendency for ecosystems to migrate towards higher latitudes and higher altitudes. The greater the impediments for this migration, the greater will be the loss of biodiversity. Approximately 20 to 30% of the species whose vulnerability to climate change has been studied will be at greater risk of extinction if the average world temperature exceeds 1.5 to 2.5°C (UNEP, 2009a).

The increase in the surface temperature and the acidification of the oceans will have very negative consequences for marine ecosystems and in particular for coral reefs. The coupling with human pressures and climate change impacts is likely to be devastating if appropriate measures are not taken. A report published in 2011 assesses the status and threats to the world's coral reefs from a wide range of human activities and from climate change (Burke, 2011). The main conclusion is that about 75% are threatened, putting at risk the livelihood of populations in many countries that depend on these ecosystems. The threat is the highest in Southeast Asia due to over-fishing and destructive fishing. Climate change adds to the risk because it induces thermal stress and reduces calcification rates through ocean acidification. The report projects that in 2050 only about 15% of the world's reefs will be in areas where calcium carbonate levels are adequate for coral growth.

Coastal areas and their ecosystems are especially vulnerable because of the combination of impacts from climate change and human pressures caused by increasing population density, land occupation, and changing land use. Mean sea level rise will cause greater erosion in the low lying coastal areas and increase the risk of inundation in river deltas, river estuaries, and coastal wetlands, negatively affecting the coastal ecosystems, in particular biodiversity-rich mangrove forests.

Small islands are particularly vulnerable to climate change owing to the greater relative importance of their coastal areas, limited natural resources, particularly water, and a generally greater exposure to extreme weather and climate events that are likely to become more intense. A particularly dramatic situation is that of the Small Island States that have a very low maximum altitude, of about 2 m, like the Maldives in the Indian Ocean and the Marshall, Tuvalu, and Kiribati Islands in the Pacific Ocean. The risks from sea level rise threaten their very existence. If the sea level continues to rise, which is very probable, their populations will have to migrate. Meanwhile, many of these islands have problems with inundation, coastal erosion, coral bleaching, and water resources, resulting from salinization of coastal aquifers. In 2009, 2 700 islanders from the Carteret Islands, Papua New Guinea, small low-lying coral atolls in the Pacific, were relocated in Bougainville to avoid the more frequent flooding from storm surges and large tides that ruined the fresh water supply and the banana and taro crops.

The health of many hundreds of millions of people is likely to be affected negatively by climate change. There will be an increase in the mortality and morbidity associated with more intense heat waves, floods, tropical cyclones, forest fires, droughts, decreasing water quality in water-stressed regions, increased ozone in the troposphere, and an increase in the area of regions affected by vector-borne diseases, such as malaria and dengue. The impacts on tourism will probably be significant, although they are still poorly studied and remain uncertain. The growing number of extreme weather and climate events will increasingly affect the insurance sector. The negative impacts on industry, urban areas, and society in general are very likely to increase as the global average temperature continues to rise.

Anthropogenic climate change is already having significant impacts, generally negative, on various human and social systems. In some regions, especially in drylands, the increase in global average temperature and changes in the precipitation regime are having adverse effects on agriculture, forests, and biodiversity. In the coastal areas, the risks of inundation and erosion are greater, and millions of people are already suffering the consequences. There are signs that climate change is beginning to create or to contribute to the emergence of critical situations in regions that are particularly unstable and vulnerable.

The dramatic conflict in Darfur that led to the deaths of hundreds of thousands of people and more than two million refugees is the first example of a possible link between climate change and humanitarian crises. The precipitation decrease of around 30% observed in that region of Africa over 40 years, is very probably due to a disturbance in the monsoon cycle caused by an increase in sea surface temperatures in the Indian Ocean, which in turn results from anthropogenic climate change. With less rain, drought and desertification increased, water resources became scarcer, and soils were unable to support both the herds of nomadic pastoralists and agricultural production by settled farmers. The violence began during a drought in February 2003 and turned into a serious ethnic conflict. Ban Ki-moon, the United Nations Secretary-General defended the relation between climate change and the Darfur conflict in an article published on 16 June in the Washington Post.

3.40 Mitigation and Adaptation to Climate Change

Faced with these scenarios of increasing risks, what are the possible responses? There are essentially two types: mitigation and adaptation. The first deals with the causes of anthropogenic climate change. Its objective is to stabilize the concentration of greenhouse gases in the atmosphere by reducing emissions and by developing potential sinks for these gases, now and in the future. Adaptation is a process of adjustment of human and natural systems to the impacts of climate change, aimed at reducing vulnerability against actual or expected effects. Adaptation moderates harm or exploits beneficial opportunities. It can be autonomous when triggered naturally, as in an organism or ecosystem that adapts itself to a changing environment, or planned, when it is the result of a deliberate policy decision.

It is important to realize that the two responses to climate change are complementary and that both involve costs. As regards adaptation, we have to consider the costs of planning and implementing adaptation measures that minimize the adverse effects of climate change. Mitigation clearly involves costs, which are heavily dependent on the extent to which greenhouse gas emissions can be reduced and the timescale required for those reductions. By significantly decreasing emissions in the short term, the mitigation costs increase and the adaptation costs in the future decrease. On the other hand, by reducing emissions weakly in the short term, the mitigation costs become lower, but the adaptation costs in the future become higher.

The Stern Review (Stern, 2006) argues that the long term benefits of strong and early mitigation considerably outweigh the costs. It estimates that the cost of mitigation actions that avoid the worst future impacts of climate change can be limited to around 1% of global annual GDP. On the other hand if we do not mitigate now, the overall costs will be equivalent in the future to losing 5 to 20% of global GDP each year. Climate change is an example of a negative economic externality with very specific characteristics. The full costs of emissions in terms of climate change impacts are not generally borne by the emitters, but by future generations. It is therefore an ethical issue of intergenerational solidarity, to the extent that later generations will experience most of the likely negative impacts. It is also an economic issue, given the scale of the damage and costs that it will impose on society (Santos, 2009).

The major difficulty with mitigation is that it is not enough to lower the emissions in some countries; emissions must be reduced on a global scale in order to stabilize the atmospheric concentration of greenhouse gases. According to the current understanding of the global carbon cycle, the climate change effects of CO_2 emissions to the atmosphere will persist for many hundreds and even thousand of years (Archer, 2010). Recently, the average annual emissions of CO_2 increased from 6.4 GtC (1 GtC or 10^9 tonnes of carbon equals 3.67 $GtCO_2$) in the 1990s to 7.2 GtC in the period from 2000 to 2005 (IPCC, 2007). Thus the average annual world emission per capita is now greater than one tonne of carbon in the form of CO_2.

There are, however, profound differences between per capita emissions in developed and developing countries. For instance, the per capita average annual emissions in the USA and EU were 19 and 8.5 tCO_2, respectively, in 2008, while in Ethiopia, Mozambique, and Nepal, to mention just three less developed countries, it was only 0.1 tCO_2. In the same year China, Brazil, and India had per capita emissions of 4.6, 1.9, and 1.3 tCO_2.

The correlation between social and economic development and CO_2 emissions is a direct consequence of the strong current dependence on fossil fuels worldwide. At the present time, the country with the largest emissions is China, with about 22% of the total, followed by the USA with about 20%. In the period from 2000 to 2008, the average annual increase in global emissions of CO_2 was 3.4%. In 2008, there was a slowdown, and in 2009 a decrease by about 3% due to the financial and economic global crisis. The significant decrease in CO_2 emissions from the more industrialized countries completely nullified the increase in emissions from countries with emerging economies, such as China and India. This situation reveals the profound contradiction between our current paradigm of development based on the intensive use of fossil fuels and the need to combat climate change.

The United Nations Framework Convention on Climate Change (UNFCCC) entered into force on 21 March 1994 and was ratified by 193 countries at the end of 2009. As stated in Article 2, its main goal is to achieve the stabilization of greenhouse gas concentrations in the atmosphere at a level that would prevent dangerous anthropogenic interference with the climate system, without, however, defining the meaning of dangerous. What exactly constitutes dangerous climate change must be decided by value judgments of stakeholders and policy makers on the basis of the risks people are willing to face or avoid.

There is some scientific consensus that, if the increase in global average surface temperature relative to pre-industrial times is kept below 2°C, the disruption of ecosystems and societies will be minimal. Above this threshold, the risks of reaching various tipping points that trigger dangerous climate change become very large. Currently the 2°C limit is well accepted by most of the international community, but the more vulnerable developing countries have been advocating without much success a lower limit of 1.5°C. It is important to recognize that the stabilization of the global average surface temperature will only occur after stabilization of greenhouse gas concentrations. But this latter stabilization can only take place after the global annual emissions decrease substantially relative to current values, after reaching a peak. The later the peak occurs, the greater will be the increase in global average surface temperature above the pre-industrial level.

What is the equivalent CO_2 stabilization concentration of greenhouse gases that would lead to an increase in the average global surface temperature lower than 2°C? The answer to this question depends on the sensitivity of the climate system, defined as the equilibrium change in average global surface temperature that would result from a doubling of the atmospheric equivalent CO_2 concentration of greenhouse gases relative to the pre-industrial value. The climate sensitivity of our atmosphere is likely to be in the range of 2 to 4.5°C, with a best estimate of 3°C (IPCC, 2007). Current values for climate sensitivity indicate that, to keep the 2°C target within reach, the CO_2 concentration at stabilization must be in the range 350–400 ppmv and the equivalent CO_2 concentration at stabilization in the range of 445–490 ppmv. To achieve these goals, the peak year for CO_2 emissions should occur between 2015 and 2020 and annual greenhouse gas emissions should be reduced by at least 50% in 2050 relative to 1990. This is a gigantic challenge. Note that the average global atmospheric CO_2 concentration in 2010 was 389 ppmv and the equivalent CO_2 concentration about 437 ppmv.

3.41 The Kyoto Protocol (2005) and the Copenhagen (2009) and Cancun (2010) Conferences of Parties

The Kyoto Protocol entered into force on 16 February 2005, after eight long years of negotiations, and has been ratified by 191 countries in 2010. All developed countries, except the USA, ratified the protocol aimed at reducing their greenhouse emissions by 5.2% by the end of 2012 relative to 1990. This mitigation is manifestly insufficient to control climate change. However, the Kyoto Protocol is a good starting point for the implementation of a future binding and comprehensive agreement involving all parties to the UNFCCC to prevent dangerous interference with the climate system, through common but differentiated mitigation responsibilities.

International negotiations to draw up a United Nations agreement on climate change for the period after 2012, when the first commitment period of the Kyoto Protocol expires, started at the end of 2007. It was expected that an agreement could be reached at the COP15 (15th Conference of the Parties to the UNFCCC) in Copenhagen in December 2009. The conference would agree on procedures for codifying emission mitigation contributions by developed and developing countries and for reviewing and updating them. In addition, it was expected that developed countries would reach a binding agreement to provide financial assistance for mitigation and adaptation to climate change in developing countries, particularly as regards adaptation in the poorest and more vulnerable ones. Furthermore, there was hope for a framework agreement to reduce emissions from deforestation and forest degradation and to promote conservation, sustainable development and the enhancement of carbon sinks in the forests of developing countries.

Before the conference, the USA announced provisional mitigation targets in line with legislation that has been blocked by the Senate: cuts of 17%, 42%, and 83% below 2005 in 2020, 2030, and 2050. These commitments represent a remarkable advance relative to the position of the preceding US administration. However, the short-term mitigation targets are very modest compared with those of the EU, since they represent only about 4% reduction in 2020 relative to the 1990 baseline. According to current estimates, the 2°C target can only be reached if developed countries reduce their total emissions between 25 and 40% by 2020 relative to 1990. Before the conference, China announced a reduction in the carbon intensity of its GDP between 40 and 45% by 2020 relative to 2005. This contribution is most welcome, but it does not necessarily imply a cut in emissions. There is still no commitment from China about the time when its greenhouse gas emissions will reach a peak. Thus, the 2°C target will be impossible to reach, since it requires a decline in global annual emissions of 50% by 2050 relative to 1990. India and Brazil also made significant mitigation pledges. However, the net result of all these mitigation pledges corresponds only to about 60% of what would be required to achieve the 2°C target.

The outcome of the Copenhagen conference was far below the expectations. On 18 December, the last day of the conference, the USA, China, India, Brazil and South Africa, which are some of the biggest carbon emitters, agreed on a document called the Copenhagen Accord which recognizes the urgent need to combat climate change and that deep cuts in global emissions are required, in agreement with the scientific consensus. However, the Accord is not legally binding and does not commit countries to agree on a successor to the Kyoto Protocol. The plenary session of COP 15 took note of the Accord but failed to endorse it. Negotiations to reach a new binding agreement will continue in the years ahead but the Copenhagen conference exposed deep and increasing divisions between three main groups of countries: developed countries, emerging economies, such as China, India and Brazil, and least developed countries especially vulnerable to climate change. China was unwilling to commit itself to a date to bend its emissions, which makes it impossible to plan the global peak.

Problem-solving possibilities are closely tied up with the mechanisms available to implement them. In the case of climate change, the world needs a binding agreement on all parties and a universally recognized higher authority to oversee its implementation. Can this authority be structurally based on the concept of nation-state sovereignty, as in the United Nations? Copenhagen showed that it is increasingly difficult to reach an agreement since the required mitigation and adaptation actions do not result from simply adding the sovereign national interests of each country, especially those of an economic and short-term nature, which are politically the most pressing. The unanimity requirement and the grassroots procedures produced disorganization and the emergence of a pledge-and-review system involving few countries. In the medium and long term of a rational development path, we would need new forms of world governance that enable the application of supranational solutions to problems of sustainable development, including climate change, in which irreconcilable nation-state interests are at odds. However, at present, it seems rather unlikely that we will follow that rational path.

The COP 16, held in Cancun in December 2010 did not achieve significant progress relative to the Copenhagen Conference. The decision of continuing the Kyoto Protocol was again put off for another year. Nevertheless the Cancun Conference helped to restore some confidence in the ability and commitment of the world community of nations to address the challenges of climate change. Since that time there has been no real progress. The next Conference of Parties to take place in Durban, beginning on 28 November 2011 is very unlikely to extend the Kyoto Protocol or to open the way to another international binding agreement on mitigation.

In spite of the difficulties in reaching a binding and comprehensive agreement on climate change, particularly within the context of the UNFCCC, there are encouraging signs of an irreversible movement in developed and developing countries, particularly in the emerging economies, towards a low-carbon economy characterized by a minimal output of greenhouse gases to the atmosphere, especially CO_2. Most developing countries, which are not heavily industrialized or populated, are already low-carbon economies. To assure sustainable development and especially to avoid dangerous climate change the carbon intensive countries should become low-carbon societies where the amount of CO_2 released to the atmosphere balances an equivalent amount sequestered or offset. To achieve this aim it is necessary to develop technologies for energy generation and for the production of materials, infrastructures, devices and services in all socio-economic sectors, including industry, agriculture and transportation, with minimal greenhouse gas emissions. Furthermore, it is necessary that energy, materials, infrastructures, devices and services are efficiently used and that the resulting wastes are disposed of or recycled with minimal greenhouse emissions. To make the transition from a carbon intensive to a low-carbon society economically viable it is indispensable to set a price on CO_2 emissions through a carbon tax, emissions trading, or both. We are still far away from a global carbon market although there is already considerable experience with emissions trading and carbon taxes in the EU and USA. Such a market would have the advantage of establishing a single carbon cost, which creates equitable access to the prevailing lower cost mitigation technologies and opportunities. The main strategies for a transition to a low-carbon economy involve primarily the efficient use of energy and the fast development of renewable energy sources. Another possibility emerging in the EU and USA is carbon capture and storage, a technology which allows the use of fossil fuels, in particular coal, with practically no CO_2 emissions into the atmosphere. Will it be possible to make a transition at the global level to a low carbon economy in time to avoid a dangerous interference with the climate system? That is a question that will remain unanswered for the next few decades, but whose answer depends to a very large extent on the attitude of each one of us toward the problem of climate change.

Finally, there is geoengineering or climate engineering, which is a more controversial approach to carbon neutrality. It involves large-scale engineering on our environment in order to combat or counteract the effects of climate change. Geoengineering should not be a substitute for mitigation and most of its technologies are likely to have side effects that cause significant environmental damage. One form of geoengineering is to create or enhance CO_2 sinks, for instance by ocean iron fertilization to stimulate phytoplankton blooms. However, this is potentially very damaging to the marine ecosystems. A more promising technology, with unlikely side effects, is CO_2 removal from the atmosphere by means of scrubbing processes, such as those that take place in devices made up of a resin that absorbs CO_2, called artificial trees. Another form of geoengineering is solar radiation management with technologies that enhance the Earth albedo or that reduce the incoming sunlight using lenses or sunshades in space. By injecting into the stratosphere aerosol precursors, such as sulphide gases, or by enhancing cloud reflectivity by spraying seawater into the atmosphere we can change the Earth's albedo. The impacts of these technologies on the environment are not well known and very likely to be damaging. Significantly, there is a growing political and economic lobby in the more industrialized countries that seeks to present geoengineering as a natural solution to the climate change challenge in the context of the Anthropocene concept.

It is important to realize that it is already impossible to avoid some anthropogenic climate change in the 21st century. What we can and should do is to implement mitigation policies and measures capable of preventing dangerous intensification of climate change during the current century and beyond. Meanwhile, we have to adapt to the impacts of the already inevitable climate change, seeking to minimize its most adverse effects worldwide. Developed countries should help the most vulnerable developing countries to adapt, since their adaptive capacity is low.

3.42 Uncertainties and Scepticism Regarding Global Warming

There are no absolute truths in science, only results and conclusions that allow us to describe and predict phenomena with a higher or lower degree of probability and accuracy. There is always the possibility that new unexpected discrepancies and new previously unknown mechanisms emerge from further investigations. In the case of the well established theories such as gravitation, thermodynamics, quantum mechanics, the evolution of species, and the greenhouse effect in the Earth's atmosphere, the probability of reaching an excellent approximation to the description and prediction of phenomena is very high, so from an operational point of view, we can trust them. When a scientific theory leads to successful technologies, the uncertainties in the theory are usually forgotten and retain only an academic interest. But uncertainties tend to be greater when we try to describe and interpret the behaviour of a complex system, such as the climate system, and attempt to project its future evolution.

In the study of climate and the construction of future climate scenarios, there are various sources of uncertainty: limited knowledge of past climates (in particular the evolution of the average global surface temperature), but also of the physical, chemical, biological, and geological phenomena that occur in the system, and of the external and internal forcing responsible for climate variability. We must also recognize the impossibility of constructing models that faithfully simulate the behaviour of the climate system. Approximations are required to represent the interactions of its various components and parameterizations to simulate phenomena that have spatial and temporal resolutions unattainable with the models. Despite the uncertainties associated with such procedures, it is safe to say that the increase in the concentration of atmospheric greenhouse gases resulting from anthropogenic emissions is causing a climate change that will intensify throughout this century, unless such emissions are substantially reduced.

The reason for this confidence stems from the fact that the greenhouse effect is based on the fundamental laws of physics. The essential part of the explanation of what is happening is simple. Consider the case of CO_2. Since it is a greenhouse gas, its molecules have the capacity to absorb and emit infrared radiation. When the concentration of CO_2 in the troposphere increases, the infrared radiation that this gas emits occurs on average at a higher altitude. However, given that the temperature of the troposphere decreases with altitude, the CO_2 then emits more in regions with lower temperature. In accordance with the theory of radiation, the total amount of infrared radiation absorbed and emitted by a gas decreases with temperature. Hence, by increasing its concentration, the CO_2 emits less infrared radiation into outer space, and this corresponds to a negative radiative forcing. This would produce an energy imbalance between the incoming solar radiation absorbed by the Earth and the infrared radiation emitted by Earth. There would be more energy absorbed than emitted, which contradicts energy conservation, one of the most fundamental principles of physics. The balance is re-established by an increase in the temperature of the atmosphere at the Earth's surface, which increases the flux of infrared radiation emitted by the CO_2.

This is the essential reason why an increase in the concentration of greenhouse gases in the troposphere raises its average global temperature. The mechanism is very well understood and contributes about $1°C$ to the climate sensitivity, which has a value in the range of 2 to $4.5°C$ (IPCC, 2009). The remaining part results from feedback in the climate system, such as water vapour, ice albedo, cloud, and lapse rate feedback mechanisms. Most of these feedback mechanisms are positive, but some may be negative. There is still some uncertainty about their magnitude, particularly cloud feedback. Water vapour feedback amplifies global warming, since a warmer atmosphere contains more water vapour, which is a greenhouse gas. A particular worrying positive feedback is the possibility that the oceans and terrestrial systems may have started to lose part of their ability to sequester CO_2 (Le Quéré, 2009).

The identification of changes in the climate records of the last 150 years and the discovery that their cause is mainly anthropogenic are probably two of the scientific propositions that have been most comprehensively analysed, scrutinised, debated, criticised, and opposed in the history of science. Analysis, scrutiny, and criticism are fundamental processes of the scientific method. They constitute the essential part of the peer reviews that are a prerequisite for publication in scientific journals. Only after going through this process, and once the paper is published, can a scientific proposition, conjecture, or theory begin to gain credibility.

To admit that climate change is mainly due to certain human activities related principally with energy, which is an essential sector for sustaining economic growth, carries big responsibilities for the future and creates a huge challenge that will be very difficult to meet. Powerful lobbies, especially those associated with fossil fuels, have tried to influence the scientific process. Climate change became an issue with very high media visibility. A strenuous search for alternative explanations for the observed increase in global average surface temperature was set in motion, and the climate change debate intensified.

One of the paths most vigorously explored is the influence that the Sun's luminosity variability and solar activity cycles may have had on the Earth's climate over the last 150 years. The Sun gives rise to the main external forcing on the Earth's climate. Its luminosity — the total radiant power emitted by the star — can be determined through the solar constant, which has been accurately measured with radiometers on board satellites since the end of the 1970s. It is well known that the eleven-year cycle of solar activity causes variations in the solar constant of a few tenths of one percent and has measurable impacts in the stratosphere, but that it has very little effect on the troposphere. Although there are indications that long period variations in the solar constant have influenced the climate, there is no credible scientific basis for attributing the recent global warming observed since the 1970s to variations in the Sun's luminosity or in solar activity (Benestad, 2002).

According to recent suggestions (Svensmark, 2007), the variation in the flux of cosmic rays could be the cause of global warming, since they may have an effect on the Earth's albedo through a link with low altitude cloud formation. The nature of this link is still largely unknown, but there has been no statistically significant correlation between the observed flux of cosmic rays and global average surface temperature over the last 30 years. Climate research should be pursued vigorously, in particular to investigate the conjectured cosmic climate forcing, and other mechanisms that may exist.

Currently, there is a strong scientific consensus that the Earth's climate is changing and that some human activities are largely responsible for it. This consensus has the support of the Academies of Science from 19 countries and of many other scientific organizations that do climate research. A recent sociology of science study concluded that, of 928 research papers on climate change published between 1993 and 2003 in peer review scientific journals, not one contradicted the thesis that some human activities are interfering with the climate system (Oreskes, 2004). In a more recent survey, 97.5% of climatologists that actively publish research papers on climate change agreed that human activity is a significant factor in changing mean global temperatures (Doran, 2009). This finding indicates that only a very small minority of climate science experts are sceptical about anthropogenic climate change. There is also a powerful sceptical movement that denies climate change and any human influence on climate. This movement will very likely become stronger as the negotiations to reach a global agreement on mitigation proceed and an increasing number of countries engage decisively in the shift towards low carbon economies.

The history of the identification and interpretation of anthropogenic climate change and of the efforts of the sceptical movements to deny its existence and prevent mitigation actions is an outstanding study case in the sociology of science. It will be viewed in the future as one of the most important tests on the role played by modern science in contemporary societies. The media are quite easy to manipulate and frequently portray the denial of anthropogenic climate change by sceptics as based on science. Many people are confused and tend to see science as just another point of view. They do not believe, as scientists do, that knowledge based on the scientific method is not a question of belief and faith.

Conservative think tanks in the USA, such as the Heritage Foundation, the Competitive Enterprise Institute, the Hoover Institute for War, Revolution, and Peace, the National Centre for Public Policy Research, the George C. Marshall Institute, the Cato Institute, and the Heartland Institute (Johansen, 2002) are responsible for a large part of the systematic effort to deny anthropogenic climate change. Their publications are mostly books, since their climate change denial papers are rejected by peer review scientific journals. The US administration adopted an attitude of denial and scepticism over the matter. An example of this attitude is a memorandum from the White House, inadvertently leaked to the press in 1990, proposing that the best way to deal with public concern over global warming was to 'to raise the many uncertainties' (New York Times, 1990).

Opposition to strong mitigation measures developed in the USA in the 1990s because of their negative potential impact on economic growth, the international competitiveness of the economy, and employment. At the same time climate change was becoming increasingly obvious to the large majority of US climatologists. The Byrd–Hagel resolution of the Senate, agreed by a vote of 95 to 0 on 25 July 1997, rejected ratification of any type of international climate change agreement with mandatory reductions of greenhouse gases emissions that failed also to set limits for developing countries. The resolution was passed a few months before the COP3 of the UNFCCC where the Kyoto Protocol was approved. This prevented the US government from ratifying the Protocol even before its approval in Kyoto. One of the most prominent sceptics, Fred S. Singer, professor emeritus of environmental sciences at the University of Virginia and president of the Science and Environmental Project Policy, stated before the House Small Business Committee that (Singer, 1998): "Climate science does not support the Kyoto Protocol and its emission controls on carbon dioxide. The climate is not warming and the models used to predict a future warming have not been validated. In any case, a warmer climate would be generally beneficial for agriculture and other human activities."

The Global Climate Coalition, formed in 1989 and generously financed by big business, especially from the oil, coal, and automobile sectors, was one of the main promoters of scepticism and made a concerted effort to convince political leaders that there were no sufficient reasons for the mitigation of climate change. One of its principal objectives was to discredit the IPCC, bringing into doubt the scientific integrity of the authors of its reports and making the UNFCCC negotiations as difficult as possible. However, by the end of 1990, some of the corporations that supported the sceptics began to feel isolated.

The first economic sector to become very worried about climate change was insurance, alarmed by the damages resulting from the higher number of extreme weather and climate events during the 1990s. Among the large petroleum companies, British Petroleum was the first to recognise the role played by fossil fuels in global warming, through declarations made by its president Sir E.J.P. Browne, in May 1997. In the following year, the company pledged to reduce its emissions of greenhouse gases to 10% below the levels of 1990, by 2010. Following this new tendency, various large companies acknowledged that climate change was not a fictitious problem and abandoned the Global Climate Coalition. Meanwhile the George W. Bush administration continued to support an attitude of denial.

In spite of the fact that the Obama administration recognizes the need for mitigation and is willing to enact legislation to this effect, we are witnessing in the USA and in other developed countries an upsurge of the sceptical movement. A poll conducted by the Pew Research Center found that the percentage of Americans who believe that there is evidence that global warming is happening shrank from 71% in 2008 to 57% in 2009. Human-induced climate change has become a partisan issue with more sceptics among republicans than among democrats. The reason for this change is probably the financial and economic crisis of 2008–2009 and the growing perception that mitigation of climate change is likely to have a negative impact on employment and economic growth in the short term.

Most conservatives do not support strong government regulations to reduce greenhouse gas emissions and a government-led effort to accelerate low carbon technologies. Their answer to climate change is adaptation. However, future adaptation to unmitigated global warming is likely to require even stronger government intervention. It is important to realize that global climate change is not a zero-sum game where the gains of some countries balance the losses of the others. As climate change intensifies unabatedly, all countries become vulnerable either directly or indirectly to its adverse impacts.

In spite of the warnings and the growing robustness of climate change science, it is very likely that the strength and visibility of the sceptical movement in the media, particularly on the internet and in the social network services, will increase. In any case, in view of the current paradigm of development, the preoccupations with assuring robust economic growth are very likely to prevail over strong climate change mitigation measures.

3.43 The Growth of the Human Population and the Malthus Theory

Let us leave the question of climate change for a while and address one of the most central contemporary issues as regards world development and the environment. The significant increase in the world population is one of the main challenges facing humanity and a crucial factor for the sustainability of its future development. To consider that the growth in the human population may constitute a problem is a relatively recent point of view in the history of civilizations. The abstract concept of world population as a mere aggregate of people whose growth can be discussed and controlled to some extent may be unacceptable in a religious perspective of human life, especially in some sectors of Christianity and Islam.

In 1790, the economist Thomas Malthus (1766–1834), the first professor of political economy in England, and probably in the world, proposed the theory that agriculture severely limits population growth, because population increases in a geometric ratio, while the means of subsistence increase in an arithmetic ratio (Malthus,

1798). He defended the need to control population growth, arguing that otherwise, increasing populations would inevitably lead to living conditions on the edge of survival, poverty, misery, and eventually collapse.

The Malthus essay was above all a critical reply to the points of view of Marie Jean Antoine Nicola de Caritat, Marquis de Condorcet (1743–1794), who had defended an optimistic vision of social progress in an earlier book. Condorcet was a philosopher, mathematician, and political scientist, a remarkable figure of the European Enlightenment and of the French Revolution in the second half of the 18th century. He supported a liberal economy, free and egalitarian public education, and women's rights.

Later Friedrich Engels and Karl Marx strongly opposed Malthus' theory. They considered it a reactionary doctrine that ignored the contribution of science to improving agricultural production, and claimed that new social systems can achieve a balance between population and means of production, thus dispelling the problem of overpopulation. They argued against the inevitability of deprivation and misery for the majority of the population under any social system. During the 19th and 20th centuries, various authors criticised the theories of Malthus, using the argument that an increase in population stimulates the technological development required to sustain it (Boserup, 1981).

Although it is very difficult to reconstruct the evolution of the world population, there are reliable indications that it grew slowly through prehistory and through the first millennia of history. A finer analysis reveals that the world population probably actually decreased during some short periods. One of the most notorious and well-documented cases happened in the 14th century when a large part of the population of Eurasia died from bubonic plague. Although the growth of the human population appears to be relatively steady and smooth, local populations are less stable and usually exhibit significant fluctuations. There are many examples of local populations that have alternated between growth and decline, and which have sometimes been on the verge of extinction, or even gone extinct.

3.44 The Collapse of Civilizations

The Indus valley civilization flourished between 2 600 and 1 800 BC, but collapsed soon after, due to a combination of various factors, the most important probably being overexploitation of natural resources and soil salinization caused by intensive irrigation. It is likely that, around that time, Indo-Aryan peoples from Central Asia migrated to the east and south, causing a transition to a new culture in the Asian subcontinent.

The Mayas give us another example of a relatively rapid civilization collapse, associated with a significant decrease in population. During the so-called Classic Period, beginning in 250 AD, they developed a remarkable civilization in a large number of cities with grandiose temples, sophisticated religious practices and forms of art, a writing system, numerals, and a keen interest in astronomical observations.

However, around the end of the 9th century, the Mayas abandoned their cities and temples in the relatively higher central region and the population decreased drastically. The causes for this collapse are not precisely known, but there were probably several converging factors: severe droughts, problems with overexploitation of natural resources, especially deforestation and soil degradation, continuous warring between kingdoms, discredit and revolt against the political and religious structures, and eventually chaos. Curiously, the crisis did not much affect the Mayan cities on the plains in the northern region of Yucatan, which continued to flourish, until they finally entered a long period of decline up to the time of the invasions by Spanish colonizers in 1511.

Another interesting and significant example, although on a much smaller scale, was the near collapse of the population on Easter Island, an isolated island in the Pacific, 2075 km from the Chilean coast. The first settlement was made by Polynesians coming from the west between the 5th and 12th centuries (Bahn, 1992), and from that point on the population remained isolated from the world until the arrival of the first European navigators in 1722. Other estimates indicate that the arrival of the Polynesians only happened later in the 13th century (Hunt, 2006).

Whenever it was, the fact is that the Polynesians developed a relatively advanced and complex civilization, with a still undecipherable writing, that led to the construction of 800 to 1000 statues called Moai sculpted from volcanic rock, with a height of between 2 and 10 m. The analysis of pollen deposited in small bogs and lakes indicates that dense forests covered the island, with various species of trees, including a very tall palm that is today extinct. Moreover, archaeological investigations show that the inhabitants fed themselves mainly on dolphins, which could only be caught in the high seas using canoes made with relatively large trees, but also on marine birds, and rats brought in boats by the colonizers. The plentiful food supply allowed for an increase in population, which probably reached 7000 to 10000 people.

When the Dutch navigator Jacob Roggeveen rediscovered the island on Easter day 1722, he registered in the ship's diary that it was covered with dry or burned meadows and that the inhabitants lived on the verge of survival and only had a few rudimentary canoes. None of this was compatible with a society capable of building the Moai. According to reports from other 18th century navigators, the population was around 2000 people. Although the details of what happened remain unknown, the records we have are sufficient to conclude that the forests were cut down to obtain land for crops, to build canoes, to make ropes to transport and erect the statues, and to use for fire.

With deforestation, soil erosion and degradation accelerated and many springs probably dried up. The marine birds left the main island and hid in the small peripheral islets, and most of the land birds disappeared. Rats multiplied alarmingly and ate the seeds of the palm trees, thus impeding regeneration. Many species became extinct. The Moai statues were ever more numerous and became larger, suggesting an untenable competition between the clans. Food products became scarce, making it impossible to sustain a complex and organized society. The political, administrative, and religious structures were probably contested and substituted by warring factions, and the population decreased by 70–80%. In the 18th century, visiting Europeans witnessed cannibalistic practices and the pulling down of the Moai statues by rival groups, the last one being pushed over and vandalized in 1894. The main conclusion is that the island population did not vanish, but that their culture suffered a profound crisis and collapsed, while the quality of life was deeply degraded.

With the arrival of Europeans in greater numbers, problems were aggravated. The population continued to decrease, this time due to diseases transmitted by the people who had established a colony on the island in the 19th century and to the savage deportation of islanders by slave traders to do forced labour in Peru and Chile. During the 20th century, the population started to increase again, reaching more than 4 000 people, and Easter Island has since become an attractive tourist destination due to the remains of its still mysterious civilization.

There are many other examples of populations that decreased violently when other peoples confronted or colonized them. In Central America, the indigenous peoples decreased by 80–90% during the 16th century with the arrival of the first Europeans, owing to the conquest and epidemics. Another example is Tasmania, where European colonists decimated the island's population at the beginning of the 19th century (Wrigley, 1969).

Throughout humanity's history, there have been many cases of more or less isolated societies whose culture developed, flourished, and eventually collapsed (Diamond, 2005). This type of problem is now of a quite different nature, as we edge ever closer to a global society. Today, the risk is a global collapse.

3.45 The Evolution of the World's Population

During the first millennium, the population of the world grew very slowly, and by the end, it probably reached around 300 million people. However, halfway through the next millennium, the increase began to accelerate, doubling in 700 years, and again after only 150 years, to reach 1 200 million in 1850. It is very difficult to determine in detail the reasons for this acceleration in the various regions of the world, but one of the most important was greater food availability and security, due in part to the dissemination of new crops and agricultural practices around the world by Europeans. The development of modern medicine, better hygiene, and better control of infections and epidemics significantly decreased the mortality rate, and in some regions, there was an increase in fertility.

Halfway through the 1920s, the world population reached 2 000 million people, and in 1974, just about 40 years later, it had doubled again, reaching 4 000 million people. In the last 40 years it has continued to increase, reaching approximately 6 830 million people in 2010. In the years around 1965, the world's population growth rate in percent reached a maximum estimated as slightly above 2.1% per year, but it then fell to the current value of about 1.2% per year. This peak in the growth rate caused a maximum in the annual world population increase of about 88 million in 1989. This fell to about 75 million in 2003 and then rose only slightly to 77 million in 2009.

The reduction is due to a diminishing global fertility rate, defined as the average number of children that would be born per woman if all women lived to the end of their childbearing years and bore children according to a given fertility rate at each age. It fell from about 5 in 1950–1955 to about 2.7 at the turn of the century (Cohen 2003). Global population projections indicate that the world total fertility rate will eventually fall below the replacement level fertility, defined as the level at which the world population replaces itself. With no early deaths in life, the replacement rate would be just slightly higher than 2.0 because of the excess of boy over girl births in human populations. Since child mortality is higher in developing countries, the replacement fertility rate is higher there. In developed countries, it is about 2.1 and in the least developed countries it can reach 3.0. The global average is presently 2.33.

The decrease in the world's population growth rate is a very important demographic fact since it may be a sign of a decisive contribution to a future sustainable development. Nevertheless, at present, raising the population by more than 70 million people every year brings with it extremely difficult challenges in terms of food production, water resources, energy generation, education, health, housing, job creation, and security.

A finer analysis reveals that during the 20th century the world's population growth migrated from the developed to the developing countries. Europe and the Americas were the regions with the highest population growth rates from 1750 to 1950. However, since that time the greatest increases have been in Africa, the Middle East, and Asia. This relocation resulted from a demographic transition process, which ideally happens in four stages.

Initially, a country has a high average birth rate and a high average death rate and these rates are nearly equal, which implies that the average population growth is near zero. At stage two, sometimes called the mortality transition, improved health and living conditions lower the average death rate while the birth rate remains high or possibly rises somewhat, thus causing a rapid increase in the population. In the third stage, also known as the fertility transition, while the death rate remains low, the average birth rate falls mainly because of changing lifestyles and increasing expectations for better living conditions. As a result, population growth slows down. In the final stage, both the average birth rate and the average death rate are low and nearly equal, leading to a very low or even negative final population growth rate. At the end of the process, we have a country with a stable population, though much larger than it was before the demographic transition.

During the last two centuries, the majority of contemporary developed countries went through demographic transitions approximately similar to the model described. Life expectancy doubled, reaching about 70 years, and birth rates fell by a factor of two or more. On the other hand, the majority of present-day developing countries are in the second or third stages of a demographic transition. A frequently argued hypothesis is that social and economic development in a given country necessarily leads to a demographic transition. Experience gained since World War II shows that cultural, religious, and geopolitical factors have a decisive influence on the viability, duration, and characteristics of the transition. In spite of this variability, a well-established link exists between living standards and the fertility transition.

The total fertility rate, sometimes just called the fertility rate, now begins to drop at an annual income per capita of US \$ 1 000–2 000 and falls until it reaches the replacement level at a per capita income of US \$ 4 000–10 000. However, there are some countries that do not follow this pattern. Currently the average fertility rate in developed countries is about 1.6, while in developing countries it is 3.1.

3.46 What Are the Limits for the Human Population?

An important demographic question is to know how the world population will evolve up to the end of this century and beyond. There are many possible ways to address this question. Let us start with the most abstract. If it were possible to find a mathematical function capable of satisfactorily reproducing the past behaviour of the world's population, its form could give us an indication of future behaviour. Contrary to common belief, the world's population growth has been faster than exponential during some periods. In the exponential curve, the growth rate in percent is constant, and this does not accurately represent what has happened since the mid-18th century.

In fact, the growth rate has been approximately proportional to the population size. This means that when the population doubles, its relative growth rate in percent also doubles. The mathematical function that satisfies this equation is the hyperbolic function, which has an asymptote and therefore becomes infinite after a finite amount of time. Known appropriately as the doomsday curve, it is obviously not a good representation for the future behaviour of the world population. Actually, we have the good news that the annual growth rate has been decreasing since the 1960s. The fact that the doomsday curve has been a reasonable approximation from the beginning of the industrial revolution up to about the middle of the 20th century is an unequivocal sign of unsustainable population growth during that period.

It is clearly impossible to make an accurate prediction of the future world population because of the many independent and unpredictable variables that contribute to its evolution. Extrapolation of mathematical curves fails because of the complexity of the problem. Nevertheless, we can make estimates based on the average trends of birth rates and death rates observed in various regions of the world over the past few years. Clearly, the further we set the horizon of the projection, the more uncertain it will be.

However, we may formulate some reliable statements about the future. The first obvious one is that the world population cannot grow indefinitely, putting aside highly speculative theories about colonising the Solar System or even beyond the Solar System. Consequently, in the long term, the world's population growth rate in percent will tend inevitably to an average value near zero or negative. Furthermore, we can safely say that it will reach zero for the first time in a period shorter than one hundred years. If the present annual growth rate remains constant for 100 years, the world population would reach 21 billion, a value that is very probably incompatible with our current high expectations of living standards and with the availability of natural resources.

There are also clear implications that result from a long-term average growth rate close to zero. In this situation, the long-term average number of births must equal the long-term average number of deaths. This means a decreasing birth rate, since life expectancy is increasing due to a generalized improvement in health care. Inevitably, there will be a strong trend toward ageing of the world's population. The phenomenon is already occurring in the developed countries.

Another way to address the future evolution of the world's population is to consider the human carrying capacity of the Earth. In biology, the carrying capacity of a species in a given habitat is the population size of the species that the environment can sustain indefinitely with its resources. Below carrying capacity, populations tend to increase, while above, they tend to decrease. The regulating factors that maintain a population in equilibrium with its habitat are very varied and variable with biological species, but frequently the most important are space, solar light, food, and predators.

A wider variety of potential constraints can limit the human population. The most important are food, water, energy, soils, diseases, waste disposal, pollution, loss of biological diversity, perturbation of the nitrogen and phosphorus cycles, scarcity of some nonfuel minerals, and climate change. The questions involved are much more complex than for other living organisms, because of our drive to improve well-being and the remarkable capacity to innovate and adapt to new situations, sometimes under very difficult or extreme conditions.

Moreover, in the human case the concept of a world population that the environment can sustain indefinitely requires further qualification, since we have to take into account the expectations of better living standards in an increasingly globalized world. In other words, the human carrying capacity of the Earth may increase considerably but at the expense of an unacceptable degradation in the relative well-being of a large part of humanity. If the human carrying capacity grows more slowly than the world population, then either population growth must slow down or the quality of life must decrease, or both. We are in a race between the human world population and the human carrying capacity of the Earth.

The point of view expressed by Condorcet is that human ingenuity can always increase the human carrying capacity of the Earth. On the other hand, Malthus believed wrongly that population growth rate would rapidly win the race against the rate of growth of the human carrying capacity. The various outcomes of this contest are describable in terms of very simple mathematical models using straightforward assumptions about the rates of increase of population and carrying capacity. There are essentially four modes of population change: steady increase, an approach to equilibrium through logistic growth, damped oscillations, and overshoot and collapse (Meadows, 1992). These schematic models are not supposed to accurately represent future scenarios of the world population, but they are useful to understand the consequences of the range of possible relations between the human economy and the carrying capacity of the Earth. In qualitative terms, we may ask which model gives the outcome that is most likely to occur.

The steady increase model, where the population carrying capacity increases indefinitely, is clearly very unlikely. In the logistic growth model, the world's population describes an S-shaped curve where the relative growth rate ultimately drops to zero and the population remains constant thereafter. This is also an improbable scenario since it is unlikely that the world population will remain approximately constant for a long period. More likely, it will decrease after reaching a peak. In the overshoot and collapse model, the world's population increases steeply, forcing a violent drop in the carrying capacity, which in turn leads to a population collapse. It is clearly the most problematic and dangerous scenario model. Although unstable, the damped oscillation model is the most probable long-term approximation to some form of sustainable development.

Currently, the crucial question for the future evolution of the world's population is to project the social, economic, security, and environmental circumstances under which the growth rate approaches zero. If the population reaches its peak in a situation of frequent and widespread crises or a truly global crisis, the chances of its size decreasing abruptly become high. On the other hand, if social, economic, and environmental conditions are relatively stable at the peak, sustainable development becomes a more likely outcome. Reaching the first peak in the Earth global population will be a decisive moment for humanity. Thus, what happens in the next few decades, up to zero population growth, is critically important for our future as humans on Earth.

The size of the world population when its growth rate reaches zero depends to a large extent on how rapidly the growth rate in the developing countries falls in the next few decades. If the fall is rapid relative to what happened in the developed countries since 1950, the convergence to an equilibrium situation will be relatively fast. However, if the decrease is slower, the convergence will take longer, and the size of the world's population will reach higher levels, with greater risks of instability.

Various experts and organizations have developed scenarios for the future evolution of the world's population based on the analysis of recent trends observed in different countries. The United Nations Population Division projections (UNPD, 2005) of the world population in 2050 are 7 680, 9 076, and 10 646 million, respectively, in the low, middle, and high scenarios. They assume a fall in birth rate from its current value of 2.6 to 2.55, 2.05, or 1.55, respectively. Despite the fall in the fertility rate, the world's population continues to grow. Indeed, in the middle scenario, it will be growing by 34 million people per year in 2050. In more recent projections the world's population in the middle scenario reaches 9 150 million by 2050 (UNDESA, 2008). It is becoming increasingly unlikely that the world's population will be below 9 000 million people in 2050. However, the world's fertility rate is likely to fall below the global replacement rate before 2050.

In the near future, population changes will continue to be profoundly different around the world. In developed countries the population is very likely to remain nearly constant from 2005 to 2050, with a value of around 1 200 million. However, in the less developed countries, it is likely to increase from the current 800 million to 1 700 million in 2050, while in the other developing countries, the growth will be slower. In the same period, half the growth will take place in only 9 countries: India, Pakistan, Nigeria, the Democratic Republic of the Congo, Bangladesh, Uganda, USA, Ethiopia, and China, in decreasing order of contribution to the world's population (UNPD, 2007). The consequences of these disparities are disturbing. The global population density, which is currently 46 people per km^2, will rise to approximately 66 people per km^2 in 2050 (Cohen, 2003). Moreover, while in the developed countries the density will remain at an approximately stable level, in the developing countries, it will tend to increase to 93 people per km^2 in 2050. This situation will cause serious problems for the sustainability of ecosystem services and food security, and it will raise political instability and the risk of conflict.

Another consequence of the increasing demographic disparities is the growing migratory pressures from the less developed countries to the industrialized countries. It is impossible to anticipate border crossing migratory movements in any quantitative way, given that they often constitute an answer, sometimes desperate, to unpredictable and rapidly evolving situations of conflict, hunger, persecution, oppression, poverty, and unemployment. However, it is very likely that the pressure to migrate from the less developed countries will increase substantially as the population continues to rise.

Population control measures can artificially change the population growth of a country. During the 1960s and 1970s, the population control movement was actively promoting reproductive health and family planning programs. Some countries, such as China, India, and Iran, have successfully reduced their birth rate with family planning policies. In China, according to government estimates, approximately 36% of the population is currently subject to the one-child policy. On other hand, some developed countries with very low fertility rates have implemented policies to increase the birth rate, such as France in the middle of last century, and Italy more recently.

At the international level, the most important organization for funding family planning and reproductive health programs is the United Nations Population Fund, which works in partnership with other United Nations agencies to attain the Millennium Development Goals. Its stated objective is "to support countries in using population data for policies and programs to reduce poverty and to ensure that every pregnancy is wanted, every birth is safe, every young person is free of HIV/AIDS, and every girl and woman is treated with dignity and respect" (UNPF, 2010).

As already noted the fall in the world fertility rate and greater life expectancy increases the average age of the population, especially in developed countries. The average world life expectancy was 47 years in 1950–1955, increased to 65 in 2000–2005, was about 67 in 2010 and will probably reach 75 in 2045–2050. In the developed countries, the increase will be greater, to around 82 years. The ratio of people not in the labour force to those that are will increase. This dependency ratio is usually defined as the ratio of the number of people under the age of 15 and over the age of 65 to the number of people with ages between 15 and 65. In OECD countries, it is expected to grow in percentage terms from 50% in 1990 to 77% in

2050. On average, for every 100 people working, there will be 77 not working. The trend will become increasingly global as we approach the end of the century. It will profoundly change the structure of societies and create new social and economic problems that will be difficult to solve.

There are attempts to project the world population beyond 2100, but the uncertainty becomes very large. Participants in an expert meeting organized by the UN (UNPD, 2004) expressed views about projections for the world population in 2300 ranging from 2 to 36 billion. This may seem too large a range to be useful, but the final outcome may actually lie outside that range. The low, medium, and zero growth scenarios go over a maximum before 2100 and then decrease, while the high scenario always increases up to 36 billion. An interesting point is that, in the medium growth scenario, the world population decreases after the first maximum at 9.2 billion in 2075 and then, following a return to replacement fertility coupled with increasing longevity, starts to increase again slowly reaching 9 billion in 2300. Another relevant projection is that, following Asia, Africa is likely to become a very important player in the world, with a share of the global population that nearly doubles from 13% in 2000 to 24% in 2300.

In conclusion, the human population is likely to level off in mid-century, very probably after 2050. Furthermore, the current expectation is that the number of people achieving higher levels of affluence, as it is currently understood, will continue to grow indefinitely before and after that time. However, this is much less certain since it would probably imply increasing social and economic inequalities, dangerous scarcity of natural resources, very damaging environmental change, and increasing security risks.

The root of the problems of sustainability has more to do with the continuation of consumerism in the industrialized countries and the expansion of this paradigm to the rest of the world, than with the increase in the population by itself. With alternative paradigms specifically aimed at sustainability, it would be possible to accommodate the world population under acceptable living conditions when it reaches its maximum. However, if resource depletion, climate change, extreme poverty, hunger, disease, terrorism, and organized crime continue and converge, the world will become increasingly unstable and the peak is likely to be reached under repeated crises. The major risk is then a sudden and large drop in the world population.

3.47 A World with Profound Social and Economic Inequalities. Extreme and Moderate Poverty. Obesity

In the last 200 years, humanity has improved its quality of life quite remarkably, particularly as regards human rights, health, housing and sanitation conditions, education, professional training, conditions of employment, access to information and communication, culture, and mobility. A far greater percentage of the world's population now has food security, better access to health, education, access to high quality water and commercial energy services, and a longer life expectancy. Nevertheless, dramatic disparities of social and economic development persist. To the majority of people that live in rich countries, the descriptions of life in the less developed countries, in situations of extreme poverty, water scarcity, overexploitation of natural resources, and environmental degradation are somewhat abstract and very difficult to imagine.

It is undeniable that we cannot exploit natural resources in a sustainable way beyond certain limits. There are two challenges associated with this proposition. The first is to identify and precisely characterise those limits. The second is to be able to use natural resources in the future in a balanced way so that the whole of humanity can benefit from them, rather than just some groups of people in some countries.

The indicator most commonly used to evaluate the level of economic progress and prosperity of a country is the gross domestic product (GDP), which is a basic measure of the overall economic output of that country in a year. In 2008, the five countries with the largest GDP were, in decreasing order, USA, China, Japan, India, and Germany with values of 14.2, 7.9, 4.3, 3.4, and 2.9 in trillions of international dollars (WB, 2008). Between countries the differences in GDP at purchasing power parity (PPP) per capita are enormous, currently reaching factors larger than 200. At the top of the ranking are countries such as USA, Norway, and the United Arab Emirates with values above 46 000 international dollars in 2009 (WB, 2010a). In the same year at the bottom of the list are countries such as the Democratic Republic of the Congo and Burundi with values below 400 international dollars. The last country in the 2009 list was the Democratic Republic of the Congo with a value of 320 international dollars (WB, 2010a).

The gross world product (GWP), defined as the sum of the GDP of all countries, has been rising steadily over the last few decades, except for 2009 because of the global financial and economic crisis. However, GDP growth is very variable from country to country, and also in time. Currently, the GDP is growing very fast in some developing economies, in particular in the so-called BRIC and BASIC countries, which includes Brazil, South Africa, Russia, India, and China. The countries of sub-Saharan Africa over the period 1975–2000 (UNDP, 2002) provide examples of slower or even negative GDP growth in developing countries. Note, however, that in the period 2003 to 2008, the average GDP of those countries have been growing at an annual rate of 6% (IMF, 2010).

There are many limitations to the use of GDP as an economic indicator. It excludes activities not provided through the market, such as household production and voluntary or unpaid work. It does not account for negative environmental externalities, the overexploitation of natural resources, or misallocated investment. Furthermore, it does not account for the wealth distribution within the country. Thus, GDP is not a reliable indicator for the standard of living and for the sustainable development of a country.

A more appropriate indicator to represent economic growth and also well-being is the genuine progress indicator (GPI). In a hypothetical situation, in which the financial gains in the production of goods and services equal the financial costs of crime and pollution associated with that production, the GPI would be zero but the

GDP positive. The determination of the GPI for various countries shows that their economies are less healthy in the long term than could have been expected (Cobb, 2001). While in the USA between 1950 and 2000 the GDP per capita increased from US $ 11 000 to US $ 33 000, the GPI per capita increased from US $ 5 000 to US $ 9 000. The growth of the difference between the GDP and the GPI from US $ 6 000 to US $ 24 000 reflects an increase in the ecological deficit and the degradation of the environment that will be left as a problem for the coming generations. Many other countries are in a similar situation.

The distribution of wealth has always been very unequal within countries, between countries, and across the total world population. Nevertheless, it is legitimate to enquire to what extent extreme poverty is acceptable in an increasingly globalized world that formally endorses the Universal Declaration of Human Rights, a fundamental constitutive document of the United Nations. What are the causes of poverty and how can we fight them? The World Bank currently defines extreme and moderate poverty as living on less than US $ 1.25 and US $ 2.00 per day, at PPP. In 2001, the number of people living in extreme and moderate poverty were 1 200 million and 2 800 million, although these estimates are highly uncertain due to the difficulty in obtaining reliable data at a local level (UNMC, 2002). In 2005, the World Bank estimated that more than 1 400 million people were living in extreme poverty (WB, 2008). More than 60% of the people under the extreme or moderate poverty line live in sub-Saharan Africa and in South Asia. In the last two decades, China accounted for most of the world's reduction in poverty.

For wealthy people in the world it is almost unthinkable to imagine living on less than US $ 1.25 per day. The fact is that it is possible to survive, but the consequences are devastating. UNICEF has estimated that in 2006 there were about 146 million undernourished children under five. About 11 million of those children die of starvation or of diseases caused by undernourishment each year (WHO, 2003). A more recent estimate gives 9.7 million (UNICEF, 2008). Almost all of these deaths occurred in developing countries, with about 80% in sub-Saharan Africa and South Asia. The fall in the infant mortality rate in the developed countries was very significant during the 20th century, but in the developing countries, the decrease has been slower, and over the last decade it has remained approximately constant. There is a strong correlation between poverty and the risk of epidemics and diseases, many of them related to the lack of access to good quality water and basic sanitation. Current estimates indicate that about one third of the annual number of deaths in the world, which stands today at around 18 million, are due to poverty-related causes.

Poverty is not an exclusive problem of developing countries. A significant part of the population in developed countries living below the poverty line suffers from malnutrition and limited access to health care. In the case of the USA, a recent report from the Census Bureau (HT, 2007) indicates that, in 2006, 36.5 million people were below the poverty line, five million more than in 2000. The only groups where incomes have grown from 2000 to 2006 were families in the highest 5% bracket of income distribution.

In a world where more than 30 million people die each year directly or indirectly from undernourishment, obesity is becoming an increasing health problem in both developed and developing countries. Weight excess and obesity are abnormal or excessive fat accumulation that presents a risk to health. They are usually assessed using the body mass index (BMI), defined as the weight in kilograms divided by the square of the height in meters. Weight excess and obesity correspond to BMI values over 25 and 30 kg/m^2, respectively. According to the World Health Organization, in 2005, about 1.6 billion adults were overweight and at least 400 million adults were obese in the world (WHO, 2010). An estimated 22 million children under five are overweight.

All these numbers have a clear tendency to increase. In the same report the World Health Organization estimates that the number of overweight and obese adults in 2015 will be about 2.3 and 0.7 billion, respectively (WHO, 2010). Weight excess and obesity lead to serious chronic diet-related diseases, such as diabetes, cardiovascular disease, hypertension, and strokes, and to some forms of cancer. Obesity accounts for 2 to 6% of the total health costs in some developed countries (WHO, 2010). According to some estimates, about 280 000 people die annually in the USA from obesity-related diseases (Allison, 1999). The obesity epidemic results mainly from changing nutritional habits that involve a higher intake of saturated fats and sugars, and from less physical activity. Rising incomes, fast-increasing urban populations, globalization of food markets, and increasing use of automated transport are the main drivers of change here.

According to the Forbes Magazine annual record of the world's wealthiest people, there were 1 125 billionaires in 2008 and their collective net worth was US $ 4.4 trillion. This value corresponded to 7.3% of the GWP. It was slightly higher than the GDP of China in the same year — a country with about 1 330 million people. The wealth of the world's seven richest people is greater than the GDP of about 40 of the most heavily indebted poor countries, which have a total population near 570 million people. There are staggering differences of wealth in our present world and these disparities are increasing. More than 80% of the world's population lives in countries where income differentials are widening (UNDP, 2007).

In terms of GDP per capita, the inequalities between the more developed and the developing countries are growing. From the mid-1970s up to 2000, the GDP increased in practically all regions, with the exception of sub-Saharan Africa. The highest rates of growth were in the richer countries of the OECD and East Asia. In the last two centuries the ratio between the GDP per capita of the richest and poorest countries has been growing alarmingly: in 1820 it was about 3, increasing to 11 in 1910, 35 in 1950, 44 in 1973, and 72 in 1992 (UNDP, 1999). The income inequalities between countries and within countries are so high, and the tendency for them to grow is so strong, that the possibility of a situation of near equality is very remote and very improbable in the foreseeable future.

Some of the most important driving forces that have been increasing the inequalities are easily identifiable. We live an age of unprecedented opportunity for the smartest, best educated, most persistent, and most cunning, almost everywhere in the world. The new opportunities result mainly from globalization, scientific research, and technological innovation and they are leading to the emergence of a global plutocracy. The low-skilled labour force is at the losing end of this world increase in income inequality. The gap between the highly educated, hard-working elite and everyone else is increasing. The growth in aggregate real wages and salaries tends to go mostly to the top earners in society.

3.48 How to Measure Inequality. The Gini Coefficient and the Human Development Index

There are various ways of measuring economic inequalities. The Gini coefficient (Gini, 1912), is one of the most frequently used. It measures the difference between a hypothetical situation where the wealth is equally distributed and the actual situation. The coefficient varies from zero, a situation in which everyone earns the same, and one where only one person earns everything. Mathematically it is the ratio of the area between the line of health equality and the Lorenz line, which represents the probability distribution of wealth, and the total area. Higher Gini coefficients indicate more unequal wealth distribution. There are large variations in the Gini coefficient in both developed and developing countries, revealing profound inequalities in some of them. Considering all countries, it varies between about 0.2 and 0.7 (UNDP, 2003). From 1990 to 1998, the Gini coefficient had the highest values, about 0.5 to 0.6, in sub-Saharan Africa, South Africa, and Latin America. In the OECD countries, the coefficient varied from 0.247 in Denmark to 0.408 in the USA (UNDP, 2001).

The inequalities within a given country diminished after World War II, but since 1970, the average tendency has been reversed, in both developed and developing countries. This was very probably a consequence of the economic policies developed and firmly established at the world level in the last few decades of the 20th century. A recent analysis made in 23 countries indicates a clear tendency towards an increase in the Gini coefficient. In the USA it increased from 0.268 to 0.430 between 1969 and 1997, while in Russia it increased from 0.289 to 0.422 between 1992 and 2007 (Bandourian, 2002; Cooper, 2009).

The human development index (HDI) provides a more comprehensive assessment of development and well-being in a given country. Invented by the Pakistani economist Mahbub ul Haq and used by the United Nations Development Programme (UNDP) in their reports since 1990, it includes three components: the average life expectancy at birth, the level of knowledge and education, assessed by the adult literacy rate and the gross enrolment ratio, and the economic situation of the country measured by the GDP per capita. It varies between zero and one, which corresponds to the highest level of human development. The HDI is generally increasing, with the exception of some sub-Saharan African countries and post-Soviet states. A decrease in HDI only occurred in four countries from 1980 to 1990, but from 1990 to 2001, this number increased to 21. In 2009, there were 38 countries with very high HDI, ranging from 0.971 to 0.902, and 24 countries with low human development, with HDI ranging from 0.499 to 0.340. Although there is a general tendency for an increase in the HDI, the differences in human development around the world are still immense.

3.49 The Reflection of Inequalities in Health

In the health sector, the inequalities between countries are growing at an alarming rate. Life expectancy, for instance, varies from less than 45 years in some sub-Saharan countries to more than 80 years in some OECD countries. The reasons for this profound disparity are well known: poverty, lack of adequate health infrastructures and medical services, lack of or inefficiency in the control of epidemics, and insufficient financing of pharmaceutical and medical research for the specific diseases of the tropical regions, where many of the developing countries are located.

The number of people infected by the AIDS epidemic increased from 0.1 million in 1980 to about 40 million in 2005 (UNAIDS/WHO, 2005), of which 25.8 million are in Africa and 7.4 million in South and Southeast Asia. Poverty is one of the key factors contributing to the propagation of the disease. Since 1981, more than 25 million people have died of AIDS, and the number of orphaned children, whose parents were victims of the epidemic, currently stands at about 13 million, the vast majority being in developing countries, especially in Africa and Asia.

Every year, about 8 million new cases of tuberculosis are detected and 5 000 of these die, almost exclusively in the developing countries of Southeast Asia, Africa, and Eastern Europe. The epidemic is spreading and becoming more dangerous, following a period of about 40 years during which the disease was in decline (WHO, 2002). An increasing number of people contract tuberculosis due to a weakening of the immune system. On the other hand, the tuberculosis bacteria reduce resistance to AIDS, establishing a strong link between the destructive powers of the two illnesses. About 15% of those infected with AIDS end up dying of tuberculosis. According to estimates by the World Health Organization, the number of people infected with tuberculosis up to 2020 will be 1 000 million, of which 150 million are likely to contract the disease and 36 million likely to die, if the fight against the disease is not reinforced (WHO, 2002).

One other disease that almost exclusively affects the developing countries is malaria, transmitted by various mosquito species of the *Anopheles* genus. It infects about 300 to 400 million people per year and causes the deaths of 1.5 million people. The malaria death rate diminished up to 1970, but has remained approximately constant since then, in part due to the increased drug resistance of the plasmodium parasites.

3.50 The Phenomenon of Urbanization

Urbanization has been one of the outstanding features of world development since the beginning of the 20th century. The percentage of the global population living in urban areas was about 2% in 1800, 13% in 1900, 30% in 1950, 47% in 2000, and over 50% since 2007 (UNH, 2011). Current estimates indicate that it will reach

5 000 million people in 2028, corresponding to about 60% of the world population, and 7 000 million in 2050.

Different forms of urbanization have evolved during this period. The dominant tendency is for the residential area to shift outwards in a process of suburbanization that leads to a network of cities. Los Angeles is the best-known example of this kind of sprawling urban area. Urban populations are increasing much faster in developing countries than in industrialized countries, but the annual growth rates are falling everywhere in the world. There is a correlation between poverty levels and urban population growth rates, although there are some exceptions. The poorer the country, the higher the urbanization rate tends to be. While in developed countries, the average value is 0.4%, in developing countries it is 2.8%, and in the least developed countries it reaches 4.6%.

In industrialized countries the percentage of the population living in urban areas has stabilized in recent years at about 75%, while it will continue to increase in developing countries to reach values of about 84% in 2030. A metropolitan area with more than 10 million people is usually defined as a megacity. The number of megacities increased from 9 in 1985 to 25 in 2005. The largest is the Greater Tokyo Area with about 34 million inhabitants. Current projections indicate a downturn in the emergence of new megacities. However, there is a tendency for the expansion and merging of highly urbanized zones to form megalopolises.

Cities are attractive because the chances of finding jobs and opportunities are generally greater there. Moreover, in developing countries there is easier access to safe drinking water, basic sanitation, health and social services and education. However, about a third of the global urban population dwell in slums under difficult, very difficult, or extreme living conditions, particularly in some of the less developed countries of Africa and Asia. World slum dwellers increase currently by about 25 million every year. The enclaves or ghettos, frequently formed due to ethnic, religious, migratory, social, or economic pressures, tend to aggravate the tensions between the majority and minority groups.

In large cities, such as Los Angeles, Rio de Janeiro, São Paulo, Johannesburg, Istanbul, New Delhi, Moscow, Rome, and Toulouse, among many others, it is increasingly common to find enclaves of residential areas of the upper-middle and upper classes, protected by ever more ostentatious and sophisticated means of security. In these protected condominiums, the quality of life is generally excellent and residents can safely use the most advanced technologies and facilities for their comfort and luxury, including shops, restaurants, and all kinds of sports and health facilities, while nearby it is common to find residential areas with substandard housing, squalor, and poverty.

The move from a rural settlement to a city offers more job opportunities, but life becomes more stressful. In general, children have easier access to education but are more easily attracted to drugs and violence on the streets. Estimates indicate that at the end of the last century, the total number of drug consumers was around 200 million, the vast majority living in cities (ACUNU, 2000). Organized crime is on the increase in many large urban areas. A significant part of it is related to the illegal global drug trade, which according to the United Nations Office of Drugs and Crime reached a total value of US $ 321 600 million in 2003, about 0.9% of the GWP in that year. Other activities include money laundering, extortion, theft, various forms of fraud, assassination, contract killing, kidnapping, prostitution, pornography, trafficking in organs and human beings, arms trafficking, computer hacking, and smuggling.

Organized crime syndicates, which operate mainly in the large cities, are well known. They include the PCC or Primeiro Comando da Capital in Brazil; the Sixth Family and the Hells Angels in Canada; the Six Triads in China; the Medellin and Cali Cartels in Colombia; the Mafia in Italy; the Yakusa in Japan; the Cosa Nostra, Latin Kings, and Aryan Brotherhood in the USA; the Juárez, Tijuana, and Gulf Cartels in Mexico, the Russian, Macedonian, Indian, and Israeli Mafias. Most have international ties in several countries, and some control a significant part of the GDP, like the Russian Mafia.

Living in large urban areas involves risks, which are in permanent mutation and vary from city to city. Human life becomes almost entirely dependent on the efficient functioning of a complex system of services that supply water, energy (electricity, gas and liquid fuels for vehicles), food products, sanitation, and waste management. The prolonged interruption of these services would have devastating consequences. Cities are voracious consumers of natural resources and depend on rural regions for food production. This dependency is increasing dangerously with the higher levels of urbanization in developing countries. The ecological footprint of the urban areas is evidently very big, several hundred times larger than the area they cover.

The security risks for both people and assets are generally much higher in urban than in rural areas, and they are growing fast with the higher incidence of terrorism, potentially involving a much larger number of victims. Urban areas with high population density are more vulnerable to natural disasters such as earthquakes, tsunamis, volcano eruptions, tropical cyclones, tornados, flooding, landslides, and epidemics. As mentioned before, atmospheric pollution is generally much higher in cities than in rural areas, reaching levels that are hazardous for human health. A large percentage of the big cities are situated in low-lying coastal regions, making them vulnerable in the long term to flooding caused by climate-change-induced sea level rise.

Cities also have very significant impacts on the environment. Most of the rainfall on urban areas is wasted, drained off into gutters and sewers, without being able to recharge groundwater aquifers or to be used by vegetation. The growth of urban areas often causes loss of good agricultural land, the fragmentation of forests, and the loss of habitats and ecosystems that are valuable in terms of biodiversity. In conclusion, increasing urbanization continues to satisfy part of the well-being expectations of hundreds of millions of people who abandon the rural areas, but creates new vulnerabilities, risks, and challenges for the urban populations, making the path to sustainability much more complex and problematic.

Cities are where most of the world GDP is generated. Currently 600 urban centres generate about 60% of the world GDP. This percentage is not likely to change much until 2025, although there will be a pronounced shift to the East. By that year, 136 new cities from developing countries, including 100 from China, are expected to reach the top 600 (McKinsey, 2011). Of the 25 cities that are expected to experience the highest GDP growth between 2007 and 2025, 21 are in the developing world (McKinsey, 2011).

We have briefly reviewed some of the most distinctive current issues in the world. In part, they are a reflection of the economics of human development since the industrial revolution. Let us therefore address some of the key aspects of that history.

3.51 David Ricardo and the First Age of Globalization

A determining feature of our world today is globalization, interpreted as an increasing interaction, transfer, networking and sharing of ideas and objectives, goods and services between people, nations, and countries on a planetary scale. It is an expanding process of global integration into a complex world system of production and exchange associated with a growing mobility of people, goods, and services. It involves practically all human activities, including trade, technology, science, information, culture, religion, health, economics, tourism, politics, government, conflicts, arms, wars, security, and terrorism. Although it is a drive to promote well-being, it may not necessarily reduce inequalities and promote sustainable development at the global level. Globalisation increasingly determines the pattern of financial and economic transactions within a country, and especially between countries, making it a crucial factor for the evolution of social and economic development at both national and global levels.

Before the 15th century, various forms of globalization were initiated, such as those that occurred in the Hellenistic and Roman Ages, the Islamic Golden Age, and the Mongol Empire. However, they were localized embryonic attempts and did not lead to the irreversible globalization initiated with the Age of Discovery. The process started with the discoveries of the Portuguese navigators along the coast of Africa during most of the 15th century, followed by the arrival of Vasco da Gama in India in 1498, the discovery of America by Christopher Columbus in 1492, and the circumnavigation of the world by Ferdinand Magellan between 1519 and 1522. The discovery of these marine routes enabled the development of trade, the scientific exploration of the new lands and seas, and the exchange of plants, animals, human populations, religions, ideas, and values on a global scale (Page, 2002).

However, the current concept of globalization is very recent, dating from the period following World War II. Some authors identify the first era of globalization with the expansion of the British Empire, or *Pax Britannica*, which followed the battle of Waterloo in 1815. Great Britain thereafter controlled a large part of world trade through its dominion over the main sea routes and overseas markets. The free trade of goods and services between countries with specific or unique export commodities, based on the exchange rate against the gold standard, intensified. There was a rapid growth of international trade and investment between the European countries and their colonies in Africa, Asia, and the Americas.

The essential features of the theory of international trade can be attributed to David Ricardo (1772–1823), a political economist descended from a Sephardic Jewish family expelled from Portugal after the establishment of the Inquisition, and born in London in 1772. At the age of 14, his father, a stockbroker, employed him full-time at the London Stock Exchange, where he later made his fortune. Meanwhile he denied the Jewish religion, married a young Quaker against the will of his parents who disinherited him, and converted to Unitarian Protestantism.

When he was 27 years old, he became interested in economics after reading the book *An Inquiry into the Nature and Causes of the Wealth of Nations* by Adam Smith (1723–1790), his compatriot and pioneer of political economics. Within a few years, he became very wealthy and one of the most important and best-known economists of the time, mainly through his practical knowledge of the way the capitalist system works. He wrote several books, which profoundly influenced the future development of economic theory, one of the most famous being *Principles of Political Economy and Taxation*.

David Ricardo was the first to clearly demonstrate with his theory of comparative advantages that, even if a country could produce everything more efficiently than another country, it would be advantageous to specialize in what it was best at producing and trading with other countries. Just like Adam Smith, David Ricardo was an opponent of protectionism for national economies, particularly for agriculture. He defended the abolition of the Corn Laws, which imposed tariffs on agricultural products.

The first modern age of globalization ended in the summer of 1914 with the most destructive war the world had ever witnessed. The major victims of World War I were precisely the countries that most benefitted from the globalized economy that was developed up to the beginning of the war. The symbiotic relationship between the world's financial centre in Britain and the dynamic economy of Germany was broken. To some extent, the war can be interpreted as a form of reaction against globalization, announced in the form of rising tariffs and tougher immigration restrictions in the years that preceded the war (Ferguson, 2008).

When the war ended, globalization suffered an almost complete collapse in the Great Depression, which began in 1929 and ran through the 1930s, with grievous and sometimes dramatic consequences in the economic, social, and political spheres. After World War II, the world had a new geopolitical configuration and the power of capitalism was concentrated in a small number of countries. The USA became the dominant economic power and was willing to lead a reform of the international economic and financial system. The process started with the Bretton Woods agreement and established the foundations for the emergence of a second age of globalization. This second age only reached its mature state and apex after the end of the Cold War, following the collapse of the Soviet Union, itself initiated by the fall of the Berlin Wall in 1989.

3.52 The Bretton Woods Agreements and the Main Protagonists: Harry Dexter White and John Maynard Keynes

In July 1944, while World War II was still raging, the US President, Franklin Roosevelt, invited representatives from 44 countries to the United Nations Monetary and Financial Conference, which lasted three weeks and took place at the Hotel Mount Washington in Bretton Woods, New Hampshire. The Conference approved a system of rules and procedures to regulate the international monetary system and created the various so-called Bretton Wood institutions, which continue to play an important role in the international monetary, financial, and trade systems. These are the International Monetary Fund (IMF) and the International Bank for Reconstruction and Development, which later became one of the five institutions of the World Bank Group. The Bretton Woods Conference also proposed an international institution called the International Labour Organization, but the US Senate did not approve it. Meanwhile a General Agreement on Tariffs and Trade (GATT) was reached in Geneva in 1947. Gradually, the GATT transformed itself into an international trade organization called the World Trade Organization, established in 1995, after seven rounds of negotiations that culminated with the Marrakech Agreement of 1994.

The Bretton Woods Conference was largely dominated by the opposing views of the USA and Great Britain on the future of the international financial and economic systems, which were personified by the economists Harry Dexter White and John Maynard Keynes. The history of the professional life of the former clearly shows the opportunities and integration capacity of the American society and the political contradictions of the time. Harry White was the son of Lithuanian immigrants, Lithuania being a part of the Soviet Union at the time. He was born in Boston in 1892, fought in France during World War I, studied economics at Harvard, became professor at the Lawrence College in Appleton, Wisconsin, and ended up accepting a position at the US Treasury Department. During World War II, he became liaison officer between the Treasury Department and the State Department on matters regarding foreign relations, whereby he participated in many meetings with delegates of the Allied Powers in World War II, including of course the Soviet Union.

White was a fervent defender of the internationalism of President Franklin D. Roosevelt and of John Maynard Keynes' economic theories. He believed that, in order to maintain peace, promote socio-economic development on a global scale, and avoid future economic depressions, it would be necessary to create institutions that could promote a stable regime of liberal international trade. White also believed that the success of these objectives depended on the possibility of cooperation between the dominant economies of the USA and the post-war Soviet Union. Following his conviction he was able to arrange for Soviet delegates to be present as observers at the Bretton Woods meetings.

John Maynard Keynes was born in 1883 in Cambridge, the son of a university economics lecturer from that city, and was undoubtedly one of the most influential figures of modern economic theory. He studied mathematics at the University of Cambridge, but later became interested in economics and began a Civil Service career in the India Office, where he was involved in the Royal Commission on Indian Currency and Finance. However, since he did not find his civil servant life sufficiently stimulating, he resigned and accepted a lectureship in economics at Cambridge. He enjoyed the company of artists and writers, such as Duncan Grant and Virginia Wolf, and was a member of the Bloomsbury Group. Furthermore, he supported the arts and was a collector of impressionist paintings. Bertrand Russell considered him the most intelligent person he had ever met.

At the start of World War I, the British Government disturbed Keynes' somewhat dilettante cultural life style, by calling him to advise the Chancellor of the Exchequer on questions of international financial and economic policy. After this appointment his professional career acquired a great notoriety and he wrote several important works, among which was the *General Theory of Labour, Interest and Money*. He defended government intervention in capitalist economies to avoid recessions, depressions, and unsustainable growth. He also proposed ways to reconcile state-led social altruism with the pursuit of individual profit through private initiative.

3.53 The Bretton Woods Institutions. The Beginning of the Second Age of Globalization and Pax Americana

In Bretton Woods, White and Keynes agreed on the essential aspects of what should be the new international financial and economic systems. However, they disagreed on one crucial aspect. White wanted the government of the USA to have control over various mechanisms and instruments of the global economy, while Keynes representing the British government was against this ascendancy. Keynes argued that the IMF should counterbalance the economic hegemony of the USA, while White considered that it should be subordinate to the economic power of the USA and promote the balanced development of international trade using the dollar as a reserve currency.

There is general recognition that Keynes conceived and formulated the terms of reference of the Bretton Woods institutions, but at the end of the conference, White's geopolitical points of view prevailed, clearly because of the undisputed economic, military, and political power of the USA. In the approved plan, the USA controls both the IMF and the World Bank, since the size of a participating country's economy and the value of its financial contribution to those institutions weight its vote. The dollar replaced the role played by the gold standard, through the establishment of fixed exchange rates for the World War II allies. The Bretton Woods agreement was the formal beginning of a transition process to the *Pax Americana*, founded on the economic and military power of the USA and heir to the *Pax Britannica* in several respects, including language, history, and culture. Britain was clearly the loser and in fact, the British parliament only ratified the agreement in December 1945 after the USA agreed to provide US $ 4.4 billion in aid to the country.

At that time, the USA was in a position of enormous advantage: it accounted for about half of world industrial production, owned 80% of global gold reserves, and was the only military power with nuclear weapons. In 1945, under the presidency of Harry Truman, Congress ratified the creation of the IMF and the World Bank. The Soviet Union, although they had been observers at Bretton Woods, did not adhere to the agreement, becoming progressively more sceptical as a very strong anticommunist feeling developed in the USA and the Cold War intensified.

Meanwhile, on 31 July 1948, Elisabeth Bentley, a defecting Soviet spy who became an informer for the Federal Bureau of Investigation (FBI), accused Harry White of espionage for the Soviet Union. After a few days, on 14 August, White was compelled to testify before the House Committee on Un-American Activities, where he repudiated the allegations. He made a heartfelt discourse on his understanding of the 'American credo', declaring his profound belief in the "freedom of religion, freedom of speech and thought, freedom of the press, freedom to criticise, in the equality of opportunities". The speech received a long applause from the audience at the hearing. However, it was White's last public appearance: he died of a heart attack two days later on 16 August.

White probably thought that he was in a position where he could help his parent country to integrate the new world financial and economic order led by the USA. He was not in fact involved in treachery against his country, although he may have passed some classified information to the Soviets. The forces unleashed by the increasing anticommunism and the growing Cold War were too strong to allow for public recognition of the decisive role played by Harry White in ensuring the success of the Bretton Woods Conference, which established the supremacy of the USA.

As for John Maynard Keynes, he maintained the prestige that he had attained with his remarkable work on modern macroeconomics and fiscal and monetary policies, but he died earlier in 1946, also due to heart problems, possibly aggravated by his deeply pledged involvement in the problems of post-war international finance. The foundations of the second globalization age, which some have also called the Golden Age of Capitalism, were now firmly established, but the two main protagonists of the process died under very different circumstances.

3.54 The Multinational Corporations and the Weakening of the Second Globalization Age

The period from the end of World War II to the early 1970s was a time of worldwide economic growth and near full employment, especially in the USA, western Europe, and some western and eastern Asian countries. Most of the countries devastated by the war, such as France, Italy, Japan, and West Germany, went through a major economic expansion. It was the beginning of the Great Acceleration referred to in Chap. 1.

International agreements that resulted from the activity of the Bretton Woods institutions considerably lowered the trade barriers. The world witnessed a rapid integration of international markets for commodities, manufacturing, labour, and capital. Globalisation accelerated in an explosive manner. Between 1950 and 1980, the number of multinational corporations grew from 7 000 to 53 000; the value of global exports increased from US $ 311 billion to US $ 5 400 billion; foreign direct investment rose from US $ 44 billion to US $ 644 billion; and tourism increased by a factor of 25 (French, 2000). The establishment of multinational corporations accelerated considerably from the beginning of the 1980s, in particular through mergers and acquisitions, and especially in the area of computers, biotechnology, and telecommunications.

Multinational corporations offer a highly functional and efficient mechanism for optimising the profit obtained from the production of goods and services. The first was the Dutch East India Company, established in 1602. It had the monopoly on Asian trade for almost 200 years, was the first company to issue stock, and lasted for about 200 years before going bankrupt due to corruption. Multinationals benefit from the possibility of delocalizing production activities to countries with cheaper labour, lower social security costs, more abundant and less tightly controlled natural resources, and less stringent environmental legislation. Furthermore, in large multinationals, production is often so remote and carefully concealed from the end user that it is very difficult to check out illegal labour and other undesirable social and environmental practices. Multinationals constitute an important component of globalization and contribute to the migration of economic power from countries to business corporations and from developed to developing countries, and they also tend to concentrate wealth in a more restricted group of very rich managers and major shareholders.

For instance, in 2005, Wal-Mart Stores had revenue greater than the GDP of Turkey, BP revenue greater than the GDP of sub-Saharan Africa, excluding South Africa, General Motors greater than Finland, and Carrefour greater than Hungary (Le Monde, 2006). The conclusions are similar when making a more appropriate comparison of GDP with the added value of corporations, defined as the sum of total wages, depreciation and amortization expenses, and profits before taxes.

One of the main examples of a giant multinational corporation is Wal-Mart Stores, founded by Sam Walton in Arkansas in 1962. Initially Walton only owned one Wal-Mart Discount City store in the small village of Rogers, but today the family controls the largest corporation in the world. In 2005, it had 6 600 stores, and was the world's biggest employer with 1.6 million employees, a revenue of US $ 315 600 million, and a net income of US $ 11 200 million. In 2008, the revenue increased to US $ 404 000 million and the net income to US $ 13 600 million.

This explosive development has resulted mainly from paying low salaries to employees, implementing innovative stock management procedures, and frequently changing suppliers to countries where the production costs are considerably lower. Wal-Mart tends to restrict its employees from joining trade unions, and to close stores and go elsewhere when the pressure to join the unions becomes too great. It has been criticised by trade unions, religious organizations, and community and environmental groups who complain about the policies and business practices of the corporation. Currently, a large part of the production is made in China, where salaries are still relatively low. In 2002, Wal-Mart acquired US $ 15 000 million of products made in China, which represented about 10% of the total volume of US imports (PBS, 2004). Wal-Mart's priority is to target the lower socio-economic groups rather than the middle class. This strategy conforms to society's current tendency to polarize in such a way that buying power is concentrated in the upper classes and the lower classes.

The second age of globalization was largely driven by worldwide development and exchange in science and technology, and by the integration of national economies into the international economy through trade and the global expansion of multinational corporations. The symbiotic relationship between the USA and China is one of the major driving forces of the current globalization model. Chinese households save a high proportion of their rising incomes, in contrast with Americans who save much less or almost nothing. China has a very high account surplus today, equivalent to more than a quarter of the US deficit. The delocalization of manufacturing facilities to sustain the insatiable consumerism of the USA provides employment to a large part of China's vast population.

However, the symbiosis between the two countries involves an underlying rivalry, since China is rapidly becoming the second world economy through strong and sustained economic growth. It is increasingly assertive on the world scene and is perceived in the USA as practising unfair competition and currency manipulation. Deteriorating political relations between the two countries are likely to run the risk of conflict.

We are in a very advanced stage of the second globalization age and in a prolonged financial and economic downturn which has a greater effect on the developed countries than on countries with emerging economies. There are some similarities between the current financial and economic world situation and the one at the end of the previous globalization age. The establishment of a parallel between them leads to the conclusion that the risk of major conflicts or wars is likely to increase. Let us therefore address the realm of conflicts.

3.55 Conflict I: The Increased Lethality of Weapons

The overwhelming majority of people prefer to live in peace. However, violent conflicts and wars are frequent in humanity's history. Human societies have a strong and perennial tendency for fighting and warfare, which very probably has its origin in the evolution of social behaviour, since our hominid ancestors, characterized by conflicts between neighbouring groups. Conflicts have sources and motivations of different nature, such as territorial, ethnic, religious, ideological, political, or access to natural resources.

In the long history of warfare, the most significant evolution has been the amazing increase in weapon lethality. Our capacity to kill our fellow human beings increased in a truly terrifying way in the 20th century. Up to the middle of the 19th century, the lethality of weapons in combat was relatively low, and the majority of deaths resulted from the inability to cure the ensuing injuries and illnesses. However, the continuous use of science and technology gave rise to ever more precise, powerful, and deadly armaments. By discovering how to produce atomic bombs at the end of World War II, the lethality of weapons reached previously unimaginable levels.

The destructive capacity of arms is comparable through the lethality index (Zuckerman, 1996), which for a sword has the conventional value of one. On this scale, a modern automatic weapon has an index of 210, a bomber aircraft an index of 480 000, and a strategic nuclear missile carrying a 25 megaton atomic bomb an index of 10 500 million. This means that one of these bombs could annihilate the whole human population, assuming that the whole population is situated in the vast area where the effects of the deflagration would be fatal.

On the other hand, since the middle of the 19th century, the dispersion of forces in the battlefield, increased mobility and communication capacity, and better medical assistance for the wounded has considerably reduced the number of casualties as a percentage of the deployed combat troops.

3.56 Conflict II: Nuclear Arsenals and the Nuclear Winter

It is estimated that, between 1945 and 2000, the five countries in the initial nuclear club, the USA, the ex-Soviet Union, France, the United Kingdom, and China, produced 128 060 atomic bombs (Norris, 2000). The race between the USA and the Soviet Union for the production of nuclear weapons that use either nuclear fission reactions or a combination of nuclear fission and fusion reactions resulted from a lack of trust and fear of being attacked. The main objective has been to reach equilibrium through a capacity of mutual assured total destruction. The nuclear arsenal reached its peak in the 1980s with 685 585 bombs in 1986 (Norris, 2000), corresponding to a destructive power equivalent to 10 tons of TNT per capita in the USA, Europe, and the former Soviet Union. Mutual assured destruction became a doctrine of military strategy based on the theory of deterrence, which corresponds to the Nash equilibrium in game theory.

At that time several articles (Crutzen, 1982; Turco, 1983) alerted to the risk that a large-scale nuclear war would cause global climate change due to emissions into the atmosphere, especially the stratosphere, of large quantities of ash, dust, and gases resulting from the explosions and subsequent fires, mainly in urban areas. That catastrophic scenario, known as nuclear winter, involves the destruction of the stratospheric ozone layer by the nitrous oxide generated by the explosions and a significant decrease in the average global surface temperature because the ash and dust reduces the amount of solar radiation that reaches the Earth's surface. The blocking of sunlight also decreases the photosynthesis leading to extreme situations in the biosphere. Furthermore, there is a sudden increase in radioactivity. All these effects combined would completely disrupt all human activities, bringing death, illness, misery, and hunger. The suffering of humanity would be enormous. The rate of species extinctions would increase dramatically and the survival of many ecosystems would become problematic, depending on the intensity of the conflict.

In the beginning of the 1990s, after the dissolution of the Soviet Union, the world nuclear arsenal began to diminish, and in 2000 reached values of around 31 000 atomic bombs. Meanwhile the nuclear club increased from the original five members to include India, Pakistan, Israel, and North Korea. South Africa developed nuclear weapons but disassembled them upon joining the Nuclear Non-Proliferation Treaty. In 2005, it was estimated that there were 29 000 atomic bombs, of which 96% were in the hands of the USA and the Russian Federation.

The element most commonly used in modern atomic bombs is plutonium, with atomic number 94. Its most important isotope is ^{239}Pu, which is fissionable, has a half-life of 24 110 years, and can be made from natural uranium. In a nuclear reactor, as already mentioned, the neutron bombardment of the uranium 238 (^{238}U) contained in the nuclear fuel rods continually produces ^{239}Pu. The latter then fissions on collision with other neutrons to produce part of the energy generated in a reactor. A sphere of ^{239}Pu weighing about 16 kg constitutes a critical mass, because it undergoes a spontaneous fission explosion. By means of neutron reflectors, a nuclear bomb can be made with just 10 kg of ^{239}Pu, forming a sphere with a radius of 5 cm. For every gram, the complete fission of ^{239}Pu produces an explosion equivalent to 20 000 tons of TNT. In addition to these properties, all the plutonium isotopes are highly toxic and radioactive. Ingestion of 0.5 g of plutonium would be fatal. Smaller amounts would have prolonged harmful effects on various organs, particularly the liver and the bones, substantially increasing the risk of cancer.

The military power of both the USA and the former Soviet Union consolidated in the second half of the 20th century through the production of large amounts of weapon-grade plutonium, containing 90% ^{239}Pu. The USA produced about 100 tons. Production peaked in the 1960s and almost died out from 1990 onwards (Samuels, 2005). A large part of this ^{239}Pu is still warehoused in various parts of the USA, principally in Amarillo, Texas, and Aiken, South Carolina. As time goes by, the helium produced by the radioactivity of ^{239}Pu alters the plutonium cores of the implosion nuclear weapons, called plutonium pits, rendering their explosive properties uncertain. The debate over how long the stored plutonium pits maintain the necessary characteristics to be used in bombs has been intense, with estimates ranging from 40 to 100 years (Dawson, 2007). In 2004, the US government proposed the production of a new nuclear weapon, called the Reliable Replacement Warhead, designed to be easier to maintain without testing, but the Obama administration ceased the program in 2009.

The balance of terror era between the two superpowers during the Cold War has ended. It was founded on the concept of deterrence and supported by intense production of ^{239}Pu. However, in the present-day fragmented world, dominated by inequalities and growing terrorism, the risk of unleashing the apocalyptic power of plutonium will probably remain, or even increase.

The fabrication of nuclear weapons is currently only an ever-decreasing step away from the production of the enriched uranium used in certain types of nuclear fission reactors. On the other hand, it is very likely that nuclear energy will continue to be a key component of the primary energy sources of many countries, especially in emerging economies such as China and India. This situation, apart from creating ambiguity and significant risks, can also be the driving force behind international geopolitical strategies.

An example was the approval into law by the US Congress, on 8 October 2008, of the USA–India nuclear agreement, which allows for the transfer of nuclear technology and nuclear fuel to India, without requiring its adherence to the Nuclear Non-Proliferation Treaty. The refusal to sign that treaty isolated India internationally with regard to civilian nuclear activities for more than three decades. That isolation has now ended, but India's military program is not subject to the Nuclear Non-Proliferation Treaty, and nor indeed is Pakistan's. The agreement does not require India to limit the production of fissile material products or the number of nuclear weapons. Proponents of the agreement argue that it will bring India closer to the USA to forge a strategic relationship that would confront the rising power of China and its alliance with Pakistan.

Currently, 189 states have ratified the Nuclear Non-proliferation Treaty. From the nine countries known to have nuclear weapons five have ratified it and four remain outside. Iran is suspected of being on the way to fabricating atomic bombs. Proliferation of nuclear weapons is clearly one of the major security threats facing the world. On 24 September 2009, the United Nations Security Council unanimously approved a resolution proposed by the USA that envisages a world without nuclear weapons, calling for an end to the spread of nuclear weapons and a reduction of global stockpiles. However, there is a very long way to go before reaching that desirable goal.

Current nuclear powers do not intend to unilaterally scrap their nuclear arsenals. It will be impossible to strengthen the non-proliferation regime and to reach complete nuclear disarmament if the nuclear weapon states do not demonstrate their joint irreversible commitment by pursuing credible disarmament steps. But in spite of all the difficulties there are also recent encouraging developments. On 8 April 2010, the USA and Russia signed the New Start Treaty that reduces the arsenal of each country to 1 550 deployed nuclear warheads, which corresponds to a cut of about 30%. The treaty was ratified by the US Senate in December and is widely recognized as a major foreign policy success of President Obama.

3.57 Conflict III: Chemical and Biological Weapons

Apart from nuclear weapons, there are also chemical and biological weapons of mass destruction. Chemists use various synthetic compounds for military purposes, sometimes derived from compounds used in industry. These include agents that attack the nervous system, the blood, the skin, the eyes, and the breathing system, causing death or temporary incapacitation. Modern chemical warfare started in World War I with the use of chlorine gas, phosgene, and mustard gas. Since that time intensive research in various countries has led to the production of more lethal chemical warfare agents, improved dispersion and dissemination procedures, and the development of more efficient means of detection and protection against chemical weapons.

On 29 April 1997, the Chemical Weapons Convention entered into force, prohibiting the development, production, stockpiling, and use of chemicals weapons, and determining the process for their destruction. Despite the fact that 188 states have already ratified the convention in 2009, only about 50% have passed legislation to prohibit the production of chemical weapons. According to various credible reports, highly lethal substances are still in use in recent conflicts in densely populated areas, such as white phosphorus in mortar shells fired by the US army in the Iraq war (Times, 2005), and by the Israeli army in the Gaza Strip war of 2008 (BBC, 2009).

Biological warfare uses bacteria, viruses, fungi, and toxins to provoke infectious diseases and poisoning liable to cause death or incapacitation of the enemy. There are records of the use of biological weapons since the beginning of history. In the 6th century BC, the Assyrians poisoned the wells of their enemies with fungi, and in the Middle Ages, during the plague, the corpses and excrement of victims were catapulted into besieged cities to spread the disease. During the Pontiac rebellion of 1763 by the native peoples of the Great Lakes region of North America, the English troops tried to infect them with measles. More recently, the Japanese used biological weapons against military and civilian personnel in World War II.

Among the weapons of mass destruction, biological weapons have a lower cost than chemical weapons, whereas nuclear arms are much more expensive in terms of production, maintenance, and launch. Nevertheless, biological weapons, despite their terrifying power of destruction, have the disadvantage that their impact is not immediate, but requires a few days to take full effect, and can have consequences that are difficult to control, even for the attacker.

The Biological Weapons Convention entered into force on 26 March 1975 and prohibits the development, production, and stockpiling of biological and toxin weapons, except for prophylactic, protective, and peaceful purposes. Currently, 171 countries have ratified this convention, but the lack of any formal regime to monitor compliance reduces its effectiveness. There are indications that several countries have active research programs in biological weapons or in biodefence that are outlawed by the convention, such as the USA, Russia, China, Iran, Syria, and North Korea.

3.58 Conflict IV: Military Budgets and the Arms Trade

Military budgets tend to mirror the inequalities in wealth and development between countries. In 2000, The USA accounted for 41.7% of the global military budget, China 3.2%, and the Russian Federation 1.3%, while the states referred to as rogue states by the USA — Cuba, Libya, Iran, Sudan, Syria, Somalia, and North Korea — altogether totalled 2.8% (SIPRI, 2003; Nielsen, 2006). In 2006 the military expenses of the USA increased to 46% of the world total and those of the NATO countries to more than two thirds of the total (SIPRI, 2006).

Currently, the USA is unquestionably the only military superpower and naturally tries very hard to preserve this status in all the regions of a rapidly and profoundly changing world. The mainstream argument is that its military supremacy is decisive for world peace. This theory generates in the USA a tendency to increase the military budget, especially after the growing involvement in conflicts during the Cold War and those that followed the terrorist attacks of September 2001. Since the end of World War II up to 2001, the USA has spent more than US $ 18 000 billion in 2004 dollars on wars (Nielsen, 2006). During this period of relative world peace, about 100 000 Americans have died in military action, 260 000 have been wounded in action, and 10 000 missing in action (Berry, 2003). From 1947 to 1989, during the Cold War, the average annual military budget was US $ 322 300 million. However, from 1990 to 2010, the average annual spending is likely to reach US $ 371 600 million in 2004 dollars (Nielsen, 2006). On 22 December 2010, the US Congress authorized a Defence Department budget of US $ 725 billion for 2011, the highest in constant dollars since 1945. The average cost per month to the USA of the current Iraq war is estimated to be US $ 8 000 million, while the occupation is costing between US $ 1 500 and 5 000 million per month (Nordhaus, 2002).

The world annual military expenditure decreased from around US $ 1 600 billion in 1985, and reached a minimum of US $ 750 billion of 2004 dollars in 1998. However, it then rose gradually to reach US $ 1 035 billion in 2004. World military expenditure in 2009 is estimated to have reached US $ 1 531 billion in current dollars, a 6% increase in real terms of the 2008 expenditure (SIPRI, 2009). This represents an increase of 49% since 2000, and corresponds to 2.7% of the GWP. There is an increasing concentration of military expenditure in a small number of countries. The 15 countries with the highest military budgets account for more than 82% of the world total. In 2009, the USA had the largest budget with 46.5% of the total, followed by China with 6.6%, France with 4.2%, the UK with 3.8%, and Russia with 3.5% (SIPRI, 2010). As regards military personnel, the USA presently has about 1 480 000 while China has 2 105 000.

The military budget in developing countries is in most cases higher than the health or education budgets, contrary to what happens in the developed countries. In Ethiopia, for instance, in the 1990s, 9.4% of the GDP was used for military expenditure, while only 1.2% and 4% were used on health and education, respectively (Nielsen, 2006). In Angola the military budget reached 21.2% of the GDP (UNDP, 2002). On the other hand, during the same period, the average military budget in a highly developed country, such as Norway, Sweden, Canada, the USA, Australia, or Japan, ranged from 2 to 5%, while the budget for health was around 7%. For education it varied between 5 and 8%. This inversion of priorities is a very serious drawback for developing countries, making the path to development much more arduous.

The arms trade is one of the most profitable forms of business in the global economy. It is led by the USA, which exports arms mainly to the developing countries, at a rate that varied between US $ 22.2 and 33.6 billion per year in the period of

1986 to 1996 (SIPRI, 2006). During the same period, importations varied between US $ 1.2 and 4.2 billion per year. The next largest arms exporters in the same period were the UK, Russia, France, Sweden, and Germany. In 2004, the first five exporting countries included the USA, Russia, and various European countries, accounting for 81% of the total global arms sales (Le Monde, 2006). In the period from 2000 to 2004, China was the largest importer of arms in the world, 95% being supplied by Russia.

Western suppliers are eager to exploit this huge market, which has an enormous development potential. At present about US $ 2.4 trillion or 4.4% of the world economy is dependent on arms production and trade. The defence industry is currently enjoying reliable and increasing revenues. This trend may be related to the fact that geopolitical tensions and both inter- and intra-state conflicts tend to increase in economic downturns.

3.59 Conflict V: The Number, Nature, and Evolution of Conflicts

The number of armed conflicts around the world is manifestly on the increase. In the five years that followed World War II, the number of conflicts grew rapidly, although they were very different in nature from the preceding war. After that period the growth continued, but at a slower rate. The Heidelberg Institute for International Conflict Research maintains a detailed database on the number, intensity, and nature of world conflicts, and classifies them as violent (including wars and crises with different levels of severity), and non-violent, including active and latent conflicts (HIIK, 2005). The total number of conflicts grew from 100 in 1945 to 278 in 2006. Of these, six were wars and 29 were severe crises characterized by the intensive use of armed violence, while 83 were crises in which violence was sporadic. The remaining 160 were non-violent conflicts, 100 active and 60 latent. The total number of high intensity armed conflicts, including wars, rose from 10 in 1945 to 35 in 2006. In 2008, 16 major conflicts were active around the world, two more than in 2007.

An important evolving trend of armed conflicts since the end of World War II is that they are happening more often within countries. The number of high intensity intra-state conflicts increased from seven in 1945 to 33 in 2003. In 2005, there were 178 intra-state and 71 inter-state conflicts. The origins of these conflicts are mostly power issues at national or international level, related to territory, autonomy and secession, ideological and religious systems, and natural resources. The importance of the latter type of motivation has been growing in the last 60 years, revealing a tendency that is very likely to strengthen in the future.

Most high intensity conflicts, including wars, occur in developing countries: since 1945, around 90% have taken place in these countries (Renner, 2001). Considering only this type of conflict in 2005, eight were in the Middle East and the Maghreb, seven were in Asia, five in Africa, and the other four in the Americas and Europe. Armed conflicts tend to seriously harm or destroy the fragile economic and social structures of the developing countries in which they occur.

Conflicts are also responsible for a large number of refugees and the relocation of large populations. Worldwide, the number of refugees grew from 2.7 million in 1972 to 9.2 million in 2004. The problem, however, is not restricted to refugees, but also to people relocated within their own country due to conflicts and crises of various natures, and to people that finally return to their country of origin. The total number of refugees decreased from 2001 to 2005, but then began to increase again. At the end of 2007, the United Nations estimated that there were 11.4 million refugees outside their countries and 26 million people displaced internally by conflict and persecution.

Meanwhile, the number of different initiatives and activities related to the growing incidence of conflicts, whether trying to resolve them in a peaceful way or inciting them by helping one side, is truly staggering. These activities have the most diverse forms, including multilateral diplomacy, bilateral negotiations, alliances, propaganda, conferences, and meetings, dissemination of true and false information, censorship, espionage, secret service initiatives, covert operations, and economic and military operations of various kinds, including trade agreements and embargoes, arms-trading, and terrorist attacks.

3.60 Conflict VI: Terrorism

Although terrorism is an extremely aggressive and violent activity, of a nature that was probably always present during the evolution of the hominids, right through prehistory and up to modern times, it has recently acquired notoriety and singular characteristics. Nowadays the concept of terrorism has very strong political and emotional overtones around the world. There are many different definitions, and this reflect the controversy in characterizing what constitutes a terrorist act and in distinguishing it from acts of unlawful violence and war. A sure sign of the disagreement is the fact that the United Nations has not yet been able to adopt a consensual definition accepted by all member states.

The word 'terrorism' was first used in France to describe the actions of the Jacobins during the French Revolution, in which Jean-Paul Marat (1743–1793), Maximilien Robespierre (1758–1794), Louis Antoine de Saint-Just (1767–1794), and Jean-Marie Collot d'Herbois (1749–1796) were important protagonists. The Jacobins themselves referred to their group as terrorists, without a depreciatory connotation. The most turbulent period of the French Revolution, in which the government of the country was held by the Comité de Salut Publique, from 5 September 1793 until 27 July 1794, was known as the Reign of Terror or simply the Terror. It was a time characterized by the intransigent defence of the revolution through brutal repression, including the systematic assassination of conspiring enemies. However, the same kind of extreme violence persisted after the coup of 27 July 1794, which marked the beginning of the Thermidor reaction to the Terror period.

A great many organizations and groups emerged in the 20th century with the aim of carrying out terrorist acts. In Europe, some of the more notable examples are the

IRA in Northern Ireland, ETA in Spain, and the Red Brigades in Italy. On the other continents, there have been many examples of groups carrying out terrorist activities, such as Sendero Luminoso in Peru, the Klu Klux Klan in the USA, the Armed Islamic Group in Algeria, the Islamic Jihad Organization in the Middle East, and the Tamil Tigers in Sri Lanka. The US Department of State keeps a list of organizations that it classifies as terrorist groups, the majority being Islamic or Communist. The attacks in New York and Washington on 11 September 2001 initiated a new age of terrorist acts on a global scale.

On 29 October 2004, Osama Bin Laden claimed that Al-Qaeda planned and executed the 9/11 attacks. Al-Qaeda, which means 'the base', is a Sunnite Islamic paramilitary organization calling for a global jihad and founded by Osama Bin Laden and a few other Islamic fundamentalists. It started as a Mujahidin organization fighting to establish an Islamic State in Afghanistan during the occupation by the armies of the former Soviet Union in the 1980s. With the withdrawal of Soviet troops, Afghanistan became a fundamentalist Islamic State, and various organizations sought to extend the fight for the implantation of radical Islam in other countries.

On 23 February 1998, the leaders of Al-Qaeda launched a fatwa creating the 'World Islamic Front for Jihad against Jews and Crusaders', for the liberation of the lands of Islam under the dominion of the Western armies, especially Jerusalem and Saudi Arabia. Terrorist acts inspired or carried out by Al-Qaeda are relatively frequent in Iraq, Afghanistan, and Pakistan, and throughout the world. Those that happened in Nairobi, Kenya, and Dar es Salaam, Tanzania, in 1989, the port of Aden, Yemen, in 2000, New York and Washington in 2001, Bali, Indonesia, in 2000 and 2005, Madrid and London, in 2004, Amman, Jordan, in 2005, Algiers, Algeria, and Rawalpindi, Pakistan, in 2007, and Yemen in 2008, are well known examples. Following the first attacks, a worldwide web of secret intelligence services from many governments that struggle to contain terrorism has discovered and aborted several other terrorist acts.

In Iraq, after the 2003 invasion by the US and allied forces, Al-Qaeda started to act in a more visible way through suicide bombings directed predominantly at the occupying armies and their collaborators, but also at the Shiites. Nowadays it is more active in Afghanistan and Pakistan. A veil of secrecy shrouds the organizational structure and short-term objectives of Al-Qaeda. It is probably a network of groups with tenuous ties, but sharing the same strong religious, ideological, and political principles and strategic objectives. Their unity stems from the common objective of a global jihad against Jews and Crusaders for the liberation of Islam from Western domination or influence, the expansion of radical Sunnite Islam around the world, and reverence of Osama Bin Laden and his fellow leaders. After a long search conducted by the US CIA, Osama Bin Laden was finally found to be living in Abbottabad in northeast Pakistan, close to Kashmir, in August 2010, and killed by a team of the US Naval Special Warfare Development Group on 2 May 2011.

A remarkable feature of present-day terrorism is the ever more frequent recourse to suicide bombings. In the last few decades, suicide attacks against military and civilian targets has increased noticeably. In the 1980s, there were about five attacks per year and they increased on average by 3% per year, while in the 1990s, this rate reached 10%. Between 2000 and 2005, there were on average about 180 suicide attacks per year. In 2005, Iraq registered an average of one suicide attack per day (Pape, 2005). However, it is a widespread phenomenon. The 472 suicide attacks that occurred from 2000 to 2004, took place in 22 countries, killing 7 000 and wounding tens of thousands (Atran, 2006).

The direct motivations for suicide attacks are mostly strategic and political. If one tries to imagine the level of frustration, despair, and submissiveness needed to perpetrate such acts, one gets a clearer picture of the profound change in values that characterizes our current age. The systematic increase in suicide attacks is a desperate form of struggle, in particular to free from occupation the motherland of the communities to which the perpetrators belong and in which they have their religious and cultural roots.

Since the end of World War II, an increasing number of territories and nations are controlled or occupied by foreign powers with overwhelming military power. The reasons for these occupations are diverse. In the case of Iraq, despite all the justifications presented for the 2003 invasion, it is very likely that it would not have happened if the country had not been rich in oil deposits. Iraq is ranked fourth in the world in terms of conventional petroleum reserves, estimated at 112×10^9 barrels. A recent study (IHS, 2007) concludes that the reserves could be greater, even twice that value, placing Iraq in third place. There is a deepening connection between the number and nature of current conflicts and the decreasing strategic reserves of natural resources. The Middle East is an outstanding example as regards petroleum and natural gas, but other regions in the world rich in natural resources are becoming active or latent theatres of conflict. That is the case of North Africa, with the most recent military entanglement of NATO being in Libya, an oil producing country.

It is difficult to see how the spread and intensification of terrorism can be stopped if the fundamental problems that drive its development are not identified and effectively addressed. Terrorism is associated with a profound transformation in the structure and hierarchy of human values. It happens when imperatives of an essentially political nature replace moral concepts, as they are usually recognized. The evolution of cruelty and terror has gone through various stages since the beginning of the 20th century. It reached a peak of inhumanity between 1914 and the end of the 1940s during the two world wars, the ensuing revolutionary repercussions, and of course under Hitler and Stalin. After this period, different forms of barbarity appeared and spread throughout the world, especially since the 1960s, with the emergence of various organizations that resort to terrorism on local, national, and global levels. Nation-States have to some extent lost their ability to prevent the spread of extreme violence, in part due to the virulence and complexity of terrorist activities and to arms proliferation and drug trafficking.

3.61 Conflict VII: A World of Fortified Sanctuaries

Are we heading for a world in which the rich countries become havens or sanctuaries, fortified against the hostility and immigration attempts of people from the poorer countries? Are we heading for a form of worldwide 'apartheid', based on fear? Are we approaching a scenario characterized by the continuing inability to resolve growing social and economic inequalities between regions and countries, and to dispel hunger, extreme poverty, and misery, especially in the least developed countries? Will the trend toward increasing conflict and terrorism persist?

In this hypothetical world, the overwhelming preoccupation of the militarily stronger and richer countries would be to maintain their sanctuary status by assuring the security and high living standards enjoyed by a significant majority of their citizens. There are alarming signs that some such structural characteristics and tendencies are already in place. Could a world where some countries are fortified havens be a feasible option given the increasing globalization, particularly as regards knowledge, information, mobility, and trade?

It is impossible today to stop the worldwide circulation of ideas and symbols that characterize our different cultures, to prevent global advertisement of the lifestyles of performing arts celebrities and sports heroes that move the crowds, or to limit the mobility of persons, goods, and services that sustain the global economy. The model of a profoundly fragmented world is intrinsically unjust. It presupposes the cultural superiority of the people in a few countries, which is contrary to human nature and has no long-lasting parallel in history. The perception of injustice is one of the strongest and most fundamental faculties of our human nature, and the struggle to overcome it eventually ends up removing all obstacles. A group, nation, or country will easily adopt as a *raison d'être* the liberation from whatever are considered to be unjust forces oppressing it.

There is a growing tendency to reinforce vigilance along borders that separate the developed countries from the rest of the world. In spite of the enormous sophistication of the means employed, efficiency is limited, and the number of illegal immigrants is increasing. Many of those that leave the north and west African coast in clandestine attempts to reach Europe drown in shipwrecks or die of exhaustion, hunger, border guard violence, or lack of medical care, or even commit suicide. The situation is comparable on the border between the USA and Mexico, but with some specific characteristics. According to the US Border Patrol (NYT, 2004), between 1998 and 2004, a total of 1 954 people died trying to cross the USA–Mexico border, but this number is likely to be an underestimate. The annual number of deaths is increasing, having doubled from 1995 to 2005, when it reached more than 500.

Meanwhile, the number of internment camps for illegal immigrants in developed countries is growing rapidly. Estimates indicate that in 2008 there were 224 internment camps in the EU where illegal immigrants await admission into or deportation from the EU. Immigration into the EU became a highly politicized and controversial issue due in part to signs of increasing immigrant crime. Some organizations are seeking to halt the proliferation of the internment camps and to improve respect for human rights in the camps (MIGREUROP, 2007). From 2001 to 2007, the government of Australia adopted an alternative policy, called the Pacific Solution, by establishing detention camps for asylum seekers on Small Island States in the Pacific, rather than on the Australian mainland. The policy was later abandoned following strong criticism. The USA and Canada, the western European countries, Japan, and Australia all try to establish 'buffer zones' in neighbouring countries to protect them from illegal immigration. They seek to establish agreements with the immigrant or transit countries to stem the migratory flow in exchange for trade facilities, but generally without much concern for the human rights of the immigrants.

The number and length of separation barriers to limit the movement of people across a certain line or border, or just to separate populations in a territory, is fast increasing. Some famous historic examples, such as the Great Wall of China and the Berlin Wall, became obsolete or were destroyed. However, there are many which are active or under construction: Israel–West Bank, Egypt–Gaza, North Korea–South Korea, China–North Korea, the Spanish enclaves of Ceuta and Melilla–Morocco, USA–Mexico, Moroccan Berm in Western Sahara, India–Bangladesh, and India–Burma, to mention just a few examples. It will be increasingly difficult to address the security concerns of people all over the world, especially in the more industrialized countries, in a globalized world where inequalities keep on growing.

3.62 Risk Society: Natural and Manufactured Risks

One of the most distinguishable features of post-modernity is the growing perception of risk, and the relentless search to identify risks, control them, and eventually nullify them. Ulrich Beck emphasizes that we live in a 'risk society' (Beck, 1992), in which the increasing preoccupation with the future and with security generates a new notion of risk. The driver in this process is the way we deal with the new threats and security problems of the post-modernistic period.

Humanity has always been subject to a wide range of very serious risks and dangers. Currently, however, there is a new category of risks, increasingly widespread and acute, arising from the great diversity and the growing complexity and intensity of applications of science and technology to promote development. These new risks, which we may call 'manufactured risks', come from human activities and are distinguished from natural risks, which are outside human control.

But the difference between natural and manufactured risks, based on the nature of their origin, is not always well defined. Frequently, human activities increase the risk associated with some natural phenomena, such as land and coastal erosion, landslides, floods, desertification, forest fires, earthquakes, and tsunamis. The construction of settlements in areas that are highly vulnerable to flooding and landslides or liable to high seismic or volcanic activity is likely to substantially increase the probability of hazards.

A significant part of the human population lives in low coastal areas with high flooding risk, including regions frequently affected by tropical cyclones and tsunamis. The nuclear accident in Fukushima, Japan, following the tsunami of 11 March

2011, is a clear example of the dangerous synergy between natural and manufactured risks. More generally, the overexploitation of natural resources tends to exacerbate some natural risks or create new ones. A wide range of human activities pollute the atmosphere, water resources, oceans, and soils, degrading the environment and generating various types of manufactured risks, especially to human health.

Manufactured risks fall into three main categories. Accidental risks associated with accidents in the various sectors of industry, transport, and trade, and in a general way with the economic activity, and conflict risks associated with the various types of human conflict, from non-violent conflict to generalized warfare. Finally, there are the environmental risks that result from overexploitation of natural resources and the pollution and degradation of the environment caused by some human activities. Accidental risks, although they may have very serious consequences for the environment, are distinguished from environmental risks, since they are associated with unexpected accidents, rather than regular consented activities with significant negative environmental impacts.

Epidemics and pandemics constitute one of the most serious natural risks in terms of the number of victims. In the 20th century, the most deadly natural disaster was the influenza pandemic of 1918–1920, which killed about 50 million people: a number of victims comparable with those caused by the black death (Potter, 2001). Next come various typhus epidemics between 1914 and 1922, responsible for the deaths of about 3 million people, the floods in China in 1931 and 1959 with around 3.7 million and 2 million deaths, and the droughts in China in 1928 and India in 1942 with 3 million and 1.5 million deaths, respectively. In this century, the most serious natural disaster was the Indian Ocean tsunami, caused by the Sumatra–Andaman earthquake on 26 December 2004, which caused more than 235 000 deaths, and the Haiti earthquake of 12 January 2010 with more than 220 000 deaths. The development of early warning systems and the improvement of disaster management procedures reduce the dangers to human life and property associated with natural disasters. Both depend largely on the advancement of scientific research and technological innovation.

Risk refers to threats to humans and to what they value. It represents the expectation of a potential negative or grievous impact of an action, event, or situation, present or future, which involves a degree of uncertainty. Sometimes it is possible to quantify the uncertainty with probabilities. In this case, a risk indicator can be defined as the multiplicative product of the value of the losses that might result from an event and the probability of that event occurring.

An example of an event with a very low probability but with potentially devastating consequences is the collision of a comet or asteroid with Earth. We have unequivocal records of such collisions, and we know that they were relatively frequent in the early phase of Earth's history. Asteroids are more dangerous than comets due to their higher density, and also because they approach the Earth more frequently. The return period of an asteroid with a 1.7 km diameter, whose impact would leave a crater with a diameter of approximately 30 km, total destruction in an area the size of France, and a global catastrophe in the biosphere, could be as low as 250 000 years.

There are various astronomical observation programmes that seek to discover and monitor near-Earth objects (NEO), in order to identify those that could have a significant probability of colliding with Earth (NASA, 2007). Currently, about 4 000 near-Earth objects have been identified, but only one of them, the asteroid (29075) 1950 DA, with an average diameter of 1.1 km, will pass very close to Earth on 16 March 2880. An eventual collision with our planet would be catastrophic, but we have plenty of time to accurately determine its trajectory, evaluate the risk involved, and if necessary implement actions to avoid the collision or mitigate its effects.

As regards manufactured risks we have the ability and the means to evaluate the risks that result from a given human activity or group of activities. Some new technologies have given rise to very serious disasters, such as the one in the Union Carbide pesticide plant in Bhopal, India, from 2–3 December 1984, and the nuclear reactor disaster in 1986 in Chernobyl, Ukraine, then part of the Soviet Union. There have been many other examples up to the present time. The perception of these disasters of a technological nature, and of the enormous and growing variety of manufactured risks, has contributed to the development of a more sceptical and critical attitude toward processes and technologies used in the industry and energy sectors. This evolution characterizes a process of reflexive modernization that gives special relevance to the precautionary principle and the concept of sustainability, and in which it is essential to maintain a constant flow of information and debate between the people involved in science, technology, and the industrial sectors and political decision-makers, stakeholders, users, and the general public.

Climate change is a good example of an environmental risk that helps to characterize the risk society. In this case, the manufactured risks result from relatively slow processes that have impacts mainly in the middle and long term. These particular impacts will be mostly negative, but some will be regionally positive, although probably only on a temporary basis. The origin of such risks derives from our current social and economic development model, particularly as regards the energy sector. It involves all citizens of the world, although in differentiated forms and with different responsibilities. The risk perception in this case, as for many other issues, depends partly on the ability of the media to communicate the results of climate change science, and partly on the way societies at national, regional, and global levels evaluate, process, and react to those messages.

Most of the information about the nature and the dangerousness of manufactured risks reaches the public through the media, in the form of news and opinions based on observations, analyses, and interpretations with very different degrees of scientific credibility. The whole communication process is vulnerable to pressures exerted by groups and networks with diverse social, economic, and political interests. Thus the prevailing perception of manufactured risks varies in time and from country to country, bearing sometimes a tenuous, adulterated, or even totally erroneous relation with scientific knowledge.

This situation is very likely to be long-lasting and to significantly worsen with time. Among the growing number of people worldwide with access to the internet, social network services, and other media, the percentage that is intellectually curious, professionally trained, used to being demanding with regard to the quality of information, and skilled in critical analysis is probably decreasing. Furthermore, the fast-increasing facility to produce and disseminate news, information, opinions, and advertisements through the internet and other social media is bound to lower their average quality and credibility. For many people, the proliferation of products with widely varying quality and sometimes with contradictory points of view is a source of confusion and contributes to devaluing the messages from robust science, which are just taken as expressing one point of view among many others.

3.63 The New Era of Global Risks

There are many human activities that modify the environment in some way and tend to degrade it over sufficiently long periods. In the past most of the impacts were only at the local or regional level. Now, human-induced environmental change happens increasingly on the global scale, either in a systemic or cumulative way. This trend is very likely to continue. The associated global environmental risks are therefore becoming a characteristic of the present and constitute an ultimate threat for the long term future.

To evaluate such risks it is essential to improve our understanding of the global terrestrial systems in which they arise. We need to characterize the threats with greater accuracy and to develop the capacity of human societies to manage, control, and minimize those risks. The impacts of human activities are often apparent only over long periods and the interactions with environmental processes and cycles frequently poorly understood. From the 18th century onwards, human activities began to have a stronger impact on the various Earth subsystems, in particular the terrestrial and marine ecosystems and the climate system. These persistent anthropogenic interferences tend to cause global environmental changes, which generate global environmental risks.

The paradigm of a systemic global environmental change is global warming, caused predominantly by CO_2 emissions. Other important examples are the destruction of the stratospheric ozone layer and changes in the Earth's albedo produced by land-use changes. Cumulative global environmental change is another type of global change arising from the accumulation of localized and regional changes, but with a wide and dense distribution worldwide. The global nature of these changes results from the fact that they appear all over the planet or that their intensity is such that they generate a problem on a global scale, including situations in which they put at risk a natural resource on a global scale. The overexploitation, depletion, and pollution of water resources (rivers, lakes, and aquifers), the loss of biodiversity, deforestation, degradation, pollution, and loss of good soils for agriculture, and the dispersion of toxic polluting agents into the environment are all examples of cumulative global environmental risks.

Each type of global environmental risk presents specific observation, monitoring, and research challenges to formulate and implement adequate measures to manage, reduce, and eventually nullify the risks on the basis of scientific knowledge. Furthermore, control and elimination of risk requires the involvement of practically all the international community of countries in order to achieve some degree of convergence in the evaluation of hazards, definition of strategies, and policy implementation. Global environmental changes are a potentially serious threat to the sustainability of our civilization, because over the long term they affect the fundamental capacities and functionalities that support life on Earth. Science and technology do not necessarily guarantee the repair or mitigation of all the damage caused by the accumulation and convergence of dangerous anthropogenic interferences with the Earth system. Some may have irreversible effects or reversible effects only over very long periods, of the order of centuries and millennia, and at very high costs to human life and the environment.

Risks associated with global financial and economic crises, such as the one that started in 2008, and world wars, in particular a generalized nuclear war, are also in the category of global risks. Globalization is rapidly integrating our societies and blending our personal and collective behaviours. Due to the increasing complexity of the system, whatever happens in one part of the world with regard to financial and economic issues, politics, or conflicts influences the rest of the world and may grow into a global crisis. It is very probable that in the future we will be frequently faced with risk, living permanently on the edge of various types of interacting global crisis.

After this very brief analysis of risk in contemporary societies, let us reconsider the issues of development, inequalities, nation-building, and current major conflicts, and the transition to a new world economic order.

3.64 Goodwill and Ineffectiveness of Aid from Developed Countries to Developing Countries

It was in Europe during the Enlightenment period that the idea of spreading the advanced civilization born out of the industrial revolution to all peoples of the world first appeared. It arose from the conviction that the European civilization was superior and thus had the responsibility to assist the rest of the world so that they could also benefit from it. Robert Owen (1771–1858) was one of the founders of socialism and an in some respects a utopian. He wrote that, if the leading powers of the world embraced the right plan, "the human race shall be perpetually well born, fed, clothed, lodged, trained, educated, employed, and recreated, locally and generally governed, and placed to enjoy life in the most rational matter on earth" (Owen, 1958).

To what extent do we still believe in this program? After World War II, the rhetoric became different. References to racial superiority and to backward and uncivilized peoples were no longer made. The main objective became decolonization and self-rule. Now in the 21st century we are entering a third phase, where economic convergence is transferring most of the world's economic power to the developing countries, particularly to the emerging economies, in a scenario of increasing scarcity of natural resources.

It was in the context of the Cold War that the Western world first organized and began to deliver what it called foreign aid for developing countries. In his inaugural address on 20 January 1949, Harry Truman announced a "bold new program for making the benefits of our scientific advances and industrial progress available for the improvement and growth of underdeveloped nations." This Point Four Program, because it was the fourth point in Truman's speech, involved the investment of millions of dollars for the development of science and technology across the world. It is estimated that in the last 50 years the West has spent US $ 2.3 \times 10^{12}$ in foreign aid to developing countries (Easterly, 2006).

This large amount of money has been very important in improving the quality of life of hundreds of millions of people. However, it has been incapable of preventing about one billion people from living in hunger today, or saving more than 150 000 children who die every year due to hunger-related causes. About ten million children die each year from diseases that are relatively easy to prevent with small amounts of money, such as malaria. However, such donations would have to reach the most needy families directly so that they could take care of the health and education of their children.

There are several major reasons why foreign aid has not been more effective. Aid plans are generally too simplistic and inadequate for the social, economic, political, cultural, and religious complexities of societies in the developing countries, very different from those of the more industrialized societies. The West tends to project its own goals and models of economic development on poor countries, without due consideration for the prior need to support institutional capacity-building to strengthen the state. The main priority is usually to create new, highly profitable business opportunities with no concern for the improvement of governance and for the ability of the state to ensure the rule of law. This imbalance favours corruption and new forms of social and economic inequities and inequalities.

3.65 Objectives of the United Nations Millennium Development

At the beginning of the new millennium, on 8 September 2000, the United Nations adopted the *Millennium Development Goals* (MDG), in a meeting that brought together the highest number of heads of state ever witnessed. All 191 member countries of the UN signed a declaration in which they promised to "eradicate poverty, promote human dignity and equality and achieve peace, democracy and environmental sustainability". The MDG comprise 8 goals and 18 targets (with 48 indicators), to be reached by 2015.

The first goal is to accelerate the eradication of poverty and hunger, halving by 2015 the 1.8 billion people living in extreme poverty in 1990 and the 842 million people who suffered from hunger in that year. The MDG Report of 2009 shows that there is globally a slight chance that the poverty goal may still be reached, but some regions will fall far short of it, in particular sub-Saharan Africa and southern Asia. As many as one billion people are likely to remain in extreme poverty in 2015. The global financial and economic crisis of 2008–2009 has slowed down the process of poverty eradication. As regards hunger, the situation is bleak. About 1.02 billion people suffered from hunger in 2009, which was more than in 1990. Over a quarter of the children in the developing countries suffer from malnutrition. There are signs of improvement in eastern Asia, but there are also signs of a worsening situation, again in sub-Saharan Africa and southern Asia.

The second MDG is to ensure that, by 2015, all the children in the world, both girls and boys, will be able to attend school and complete primary education. In 2007, about 72 million children of primary school age did not go to school. The situation is improving, especially in sub-Saharan Africa, where the increase in school enrolment was 25% in the period from 1999 to 2007. Estimates indicate that about 29 million children will still not be attending school by 2015, which implies that this MDG will also fall short of its pledge.

The third goal is to achieve gender parity in primary and secondary education by 2005, and in all levels of education by 2015. The first target was not reached in 2005. However, there was some progress. By 2007, in developing countries, 95 girls enrolled in primary school for every 100 boys, which is an improvement on the value of 91 in 1999. But unless there is a renewed commitment, the 2015 target will not be fulfilled. The gender disparity is greater in higher education and becomes even greater when one considers access to employment, especially in management and politics.

The fourth goal is to reduce by two-thirds, between 1990 and 2015, the under-five mortality rate. The number of young children dying per year declined from 12.6 million in 1990 to 9 million in 2008. However, with this rate of progress it will be impossible to reach the goal. The leading causes of mortality — pneumonia, diarrhoea, malaria, and measles — are preventable through relatively inexpensive improvements in health care services. Between 1990 and 2006, there was no progress in reducing child mortality in 27 countries, most of them in sub-Saharan Africa.

The fifth goal is to reduce by three-quarters, between 1990 and 2015, the maternal mortality rate. In 2005, more than half a million women died during pregnancy, childbirth, or within the six weeks of delivery (WHO, 2007). The difference in the risk of dying from treatable or preventable complications of pregnancy between sub-Saharan Africa and developed countries is staggering: 1 in 22 compared with 1 in 7 300. Very little progress was made in saving mothers' lives in sub-Saharan Africa and southern Asia. Globally, the maternal mortality rate decreased less than 1% annually between 1990 and 2005, making it almost impossible to reach this MDG. Another target is to achieve universal access to reproductive health. Adolescent fertility is slowly decreasing. On the other hand, the unmet need for family planning is still very high in developing countries, particularly in sub-Saharan Africa. This situation results in unwanted pregnancies and births and undermines the achievement of several other MDGs.

The sixth goal is to halt and begin to reverse the spread of HIV/AIDS and the incidence of malaria and other major diseases. More than 25 million people have died from AIDS since the identification of the epidemic in 1981. The total number of people infected with HIV rose from 29.5 million in 2001 to 33 million in 2007. However, the wider access to treatment has contributed to the first decline in the annual number of AIDS deaths since the epidemic began. Malaria and tuberculosis mortality rates are falling, but not fast enough to meet global targets.

The seventh goal is to integrate the principles of sustainable development into national policies and programs and reverse the loss of environmental resources. This is clearly an immense and highly diversified challenge. There are some quantified targets. In the case of water, the target is to halve the number of people that do not have access to safe drinking water and to basic sanitation services. As already mentioned, there are currently about 1 100 million people without access to safe drinking water and around 2 400 million without access to toilets or other forms of basic sanitation. Water scarcity presently affects almost half the world's population and this percentage is increasing. There is also a target to achieve a significant reduction of biodiversity loss by 2010 but currently the number of species threatened with extinction is rising rather rapidly. Another target is to achieve a significant improvement in the lives of at least 100 million slum dwellers by 2020. It will be very difficult to achieve this objective because the number of global slum dwellers continues to increase significantly: it grew from 776.7 million in 2000 to 827.6 million in 2010 (UNH, 2010).

Finally, the eighth goal is to form a global partnership for development. The developed countries assume the commitment to assist the special needs of the least developed countries. They also promise to develop further an open, rule-based, predictable, and non-discriminatory trading and financial system. Nevertheless, market access for most of those countries has not improved and the agricultural subsidies in developed countries overshadow the investment in development aid. The target of dealing comprehensively with the debt of developing countries has had measurable success. Furthermore, the target of making available the benefits of new technologies, particularly in the information and communications sector, is having remarkable success. The use of mobile phones and internet is increasing rapidly in almost all developing countries. As part of the partnership deal, the developing countries promise to move towards responsible and democratic forms of government and viable economies, and to guarantee human rights, security, and social solidarity.

The MDGs are a set of ambitious objectives that would significantly move the world along the road to sustainable development. The exercise of devising the MDGs was important in identifying the most pressing current development problems, finding the best solutions, and disseminating their awareness throughout the world. However, most of the MDGs are practically impossible to achieve by 2015. The tremendous effect of the 2008–2009 economic downturn in slowing the rate of progress toward the MDGs has revealed the fragility of our global system. The economic crisis, as well as high prices for food, energy, and other commodities, has dramatically increased poverty and hunger in developing countries, particularly in the least developed among them. Furthermore, the crisis has pushed tens of millions of people into unemployment or vulnerable employment. Developed countries significantly reduced foreign aid funding because of the massive investments made in their banking systems to combat the economic downturn.

There are in fact many past examples of good development and assistance programs that have been unsuccessful. A United Nations summit meeting in 1977 agreed to establish 1990 as the year when there would be universal access to safe drinking water and to basic sanitation. The reality in 1990 was very far off the proposed goal. In 2000, the MDGs revised the targets and established new ones for 2015, but these are probably also unrealistic. In 1990, the United Nations specified the year 2000 as the time limit for assuring universal primary education. Again, the system failed, and in 2000 the goal was postponed to 2015. In all cases, no individual and no institution has assumed responsibility for the failure to meet the initial goals.

Normally, development assistance relies on grandiose and ambitious plans drawn up in a top–down fashion, without a critical analysis of the specificities and conditioning factors of the countries that receive the aid. There is usually no field monitoring of the way the plans are executed and no responsibility for the failure to reach the objectives within an established timeframe.

There are, however, other aspects to bear in mind. A recent report (Campbell, 2007) concluded that the MDGs would be difficult if not impossible to achieve with the current population growth rate in the developing countries, especially in view of the very large number of people living in extreme poverty. The solution advocated is to increase the support for family planning programs.

The basic needs of the poor in developing countries are not satisfied simply because they have neither the money nor the political power to secure them. Furthermore, they cannot hold anyone responsible to meet those needs. Without effective development assistance from the rich countries, the dramatic situation of those living in extreme poverty will persist, and their number is likely to increase, especially with recurrent global financial and economic crises. It is important to realize that development assistance is necessary but insufficient to induce sustainable prosperity in developing countries. There is always the risk that foreign aid will become a process of imposing Western social and economic models on societies with histories, cultures, and religions that are very different from those of the West. That strategy leads to failure. Success on the way to sustainable prosperity in poor countries can only result from the motivation, commitment, and effort of the local populations.

Within this framework, Western countries have a unique role to play in capacity building, including human resources, organizational, and legal framework development. Furthermore, they can also help through international exchanges, exposing new ideas, new products, new methods of production, and new technologies to the developing countries. In any case, help must be well adapted to local specificities. If developed countries disregard their responsibility to involve those in need, the future world will be less free and more fragmented, unstable, and insecure.

3.66 Public Development Assistance — Promises and Reality

An important component of foreign aid is the Official Development Aid (ODA) from OECD countries belonging to the Development Assistance Committee (DAC). In accordance with a decision taken by the United Nations General Assembly in 1970, each country belonging to that OECD committee should provide 0.7% of its GDP in ODA by 1975. However, 36 years afterwards, the vast majority of OECD countries have still not reached that goal. In 2004, only five countries, Norway, Denmark, Luxemburg, Sweden, and Holland, had made contributions above the barrier of 0.7% of the GDP. In the same year, the USA reached a modest 0.17%, but in absolute terms it made the largest contribution, namely US $ 19.7×10^9. The countries belonging to the DAC in 2004 contributed an average of 0.26% of their GDP to ODA. Note that the target of 0.7% of the GDP has been repeatedly confirmed and assumed in international meetings, such as the International Conference on Financing for Development in Monterey, Mexico, in March 2002, and the World Summit on Sustainable Development in Johannesburg, in August of that same year.

It has been calculated that, if each member country of the DAC reached 0.7% of GDP in ODA, by the year 2015, the resources generated would be sufficient to achieve the MDGs. The total ODA decreased during much of the 1990s and reached a minimum in 1997. It then increased to reach a value of US $ 107.1×10^9 in 2005, which represents 0.28% of the total GDP of the countries involved. Much of the increase goes to fulfilling the geostrategic interests of the donor, such as fighting terrorism. The forecasts of the DAC indicate a tendency to reach 0.36% of the total GDP in the next few years, which is very likely to be insufficient to achieve the MDGs.

Why do the rich countries refrain from carrying out their promises of development aid to the poor countries? The reasons are complex, varying from country to country and in time. They relate mainly to the political assessment of priorities in the distribution of financial resources in the context of the cultural, social, economic, and political situation of the donor countries. The concern to reach a more equitable and supportive world is widespread, but the political priority for action in that direction at the national level is frequently low, both in developed and developing countries. Questions of national sovereignty, territory, military and economic power, politics, security, and all those that refer to the affirmation and preservation of ethnic, religious, and cultural identities in the international context, have a much higher priority. The mediocre activity of the governments in most developed countries as regards development aid is probably an accurate reflection of the opinions and priorities of their electorates. A recent opinion poll in the USA revealed that only 3% gave high priority to development aid (Time, 2007).

The private sector has also had a long experience of development assistance in the developing countries, championed mainly by NGOs and various types of foundation. This form of assistance is a great hope for the future, because it tends to be more responsible as regards achieving objectives, more effective, more innovative, more sensitive, and more adaptable to local realities. Nevertheless, despite their growing importance, they are utterly unable to substitute for the ODA deficit that stands in the way of accomplishing the MDGs by 2015.

Many defend the point of view that the social and economic development of the developing countries depends almost exclusively on the evolution of their cultural characteristics and on the quality of governance, and that the ODA is almost irrelevant. To what extent is this thesis defendable? This question leads us to consider the scope and strength of state functions.

3.67 The Scope and Strength of a State. Constructing Democratic States

The fundamentally new socio-political structure of the early agricultural civilizations is the state. The first states arose in Mesopotamia, associated with the cities that flourished there around 6 000 years ago. Then the Egyptian, Chinese, Greek, and Roman civilizations flourished, providing a range of different models. The present day modern states of the industrialized countries originated from those formed in Western Europe in the late Middle Ages, after incorporating the results of a long and often difficult experimentation with its religious, military, and economic components.

The modern state has various priority functions: to exercise sovereign authority and defend its territory, to provide and enforce law and order, to provide and protect civil liberties, security for people and assets, and property rights, to exercise taxation powers and macroeconomic management, to manage public finances, public borrowing, and international relations, to provide public health, social assistance, education, culture and scientific research, environmental protection, and so on. Essentially, according to Weber, the state is "a human community that (successfully) claims the monopoly of the legitimate use of physical force within a given territory" (Weber, 1946).

The state can take on a greater or lesser number of functional responsibilities and exercise them to the benefit or detriment of the common good. The scope of state functions is distinguishable from the institutional capacity to enforce laws or, abbreviating, the strength of the state. At one extreme, there are states that assume a vast array of functions, far beyond the minimum responsibilities of modern states, but have a much-reduced capacity to enforce the law. At the other extreme, one finds states that assume relatively limited responsibilities but have a strong institutional capacity for law enforcement. Fukuyama argues that Brazil is an example of the first tendency, while the USA is an example of the second (Fukuyama, 2004). The greater part of political history in the 20th century has to do with the definition and implementation of the scope of state functions and the strength of the state.

At the beginning of the 20th century, the scope of state functions was very limited in Great Britain, which was a model of a liberal democracy, as well as in other European states. With World War I and the Great Depression, various forms of totalitarianism developed in the Soviet Union and Germany, which sought to destroy civil society, transforming the people into mere subordinate elements for the ideological and political ends of the state. Nazism was responsible for the greatest horrors of recent human history and was defeated in World War II. As for left wing totalitarianism, it collapsed after the fall of the Berlin Wall in 1989, except in a very few states.

During the first three-quarters of the 20th century, the scope of state functions increased in practically all the democracies and non-totalitarian states. This trend led to an increasingly higher percentage of the GDP being consumed by the state sectors. In Sweden, for instance, the state sectors rose from 10% of the GDP at the beginning of the century to 70% in the 1980s (Fukuyama, 2004). The systematic increase of the social responsibilities of states was one of the main factors at the origin of the neoliberal reaction in the last few decades of the 20th century.

It was at precisely this time that a large number of countries in Europe, Latin America, Asia, and Africa began to set up democratic regimes, a movement known as the third wave of democratization (Huntington, 1991). The democratic transition frequently occurred in very difficult economic situations. To face these difficulties, the International Monetary Fund, the World Bank, and the US Treasury Department conceived of a reforming package of economic and financial policy prescriptions to promote development in the developing countries, called the Washington Consensus by John Williamson in 1989. A commitment to free-market fundamentalism based on privatization, liberalization, and price stability was a condition for receiving financial help. This program implied a weaker state with a reduced scope in the young democracies, especially in the economic sector, when in fact there was an urgent need to broaden and strengthen the capacity of the state institutions. The development aid based on the neoliberal principles of the Washington Consensus did not achieve its objectives, and in certain cases worsened the situation of the country that received it. In Africa, in the last three decades, the institutional capacity of many of the new independent states has deteriorated, despite external development assistance by the developed countries (Van de Walle, 2001).

Nowadays, emerging economies, especially China, constitute new centres of economic power that provide new forms of global cooperation and development in competition with industrialized countries. In the last 30 years, China has been the fastest-growing major economy in the world with a political regime that is very different from a Western type democracy. Meanwhile, in the majority of the developed countries democracies, national deficits and debts are growing, an increasing number of voters tend to distrust their political leaders, and many different interest groups have the ability to block change. The appeal of the Western politico-economic model to promote social and economic development is fading.

The model of the modern Western democratic state is far from being universally accepted and applicable. After World War II, decolonization gave birth to a large wave of new states which had varying degrees of success. In some cases, for example India and South Africa, the state was formed and consolidated, but in many other cases, in parts of Africa, Asia, and the Middle East, the state remained weak and practically non-existent, without the institutional capacity to exercise most of its priority functions. In other cases, the state collapsed or did not truly form, due to the arbitrary nature of its formation in terms of the ethnic and cultural diversity of the peoples in its territory, and the incapacity to successfully apply the available models of governance that enable social development and economic growth.

3.68 The Afghan State and Its Influence on World History

A particularly enlightening example is Afghanistan, situated in a mountainous region of great geostrategic importance, one of the key meeting points of European and Asian civilizations, and an important crossing point between East and West and North and South. The country comprises a wide variety of ethnic groups of which the most important, representing about 41% of the population, are the Pashtuns. In Persian, they are called Afghans, and locally, since the Islamic conquest and the conversion to Islam in the 7th century, the two words have become frequently synonymous. They are the largest ethnic group in the world with patriarchal characteristics and a segmented social structure, succinctly described by the Arab proverb: "Me against my brothers; me and my brothers against my cousins; me, my brothers, and my cousins against the rest of the world."

Afghanistan has a turbulent thousand-year history of resistance against frequent invasions. The martial abilities of the Afghans were recognized throughout history. After invading the region in the years 330 to 329 BC, Alexander the Great only managed a brief occupation due to the strength of the resistance and the harshness of the environment. He called the Afghans a leonine and brave people. Much later, they were one of the few regions who did not surrender to the domination of the British Empire, despite various attempts in three Anglo–Afghan wars in 1839–1842, 1878–1880, and 1919, motivated mainly by the rivalry between Great Britain and Russia for the control of Central Asia. In the first war, the British and Indian contingent of around 16 500 people stationed in Kabul, of which 4 500 where military personnel, was forced to retreat to India in mid-winter of 1842. The Afghan guerrillas attacked and finally destroyed the retreating force in the gorges and passes along the Kabul River valley. Only one Briton, Dr. William Brydon, survived to reach Jalalabad.

After the second Anglo–Afghan war in 1893 the British tried to weaken the Pashtuns by dividing them with a 2 600 km long frontier line that still separates Afghanistan from Pakistan today. Mortimer Durand, then Foreign Secretary of the British Government in India established the frontier, known as the Durand Line. It separates the Pashtun territory into two parts: to the east, in Pakistan, live about 26 million Pashtuns, while to the west live around 14 million. The loya jirga, name given to the grand assembly of the Pashtun people, declared the Durand Line invalid in 1949, following the independence of India and Pakistan, but without any practical effect.

The revival of tensions between Great Britain and Russia after the revolution of 1917 led to the third Anglo–Afghan war in 1919, when, for the first time in history, the Western forces used air strikes to attack the Afghans. After the end of the war, Afghanistan gradually began to win its independence and had a relatively stable period between 1933 and 1973. However, in 1978, the Communist People's Democratic Party of Afghanistan won control of the government. The country entered the orbit of the Soviet Union and became a pawn in the game of the superpowers, protagonists of the Cold War.

Six months before the invasion of Afghanistan by Soviet troops in December of 1979, the CIA began to provide secret aid to the Mujahedeen, who were opposing the pro-Soviet regime in Kabul. According to an interview given by Zbigniew Brzezinski, President Jimmy Carter's National Security Advisor, to the Nouvel Observatoire in January 1998, the USA knew that by supporting the Islamic fundamentalist movements they would very probably provoke a Soviet invasion of Afghanistan. Apparently, the intention was that this would give the USSR its Vietnam War. The tension between the USSR and the USA increased considerably after the Soviet invasion of Afghanistan. Finally, after a very violent civil war, the Soviet Union abandoned the country in chaos in 1989. On 9 November of the same year, the Berlin Wall fell.

It was in this situation that the radical Islamic leaders took control of Afghanistan. After a period of infighting between various Mujahedeen factions, the fundamentalist Sunni Islamic political movement of the Taliban, meaning the students, emerged as the dominant force and proclaimed the Islamic Emirate of Afghanistan in 1996. By the end of 2000, the Taliban controlled more than 90% of the country. They imposed harsh Sharia punishments for crimes, curtailed civil liberties and violated human rights, for instance, by forbidding girls from attending school and universities. At the same time, they systematically repressed the communists and opium production came nearly to a halt.

Between 1988 and 1989, Abdullah Azzam and Osama Bin Laden, who had been very successful in raising funds and recruiting foreign Mujahedeen for the war against the Soviets, formed the fundamentalist Islamic group of Al-Qaeda. One of its main goals was to take the jihadist cause elsewhere in the world after the Soviet withdrawal from Afghanistan. In 1996, Al-Qaeda relocated its headquarters in Afghanistan and enjoyed protection by the Taliban. A few years later, the suicide terrorist attacks against the USA on 11 September 2001, conducted by Al-Qaeda, led to a turning point in world history.

The USA responded immediately with Operation Enduring Freedom. They invaded Afghanistan, leading a Western multinational military force on 7 October of the same year, with the goal of destroying Al-Qaeda and defeating the Taliban, and thereafter helping to build a democratic Afghan state, with an institutional capacity to affirm and consolidate itself throughout the whole territory. In 2011, after 10 years, the Taliban still hold a considerable part of the power in the country, especially in rural areas, and the presence of Western troops led by NATO continues to be indispensable to maintain the stability of the Afghan government in Kabul.

Throughout its dramatic past and present history, Afghanistan remains one of the poorest and least developed countries in the world. The GDP per capita at purchasing power parity was 935 US dollars in 2009, among the 15 lowest values worldwide (IMF, 2010). About two-thirds of the population lives on less than two dollars a day. Government corruption is rampant and about one-third of an Afghan's income goes into bribes. Nearly one-third of Afghanistan's GDP comes from growing poppies for opium production. On the bright side, the country has benefited from considerable international assistance and investment since 2002. There are large unexplored gas, oil, and mineral deposits, the latter including copper, iron, gold, and lithium, a strategic metal with important applications in the energy sector, in particular to produce batteries. However, Afghanistan remains a dysfunctional state, trapped in a prolonged war.

On 1 December 2009, President Barack Obama ordered the deployment of an additional 30 000 troops over a period of six months and proposed to begin troop withdrawals 18 months from that date. A major conference on Afghanistan took place in London in January 2010. Later, the Taliban said that they would not engage in peace talks with the Kabul government until there were no more foreign troops on Afghan soil. The initial objective of completely defeating the Taliban is becoming increasingly remote. On 17 June 2011, the UN Security Council split a sanctions list on Al-Qaeda and the Taliban, sending a clear sign of reconciliation in Afghanistan to facilitate talks between the USA and the Taliban.

In addition, it is ever more difficult to define the precise objective of the Western countries in Afghanistan. History teaches us that, since World War II, it is impossible to achieve peace and development in a country under foreign military occupation, no matter how focused and sophisticated the intervention by the occupying troops. In spite of all uncertainties about the future, it is very unlikely that Afghanistan will become a modern democratic state, friendly with the West, at least not in the next few decades. More probably, the Taliban will regain control of the country after the withdrawal of most of the foreign troops.

Afghanistan has been a victim of several global geostrategic conflicts throughout its history. The current war is just one more case in the long Western learning process with regard to war and peace in a world under profound economic and power transformations. The decisive question is to define the objective of military interventions on foreign lands and the circumstances under which they are justifiable. In this respect, the invasion of Iraq in 2003 is an important case study.

3.69 The Neo-Conservatives and the Invasion of Iraq

After the attacks of 11 September 2001, the USA reoriented its foreign policy to legitimize direct intervention in ailing, collapsed, or badly governed states which they considered to be a threat to national security. An example of this new orientation is the mandate of the 2002 National Security Strategy of the USA (NSSUS, 2002), which recognizes the possibility of resorting, if necessary, to preventive wars against sovereign states, including occupation and regime change. The objective is to create democratic and market-oriented states, which may involve tearing apart the old structures, so that more people enjoy the benefits of peace, democracy, and market economies (Krasner, 2005).

This post-modern imperialism for ailing states has its foundations in the doctrine of the neo-conservative movements created in the 1960s by various authors, such as Leo Strauss, Irving Kristol, and Norman Podhoretz. According to their followers, the USA had the economic and military means necessary to develop an expansionist geopolitical strategy of 'nation-building' along the principles of neo-conservatism, considered beneficial for the countries involved. Under the Presidency of George W. Bush, the foreign military interventions of the USA were often presented and justified as a 'war on terror', whose goal was to eliminate international terrorism and terminate state sponsorship of terrorism. However, 'terror' is more a tactic than an enemy that can be defeated by war. The expression thus implies the possibility of a state of continuous warfare. Furthermore, terrorism can be an ambiguous word, since it does not have a universally accepted meaning. Under President Barack Obama, 'overseas contingency operation' replaced the term 'war on terror'.

On 5 August 2004, the government of the USA created the Office of the Coordinator for Reconstruction and Stabilization to "prevent or prepare for post-conflict situations, and to help stabilize and reconstruct societies in transition from conflict or civil strife so they can reach a sustainable path toward peace, and a market economy" (USDS, 2006). One of its functions was to coordinate the actions of reconstruction and stabilization between the civilian agencies and the armed forces in countries subject to military intervention. This process involves the United States Agency for International Development, the World Bank, and the International Monetary Fund.

The most revealing example of the application of the neo-conservative doctrine was the invasion of Iraq, beginning on 20 March 2003, with the aim of overthrowing Saddam Hussein, replacing its regime by a democracy and installing an ultra-liberal economy. Paul Bremen, President of the Coalition Provisional Authority that governed Iraq after the invasion, ordered the immediate layoff of 500 000 soldiers and civil servants, the privatization of 200 state enterprises, the removal of restrictions on foreign investment in the non-oil sector, a drastic reduction of taxes, and the end of import tariffs (Easterly, 2006). Iraq was a unique opportunity to construct a modern democratic and capitalist state in accordance with neo-conservative concepts, using a top–down approach which would serve as role model for the progressive transformation of the Middle East.

It was argued that the success of the construction of a democratic Japan after World War II could be repeated in Iraq. However, the two situations are not comparable: Japan was a homogeneous nation from the ethnic, cultural, and religious point of view. It also had a strong state with a broad institutional capacity long before war. It was relatively easy to replace the old authoritarian regime by a democracy because the transition had the support of a functional state with disciplined state workers.

The insurrection and violence in Iraq over the last eight years following the military intervention have clearly shown that the utopian ambition for Iraq was somewhat naïve and completely inaccessible. The Islamic population made up of about 60% Shiites and 40% Sunnis had previously had a relatively peaceful coexistence, despite the antagonism and age-old mistrust between the two branches of Islam. With the invasion and the transfer of power from the Sunnis to the Shiites, the equilibrium was broken and this led to violent conflict. Meanwhile, Iran dominated by the Shiites reinforced its regional power, contrary to the interests of the USA, running the risk of further religious violence in the Middle East.

Up to 2006, the Iraq war had caused the deaths of 600 000 civilians and more than 33 000 servicemen, as well as over 3 000 American troops. About 4 million Iraqis abandoned their homes, some of them taking refuge in neighbouring countries, especially Jordan and Syria. Kuwait has closed its borders to Iraqi refugees and Saudi Arabia is building a very sophisticated 900 km barrier along its desert frontier. The supply of water and electricity continues to be uncertain and the health services have been close to collapse.

On the other hand, the level of violence has decreased substantially from a peak in 2006. Iraq's oil production has finally reached pre-war levels, with 2.4 million barrels per day. The country intends to increase its production to 12 million barrels per day by 2017, which is more than the 10 million barrels per day that Saudi Arabia intends to produce in 2011. If Iraq attains this goal and is relatively stable and friendly with the West, some will say that one of the main objectives of the Iraq war has been achieved. Currently, however, the per capita income is about half of its value before the invasion and unemployment in the age bracket 15–29 has reached 43% (Time, 2011). About 40% of Iraqi professionals have left their country since 2003 (USLP, 2011).

The political situation also remains volatile and lack of security continues to be a permanent and critical preoccupation. The powerful Shiite cleric Moqtada al-Sadr says that his Mahdi Army will remain ready to fight at least until the American troops get out of Iraq. This however, is very unlikely to happen in the foreseeable future. In April 2011, there are still 46 000 US troops in Iraq (USLP, 2011). One example of the strong commitment of the USA to Iraq is its new embassy in Baghdad, built with a final cost of about US $ 736 million (The Washington Post, 2007) and inaugurated in January of 2009. It is the largest embassy in the world, installed in a massively fortified compound, covering an area equivalent to the Vatican in Rome.

Iraq is likely to continue to be a serious concern for the West for many years to come, requiring huge military spending from the USA and possibly other countries. This cost will be bearable if Iraq remains friendly or even just accommodating with regard to the West, thereby guaranteeing safe access to one of the largest conventional petroleum reserves in the world.

3.70 How to Address the Challenge of Weak, Failing, or Collapsing States?

One of the gravest risks to world security at the present time is the transnational threat that emanates from weak, failing, or collapsing states. Such states have been at the origin of a large number of international crises since the fall of the Berlin Wall in 1989. Poorly governed developing countries can generate or contribute to regional instability and conflicts, international crime, weapons proliferation, mass migration, and environmental degradation. All of these issues are likely to have very negative international implications.

The situation is particularly difficult in some developing countries of Africa and Asia, some of which are involved in internal or external conflicts. Part of the responsibility falls upon the Western countries, which in the recent or more remote past promoted the formation of a large number of countries without due consideration for their future viability. In the painful process of decolonization, the emergence of new countries and the establishment of their borders resulted mainly from the contemporaneous geopolitical strategies of the Western powers and the partitions between their zones of influence, without giving much consideration to ethnic, religious, and cultural aspects.

Strengthening the state and improving the governance in these weak, failing, or collapsing states is one of the major challenges for foreign policy in the 21st century. In this respect, the developed countries still have to find the methodology to reach the right balance between a proactive and a reactive role. The difficulties encountered in Afghanistan reveal the problems that can result from trying to promote the development of a country and to strengthen its state while subjecting it to a military occupation.

The main characteristics of a failing state are a weak central government with little practical control over much of its territory, widespread corruption and criminality, large migration movements, economic decline, and increasing poverty. Since 2005, the US think tank Fund for Peace and the Foreign Policy magazine publishes a Failed States Index that evaluates those indicators and produces an annual ranking in four categories: alert, warning, moderate, and sustainable. For 2009, the list had only 13 states under the category of sustainable, Norway being the country with the lowest failed state index. The country with the highest index was Somalia followed by Zimbabwe, Sudan, Chad, Democratic Republic of the Congo, Iraq, and Afghanistan.

Somalia has no effective central government since warlords toppled the dictator Mohamed Siad Barre in 1991. Since then anarchy and violence has engulfed a large part of the country. According to the International Committee of the Red Cross, about 40% of the population of 3.4 million people requires humanitarian assistance, but relief operations are very limited because of insecurity. No one really knows how to deal with the situation, even though the scale of human suffering for most of the population is dramatic.

Delivery of food aid shipments to Somalia is now much more expensive and risky because of the piracy that has developed off its coast in the strategic Gulf of Aden since 2005. In the two years up to November 2009, 88 boats were sequestered and about 1500 persons made hostage (El Pas, 2009). The paid ransoms amount to tens of million of US dollars and contribute to some surprising forms of economic revival along the coast. The funding of piracy operations is now relatively well-established, using capitalistic procedures with a kind of stock exchange in Haradheere, which has attracted financiers from the Somali diaspora and from nationals of other countries.

In 1994, under former President Clinton, the USA adopted a new security strategy developed in part by the national security adviser Anthony Lake, which gave high priority to the containment of what he characterized as rogue or backlash states. These are states with undemocratic regimes that promote radical ideologies and are assumed to pose a threat to international security and especially to US interests and ideals. A rogue state usually has an authoritarian regime that severely restricts human rights and is under the accusation of sponsoring terrorism. The administration of President George W. Bush considered that Afghanistan, Cuba, Iran, Iraq, Libya, North Korea, and Syria were rogue states. This terminology has been in official use almost exclusively by the USA and UK, and criticized by France, Russia, and China, among others.

One of the main current concerns in international relations is the risk of nuclear weapons proliferation, especially in the so-called rogue states. North Korea withdrew from the Nuclear Non-Proliferation Treaty in 2003 and conducted its first nuclear test on 9 October 2006, in spite of the efforts made by the USA and other countries to prevent the development of nuclear weapons. The country, governed by a kind of hereditary dictatorship, is profoundly isolated internationally and has an almost entirely state-owned economy. There are many violations of human rights and a large part of the population is undernourished or close to starvation. The first priority of the government is to assure the military power of the country, and relations with South Korea are almost permanently on the brink of war.

Another complex current risk of nuclear proliferation concerns Iran, an ascending regional power, currently with a dictatorial regime that suffocates all opposition and does not respect most human rights and civil liberties. Most of its power comes from its large oil and natural gas reserves and from the increasing influence of the Shiites in the region following the military intervention in Iraq. Iran is currently the fourth largest oil producing country, with an output of 4 172 million barrels per day in 2009. Despite repeated statements from the government that its nuclear programme is exclusively of a civilian nature, a significant group of countries believes that one of its objectives is to develop nuclear weapons, as other countries in the region have done, particularly Israel, Pakistan, and India.

In the past, the West considered that it was not necessary to question and to try to prevent Israel from becoming a nuclear power in the Middle East, creating a regional asymmetry that will be increasingly difficult to manage. Israel is determined to prevent Iran from having nuclear weapons, by military means if necessary. Meanwhile, as a last resort, the accepted international policy has been to impose sanctions on Iran and to try to isolate the regime. However, it is very uncertain that such a policy will deter Iran from enriching uranium to weapon-grade level.

In a rapidly changing world, the emerging economies have an increasingly divergent perception and involvement in the Middle East conflicts, as compared to the USA and its allies in the West. They tend to adopt more pragmatic positions dictated by their economic interests. Furthermore, they are unwilling to support indefinitely Western policies likely to exacerbate tensions and lead to armed confrontation. In the medium and long term, it will probably be impossible to prevent a country like Iran from producing nuclear weapons. Air strikes against Iran's nuclear installations would only buy time at a huge cost. Unless there is a change of course, the present tensions in the Middle East are likely to lead to recurrent critical situations and violent military conflicts.

3.71 Is There a Current Clash of Civilizations?

It is nowadays frequent to refer to a putative conflict between civilizations and exemplify it with the tension described in indefinite terms between the West and Islam. This trend in opinion has its origin in the recent thesis of Samuel P. Huntington (Huntington, 1993; Huntington, 1996), according to which the dominant cause of conflict after the Cold War results from the confrontation of different cultural identities and religions, and is not of an ideological or economic nature. Nation-states remain the leading actors on the world stage of international politics, but the dynamics of the play are the confrontations between nations and groups of nations belonging to distinct civilizations. According to Huntington, this new re-structuring of conflict results from the fact that human rights, liberal democracy, and capitalist free markets became the only effective ideological alternative after the fall of the Soviet Union. The current conflict between Islamic and non-Islamic communities and countries is viewed as a clash between two different civilizations.

One of the problems with this thesis is the difficulty in characterizing the concept of civilization in the present-day world, and in identifying the supposed different civilizations and their relation with culture and religion. At the present time, it is impossible to characterize and identify a 'Western civilization' or an 'Islamic civilization'. Nowadays we are witness to a strong interchange of experiences, ideas, values, and goals between all countries in the world, and a profound cultural and political diversity among the believers of a given religion or religious denomination, in particular the Jewish, Christian, and Islamic religions.

In a world dominated by an increasing globalization of the economy, trade, employment, communication, information, internet and social network services, tourism, and the applications of science and technology, conflicts over civilization issues are rather unlikely unless there are specific political or geostrategic factors that exacerbate the cultural and religious tensions. The battlegrounds of the future are likely to be determined primarily by the economics of natural resources and not by the fault lines between coalescing civilizations.

In the last few decades, Islamic societies have benefited from an intense process of secularization and politicization leading to a massive investment in education and professional training, economic development, the emergence of a new middle class, a growing contact and exchange with Western countries, and a remarkable growth in the means of communication and information exchange through television, the internet, and other social media. One can distinguish essentially two types of social and political discourse in contemporary Islamic societies. On the one hand, there is the proactive discourse characterized by the conviction that the Islamic religion is the unquestionable supreme truth and that Islam is capable of renewing itself and offering a strong and attractive alternative to other models of society, the Western model in particular. On the other hand, in the secularized discourse, there is less integration between the social and cultural experiences of Islam and the religious component. The other religions are accepted as alternative pathways to reach the sacred and the divine enlightenment. In the proactive Islamic discourse, an extreme tendency defends the jihad against non-believers and the establishment of Islamic states.

The proactive discourse was inspired in the 20th century by various intellectuals and activists, such as Abul-Ala Maududi (1903–1979), a Pakistani journalist, theologian, and political philosopher, and Sayyid Qutb (1906–1966), an Egyptian theologian, writer, and poet. There are also moderate and traditional tendencies that privilege the use of reason as opposed to force, and finally quietist tendencies, characterized by a separation between religion and politics. A network of political and social organizations, usually clandestine, such as Al-Qaeda, the Egyptian Islamic Jihad, and the Salafist Group for Call and Combat, among many others, represents the most visible face of the extremist positions.

Islamic societies generally react against the growing influence of the cultural, social, and political models of the West and against integration into the prevailing liberal economy, which is becoming ever more global. They resent the risk of losing their social, political, and cultural identity, which, in the case of Islam, integrates the religious component. The reactive discourse in Islamic societies is less concerned with the rebirth of codes of conduct and religious practices in their purest ancestral forms than with the need to reaffirm the essential identity of Islam in the modern nation-state. This requirement manifests itself in the use of the state instruments of power: construction and restoration of monuments and mosques; civil codes transcribed or inspired in the Sharia; rigid dress codes for women; and strict obedience to religious practices. Islamic societies do not perceive science and technology as constituting a direct risk to their cultural identity. On the contrary, there is a widespread and strong interest and investment in their development.

Fundamentalist tendencies are also present in other religions. The Haredi Jews, also known as ultra-orthodox, believe that the Torah is the supreme law of the community. Obedience to it must prevail even if it contradicts civil laws. The majority of Haredi Jews reject Zionism, claiming that Jewish political independence is reachable only through divine intervention with the coming of the Messiah. They make up about 10% of Israel's population of 7.5 million and constitute the fastest growing group in present-day Israel, strictly following the Torah commandment of 'be fruitful and multiply'. They defend segregation from non-Jewish culture and restrict the access to some characteristic services of the globalized modern society such as television, films, secular newspapers, and internet.

Recently, on 17 June 2010, an estimated 70 000 Haredi Jews staged protests in various Israeli cities against a ruling by the Supreme Court ordering the jailing of parents of Ashkenazi background for refusing to send their daughters to a school with Sephardic girls, where the girls from the two communities would have to mix. About 60% of Haredi men receive welfare subsidies for full-time Torah study. They do not integrate the labour force and are exempt from the military obligations required of all other Jewish Israelis. Their growing number is creating tensions with mainstream Israelis and the Israeli government.

Besides religious fundamentalism, there are other reasons for tension between different cultures. One of the most important is the Middle East conflict between Israel on the one side and Palestine and most Islamic countries on the other. However, the reasons for the conflict cannot be reduced to a clash of civilizations. The same conclusion applies to the wars in Iraq and Afghanistan. In the case of Iraq, it is impossible to dissociate its large oil reserves from the initial motivation for the military invasion. Troop withdrawals began in August 2010, but some troops will remain to train and advise Iraq security forces, and to protect diplomats and civilians working in Iraq. In the case of Afghanistan, the initial objectives of the war against Al-Qaeda and the Taliban have not been entirely achieved, but problems were created with the generation of regional tensions between the USA, Pakistan, and India that will be difficult to appease.

Likewise, the growing competition and tension between the USA and China is not the outcome of a clash of civilizations. China has embraced the liberal economic models of the West, and Chinese citizens strive to emulate their lifestyles. Furthermore, at present, the two countries have inseparable financial and economic bonds.

In a strongly globalized world, cultural and religious identities are unlikely to be the primary cause of violent conflicts. The main drivers are weak, failed, or dysfunctional states, territorial and military occupation of foreign land, and the quest for natural resources. The risk of conflict also increases as a result of strong economic and military growth, especially in large countries or strong regional blocs. National preservation and development of economic and military power at a national, regional, or even at a global level, and security of access to the natural resources essential to sustain the paradigm of continuous economic growth are now the main driving forces for peace and war.

3.72 The Extraordinary Israeli–Palestinian Conflict

The most extraordinary, complex, and enduring conflict of the 20th and 21st centuries is surely the Israeli–Palestinian conflict. It has deep roots in the millenary Jewish civilian and religious history, the Jewish diaspora up to the present, Islamic history, British Empire history, and the events of World War II. It can be analysed from all these different angles, and in each case it usually arouses passionate opinions and controversy. The crucial point is that it involves the creation of a new Nation-State in modern times and in a land mostly occupied by people who did not belong to the same nation. The odds against success in this endeavour were very high right from the beginning.

Beginning in 1882 with the Bilu movement, thousands of persecuted eastern European and Russian Jews sought refuge in Palestine, attracted there by the appeal of the new ideology of Zionism, founded by Theodor Herzl. This wave of immigrants, the first Aliyah, reached a land ruled by the Ottoman Empire, which had an 85% Islamic majority, a Christian minority of about 9%, and an indigenous Jewish community, called Yishuv, of about 3%, living mostly in Jerusalem, Safed, Tiberias, and Hebron, the towns of rabbinical learning. The Jewish population in Palestine went on increasing from about 24 000 in 1882 to 85 000 in 1914.

Following World War I, Great Britain administered Palestine on behalf of the League of Nations, between 1920 and 1948, a period known as the British Mandate. But after World War II Britain did not have the resources or the resolve to remain in Palestine. Meanwhile the leaders of the Yishuv were determined to bring Jewish survivors of the genocide from Europe to Palestine. They declared a revolt against the British Mandate in Palestine and refused to respect the limits on immigration imposed by the 1939 White Paper issued by the British Government under Neville Chamberlain. The White Paper determined the creation of an independent Palestinian State in 1949, governed by Palestinian Arabs and Jews in proportion to their numbers in the population.

Although David Ben-Gurion, who was to be the first prime minister of Israel, had pledged to help Britain in the war against Nazi Germany, other more radical Zionist parties launched an armed insurgency with the stated aim of driving the British out of Palestine. Most of the violence was perpetrated by the Jewish terrorist organization Irgun, responsible for the bombing of the King David Hotel in Jerusalem, on 22 July 1946. Meanwhile, the antagonism between Jews and Arabs was ever more irreconcilable and the British had no leverage over the disputing parties. At the end of the war in 1945, the Jewish population in Palestine reached about 600 000 people, while the Arabs were more than 1.2 million. On 25 February 1947, Ernest Bevin, the British Foreign Secretary, referred the Palestine question to the recently created United Nations.

An eleven-nation Special Committee on Palestine (UNSCOP) was created and recommended the partition of Palestine into a Jewish and an Arab State. Zionist activists forcefully lobbied United Nations members to secure a two-thirds majority to carry out the Partition Resolution. The US President Harry Truman recalls in his memoirs that he never "had as much pressure and propaganda aimed at the White House as [...] in this instance" (Lenczowski, 1989). The same kind of pressure on the USA by the Israeli lobbies continues unabated up to the present time.

On 27 November 1947, the UN General Assembly adopted the Partition Resolution by a vote of 33 to 13, with 10 abstentions, calling for the termination of the British Mandate for Palestine and the partition of the territory into two states, one Jewish and one Arab. After passing the resolution, a transitional period under the United Nations was to begin that would last until the establishment of the two states. The Palestinian Arabs and also the Arab world remained fiercely opposed to the partition and to Jewish statehood in the Palestinian land.

A civil war in Palestine broke out almost immediately. A few months later the establishment of the state of Israel was declared on 14 May 1948, the day when the British Mandate over Palestine expired. The new state was recognized that same night by the USA and three days later by the USSR. By that time the Jews of Palestine, now Israelis, had secured the occupation of the main towns in the Palestine coastal plain and the Galilee panhandle, driving about 250 000 Palestinians from their homes. The Palestinian Arabs were left without a state. Instantly, several Arab states, including Egypt, Iraq, Jordan, Lebanon, and Syria, attacked Israel, but they were divided and weakened by the British and French colonial rule. The war ended in March 1949 with the signing of the 1949 Armistice Agreements, but the Israeli–Palestinian conflict persists unabated up to now.

The international community continues to press for the two-state solution as the way to achieve peace, but progress in that direction is almost non-existent. Recent reports published on the 40th anniversary of the Israeli–Arab war of 1967, one by Amnesty International and the other by the United Nations Office for the Coordination of Humanitarian Affairs (OCHA, 2007), indicate that the occupation of the West Bank by Israel is progressively inhibiting the establishment of a Palestinian state. The annual increase in the number of Israeli colonists is 5.5%, which is about three times the increase in Israel's population. There are continual delays in the negotiations that constitute the remaining hope for a peaceful solution to the conflict.

The West Bank is a labyrinth of barriers where the 2.5 million resident Palestinians have great difficulty in moving around. According to the United Nations Relief and Works Agency for Palestine Refugees, there are more than 50 Palestinian refugee camps spread throughout various neighbouring countries of Israel. The number of refugees rose from 914 000 in 1950 to 4.3 million in 2005, and continues to rise. The populations of these camps live in extremely precarious conditions of poverty, with high unemployment and very little hope for the future. It is not surprising that, as the years and the successive generations go by, these camps offer a fertile ground for the development of extremist Islamic movements.

Regarding Gaza, Israel and Egypt imposed a blockade since the Hamas formed a government there and took control of the strip in June 2007, after winning the 2006 Palestinian legislative elections. Israel maintains that the blockade is necessary to limit Palestinian rocket attacks on its towns. Meanwhile the humanitarian situation in Gaza is worsening. Recently, on 31 May 2010, the United Nations Security Council condemned the acts resulting in civilian deaths during an Israeli operation against a Gaza-bound aid ship convoy. It also stressed that the situation in Gaza is not sustainable and reiterated its concern regarding humanitarian aspects.

In the absence of any sort of sustained negotiation between the parties, the Israeli–Palestinian conflict is likely to continue its slow evolution until it reaches extreme situations. Meanwhile, Israel fights continually for its security and against the risk of becoming increasingly isolated internationally. The very tense relations between the present Israeli and Iranian governments, particularly in view of Iran's nuclear programme and the support Iran gives to the Hamas and Hezbollah groups, constitute a dangerous threat not only to those countries but also to the rest of the world. Israel's deep preoccupation with the 2011 uprisings in Arab countries governed by corrupt and dictatorial regimes, especially Egypt and Syria, reveals very clearly that in the middle to long term it will be increasingly difficult and risky to avoid serious negotiations for a two-state solution to the Israeli–Palestinian conflict.

In spite of all the conflicts, Israel is largely a successful Nation-State, particularly as regards its internal development. It is an industrialized country with a fully democratic regime and the majority of its citizens share with the West the same type of culture and lifestyles. It enjoys a very close relationship with the USA, where the Israeli lobby is very powerful. Clearly, Israel and its Islamic neighbours have different cultures and religions. However, the conflict between them is very far from a clash of civilizations. The conflict has to do primarily with territory and the desire to somehow reverse history up to 1882.

The first reaction of the Arab countries in 1948, after Israel declared independence, was to stress the illegality of the territorial occupation. Now some Arab countries have full diplomatic relations with Israel, such as Jordan and Egypt. However, there are still 22 United Nations member states that do not recognize Israel as a state: Afghanistan, Algeria, Bahrain, Bangladesh, Brunei, Chad, Cuba, Indonesia, Iran, Iraq, Kuwait, Lebanon, Libya, Malaysia, North Korea, Pakistan, Saudi Arabia, Somalia, Sudan, Syria, the United Arab Emirates, and Yemen.

3.73 The Arab Uprisings from 17 December 2010

A 26 year old Tunisian Tarek al-Tayyib Mohamed Bouazizi, commonly known as Mohamed Bouazizi, lived on the outskirts of Sidi Bouzid, a provincial town with a population of about 40 000 people in the centre of Tunisia, a world apart from the expensive coastal touristic resorts and towns where the Tunisian elite lives. Sidi Bouzid has an unemployment rate of about 30% and suffers from strong and widespread corruption and nepotism, like many other towns in the developing world. Mohamed Bouazizi was educated in a one room country school in Sidi Salah and later left school at the age of 19 to work full time.

He supported his mother and five younger siblings, including the expenses of one of his sisters to attend university. He became a street vendor, taking his wooden cart on foot over two kilometers to the market in the souk, where he sold fruit and vegetables, earning about 150 US $ per month. He did not have a permit to sell from his cart, but according to Hamdi Lazhar, head of Sidi Bouzid's state office for employment and independent work, no permit is needed for that (The National, 2011).

On 17 December 2010, Faida Hamdy, a 45 year old municipal office woman confiscated his electronic scales. They swore at each other and apparently she slapped him in the face. Bouazizi was publicly humiliated. He tried to retrieve his scales in the municipal building but to no avail. He then walked to the Governor's office and demanded an audience but was refused. No one listened to his grievances. He then acquired a can of gasoline from a nearby gas station doused and set himself on fire in front of the Governor's main gate, less than an hour after the violent discussion. On 4 January, eighteen days after his self-immolation, he died in hospital in Ben Arous, after being visited by the then president Zine El Abidine Ben Ali.

The protests in Sidi Bouzid started soon after Bouazizi immolated himself, and spread rapidly. The next day the police started beating the protesters and firing gas. According to Bilal Zaydi (The New York Times, 2011) relatives and friends of Bouazizi threw coins at the gate of the Governor's office saying: "Here is your bribe." The death of Mohamed Bouazizi reverberated through the Arab world, helped by Facebook and Al Jazeera. His grievances were similar to those of many millions of young people, victims of unemployment, corruption, petty bureaucratic tyranny, autocracy, and the absence of some very important human rights. His death became a catalyst for the Tunisian Revolution that led to the end of the Ben Ali's 23 year autocratic and corrupt regime. In fact Ben Ali resigned from the presidency and left Tunisia on 14 January just ten days after Bouazizi's death. The country was in a far from equilibrium situation, vulnerable to an isolated incident of this kind that eloquently exposed the profound injustices and stimulated the imagination, confidence, and courage of the people.

The rapid success of Tunisia's protests and revolution sparked similar protests right across the Arab world from North Africa to the Middle East. The revolutionary wave of demonstrations and protests spread to Egypt, leading to the resignation of President Hosni Mubarak on 11 February 2011, after weeks of protests in Cairo and other cities. Mubarak had ruled for about 30 years, suppressing dissent and jailing his opponents. In May 2011 there is an ongoing civil war in Libya involving a NATO air intervention, and civil uprisings in Bahrain, Syria, and Yemen and there have been protests in many other countries in the region, including Algeria, Iran, Iraq, Jordan, Kuwait, Lebanon, Morocco, Oman, and Saudi Arabia.

One of the most important driving forces behind the Arab uprisings was demography, coupled with high and widespread unemployment. About 60% of the population of the countries involved are under 30, reaching 74% in Yemen (Time, 2011). Furthermore the percentage of people aged between 15 and 29 not working and not going to high school or university is very high, ranging from 25 to 40%, except in Qatar and the United Arab Emirates where it is 13% and 16%, respectively (Time, 2011). This young generation watches satellite TV networks, uses Facebook and internet, and realises that there is no fundamental reason why they should not have freedom, human rights, civil liberties, the rule of law, proper political representation, and better opportunities for employment and economic development of the kind they see in many democratic countries, instead of being governed by corrupt despots, bent on purely selfish profiteering for themselves, their families, and their friends.

The 2010–2011 protests in North Africa and the Middle East, also known as the Arab Spring, constituted a post-Islamic uprising driven by technology and demography. It was not driven by young Muslims seeking to re-establish some form of authentic Islamic identity to escape Western influence. The motives of the demonstrations and protests are universal values shared by a majority of people in the West and in most countries of the world. However, there is also a strong element of justified resentment against the West, since Western governments acted as if the only option to prevent radical Islamism was to support corrupt and despotic regimes.

It is still too early to know the outcome of the Arab Spring, in other words, whether it will manage to establish a new sense of national identity based on the idea of citizenship and on the defence of civil liberties and human rights, or whether it will become entangled in conflicts arising from tensions between secularist and Islamist tendencies or between a variety of clans, sects, ethnicities, and religions. The uprisings have shown that there is a third way, distinct from Islamic political regimes or Western-backed despots. They go against the idea of a clash of civilizations and reflect a growing global movement toward a cultural convergence.

On 17 March 2001 the UN Security Council adopted a resolution on Libya to establish a no-fly zone and use all means necessary short of land occupation to protect civilians from the attacks of the Kaddafi forces. This resolution formed the legal basis for a NATO military intervention. Significantly the four BRIC countries plus Germany abstained in the Security Council vote. In June 2011, the military intervention is still going on. It is proving to be more expensive than expected for countries currently under extreme fiscal pressure and has generated strong tensions within NATO. Meanwhile in Syria the forces of President Bashar al-Assad are brutally repressing an enduring uprising against his dictatorial regime.

3.74 The Financial and Economic Crisis of 2008–2009 and the Transition to a New Economic Order

The time between the end of World War II and 2007 was an exceptional period of sustained social development and economic growth throughout the world. This was the time of the Great Acceleration after the war and the golden age of the second globalization. There were economic downturns and financial crises, but they were relatively weak, such as Mexico in 1994, Eastern Asia in 1997, Russia in 1998, and Argentina in 2001, all relatively localised and much less pronounced than the Great Depression of 1939. After the end of the dot-com bubble in 2000, world economic growth became very strong. The real annual growth of the GDP around the world increased steadily from 2.28% in 2001 to reach 5.06% in 2006 and 5.17% in 2007 (IMF, 2010).

This was also a period of strong economic convergence. In the five years from 2001 to 2005, the average growth of the GDP per capita in the emerging economies was 7%, while in the developed countries it was only 2.3% (The Economist, 2006). However, signs of disequilibrium in world markets started to appear in 2007. World food prices, especially for rice, wheat, and corn, grew abruptly in 2007–2008, causing social and political instability in several developing countries and dramatically increasing the total number of people living in hunger to more than one billion. The main causes for the food price crisis were the growing use of biofuels, in particular ethanol made from corn in the USA, and an increasing demand for more varied diets in emerging economies, especially in Asia, particularly as regards meat consumption. Other root causes may have been structural changes in trade and agricultural production and lower agricultural productivity due to greater climate variability associated with climate change.

Shortly before this, in the USA, a speculative bubble (some prefer to call it an economic bubble) increased real home prices by about 90% from 1997 to 2006. Late in the same year, housing prices started to drop moderately, making refinancing more difficult. Default and foreclosures increased dramatically, leading to the collapse of housing prices and of the subprime mortgage market during 2007–2008. The crash severely affected US and European banks, which lost about US $ 2.8 trillion on toxic assets and bad loans from 2007 to 2010. Several major financial institutions, such as Lehman Brothers, Fannie Mae, Freddie Mac, and AIG went bankrupt, underwent acquisition under duress, or were taken over by the government. To prevent the global financial system from collapsing, the US Federal Reserve, the European Central Bank, and other central banks injected about US $ 2.5 trillion into the credit markets.

It is important to note that during the downfall of the housing market, the price of oil started to increase vigorously. The price of the oil barrel jumped from about US $ 60 in early 2007 to US $ 148 on 11 July 2008, contributing to the slowdown of the economy in the oil importing developed countries. Other commodities also saw price increases. Coal went up from US $ 50 per metric ton to US $ 170, and copper reached US $ 7 000 per ton. These price increases were partly due to the flow of speculative capital from the housing to the commodity markets. It is likely that some of this flow was also partially responsible for the food price crisis at the end of 2007. After the financial crash of 2008, the price of oil jumped back to nearly 40 US dollars in January 2009.

The financial and economic crisis of 2008–2009 was the worst since the Great Depression of 1929. Its underlying cause was the association of very low interest rates in the USA and Europe, unprecedented levels of liquidity, and unregulated financial innovation and complexity in the markets. The Chairman of the US Federal Reserve, Alan Greenspan, introduced the low interest rate policy following the dot-com bust and the 9/11 attacks in 2001. The gigantic increase in liquidity had to do with structural imbalances in world markets. Enormous financial surpluses had accumulated outside the developed countries, especially in China and the oil producing countries of the Middle East. A particularly good example is the symbiotic relationship developed between China and the USA through the accumulation of huge currency reserves in China, obtained by exporting manufacturing capacity to satisfy the insatiable consumerism of the USA, which Niall Ferguson calls Chimerica (Ferguson, 2008). With these reserves, China buys US government securities and functions as the USA's banker.

The third important aspect is that the regulatory framework became insufficient to cope with the development of innovation, particularly in non-bank financial institutions and financial products, such as derivatives. Because of the crisis, the real GDP world growth declined to 2.99% in 2008 and to −1.05% in 2009 (IMF, 2010). The decrease in international trade in 2008–2009 was one of the most severe on record. Probably the most significant aspect of the crisis is that it affected the industrialized countries much more than the emerging economies. China's financial system plays a relatively small role in its economy and had no exposure to the toxic assets that contributed to the downfall in the USA and Europe.

In this context, it is important to recognize the distinction between markets that correspond to real economic activities, such as the commodities markets, and the derivatives markets that correspond to purely financial operations. Only 5% of current daily transactions correspond to commercial operations in goods and services. The remaining 95% are speculative capital movements (Jacque, 2010). The New York Stock Exchange, which is the largest in the world, and the London Stock Exchange are the most important specialists in the derivatives markets. After the crisis, the US and European governments decided to tighten the regulation of the financial markets through tougher risk management and surveillance of financial operations, but the major banks have been fiercely resisting such moves. It is a losing game, because market investors are much bolder, more imaginative, and more determined than the regulatory officers of state financial institutions. This imbalance lies at the heart of the current unsustainability of the global financial system.

In 2010, the world economic recovery was advancing due to the robust growth in the emerging economies. They provided much of the global growth mainly through domestic demand-led recoveries in China, India, and Brazil. According to the World Bank, developing economies are expected to have an annual growth between 5.7 and 6.2% in 2010 and 2011, while high-income countries are projected to grow between 2.1 and 2.3% in 2010 and between 1.9 and 2.4% in 2011 (WB, 2010). While growth is moving to the developing countries, debt is moving away from them into developed countries. The direction of capital flow is currently from East to West instead of West to East as it was for centuries.

Another important aspect of the financial and economic crisis of 2008–2009 was the emergence of state-directed capitalism. To lessen the pain of the current global recession, and to revive ailing economies, developed countries have resorted to state interventionism in ways similar to those practised by emerging economies. The neo-liberal interpretation of the free market has now receded, at least temporarily, and in its place has emerged a last-ditch tendency toward new forms of state capitalism. Economic decisions are now much more dependent on government policy. This is probably a long-lasting trend which results from the growing financial and economic instability at the global level, and which may remain after the crisis.

An immediate consequence of the global recession was the financial crisis in Iceland in 2008, followed by Dubai in 2009, and by several European countries in 2010 and 2011. Among them Greece, Ireland, Portugal, Spain, the UK, and other EU countries have developed rising government deficits and debt levels in recent years, triggered a sovereign debt crisis in some of them. The risk of some eurozone countries defaulting on some of their sovereign debt has been avoided up to May 2011 through massive loans obtained after government approval of strict austerity measures, including fiscal adjustments and structural reforms. Greece, Ireland, and Portugal have received large loans from the EU countries and the IMF, but conditional on harsh austerity programs. Other EU countries may follow, but the bailouts will probably be insufficient to solve the debt problem, because they are taking too hard a toll on economic activity, thereby suppressing economic growth. A possible way out is for the breakup of the eurozone area. An important point is that not all eurozone countries were fully prepared to participate in the single currency experiment. The other is that competition between the US dollar and the euro in a period of high public debt in their respective countries is drawing international capital away from the euro.

The bailout strategy that already rescued three peripheral eurozone countries is proving insufficient, up to June 2011, to solve the sovereign debt crisis of those countries, because their political systems are on the verge of being unable to implement the reforms required in return. A parallel dynamic is unfolding in the USA, where there is no political agreement on how to deal with the debt ceiling and tackle the deficit, because that would cut expectations for the nation's economic growth.

According to the IMF, gross public debt in the high-income countries rose from about 75% of GDP in 2007 to 110% in 2015. However, in emerging economies it is below 40% and projected to remain stable in the future. To sucessfully address the consequences of the developing imbalances in the world financial and economic systems would require new approaches to politics.

Another consequence of the financial and economic crisis has been the growth of unemployment, especially in the more industrialized countries. In the EU, unemployment rose from 7.8% of the workforce in 2008 to 10% in May 2010, corresponding to about 23 million people and a job loss of 3.9 million. Among the young, the jobless rate in 2010 is on average close to 20%, reaching 40% in Spain. In the USA, the unemployment rate at the end of 2009 rose above 10% for the first time since the 1980s. About 45% of the jobless are long-term unemployed, pointing to structural unemployment problems.

In fact, the current globalization process involves structural imbalances that played a crucial role in the financial and economic crisis of 2008–2009. The integration of the world's economy involves primarily commodities, capital, and labour, which are the three main factors of production. Multinational corporations have taken full advantage of the fact that the cost of doing the same job in different countries varies significantly. Outsourcing in emerging markets can substantially lower the labour costs. Until recently, this outsourcing was mostly limited to unskilled work. However high-quality and high-technology labour is now available in emerging economies as a result of a combination of improvements in education and professional training, infrastructure development, advances in technology and production techniques, and the spread of digital standards worldwide. This specialized labour is increasingly available in Brazil, China, and India, among other developing countries, and its productivity often exceeds that of the workers in the high-income countries. In the future, it will tend to dwarf the equivalent workforce of the more industrialized countries.

Much of the production from this new labour force is for export and not for domestic consumption. This trend increases structural unemployment in the more industrialized countries and generates deep imbalances in the globalization process. Labour markets in developed countries could regain competitiveness through currency revaluations in emerging economy countries, but that would slow down the fundamental process of economic convergence between developed and developing countries. Furthermore, the world GDP growth and especially the fast growing GDP of the emerging economies require an increasing supply of commodities. Without a readjustment in currency values, the price of commodities becomes too high in emerging economy countries and too low in developed countries, leading to financial tensions in commodity markets. Consumers in developed countries tend to increase their consumerism while those in developing countries consume less than they otherwise would.

These imbalances increase commodity prices since they are set in unconstrained global markets, while the currency values are subjected to foreign exchange control and interventions, particularly in some countries such as India and China. Developed countries should consume less, increase their savings, and adhere to stronger fiscal discipline. However, voters in developed country democracies prefer profligate policies, where governments spend more on social programs without increasing taxes, forcing them to incur even bigger deficits. The pressures from an ageing population aggravate the problem.

On the other side, developing countries, particularly the emerging economies, generally have low-cost labour, high exports, low debts, large currency reserves, and large current account surpluses, and provide more capital to developed countries than they receive from them. The two blocks display profound differences as regards demography, social, economic, and cultural perspectives of development, and quality of life. There are many constraints that prevent a simple transfer of the Western development model and way of life to the developing countries. The developed countries try to preserve the model while the developing countries aspire to reach an equivalent one.

These tendencies are likely to maintain or exacerbate the structural imbalances in the globalization process, increasing the risk of massive changes in commodity prices, rapid currency shifts, and large increases in interest rates or even defaults for heavily indebted countries. In the medium to long term, further globalization is likely to prevail through new development models shared by the powerful market states, which include the USA, EU, Australia, Canada, Japan, Russia, Brazil, China, and India, among others. Probably this convergence and integration will involve repeated financial, economic, and environmental crises. In spite of this process, significant inequalities, extreme poverty, and hunger are very likely to continue to exist throughout the world, particularly in the least developed countries.

The economic and financial crisis of 2008–2009 has damaged the Western economies, while leaving emerging economies almost unscathed. Psychologically this has very strong implications in the developing world and diminishes the prestige of the West. The rising importance of the group of 20 leading economies reflects the changing balance of power. Currently the leading emerging economies are China, India, Russia, and Brazil, responsible for about two fifths of the total GDP of the developing countries. Estimates of the year when China's economy will become bigger than the USA's have been shifting continuously from the 2040s to the 2020s. After the overtake, the largest economies in the world will likely be, in decreasing order: China, USA, India, Japan, Brazil, and Russia. The group of developing countries already consumes more than half of the world's primary energy and holds the majority of global reserves of foreign currency.

In spite of this new trend, some developing countries had negative GDP growth rates per capita in recent decades. The average value of the growth rate in the period from 1980 to 2002 was −5.8% in Sierra Leone, −5% in the Democratic Republic of the Congo, −3.9% in Liberia, −1.8% in Zambia, and −1.6% in Nigeria, just to mention a few examples (Easterly, 2006). Many of the poor countries continue to struggle against extreme poverty, malnutrition, and serious lack of health assistance, good quality drinking water, and basic sanitation.

After the Asian crisis of 1997 most of the emerging economies, until then highly dependent on the IMF, went from being debtors to creditors, and now have large foreign exchange reserves. Consequently, the financial role played by the IMF in the emerging economies and in other developing countries has become more limited. It is now playing an increasing role in some debt ridden developed countries, such as Greece, Ireland, and Portugal. The World Bank, often accused of defending the interests of the USA, is going through a transformation period and adopting a more pragmatic and less arrogant style of intervention.

The 2009 G20 Summit in London indicated that the financial and economic crisis of 2008–2009 revealed the need to thoroughly revise the role of the Bretton Woods Institutions, particularly the IMF. On 19 September 2006, a reformation of the IMF took effect, in which the percentage of the total number of votes of China, Mexico, South Korea, and Turkey were raised from 2.94%, 1.2%, 0.76%, and 0.45% to 3.66%, 1.43%, 1.33%, and 0.55%, respectively. However, these changes are insufficient to accurately reflect the recent changes in the balance of financial and economic power. For example, the USA's GDP was 2.9 times higher than China's in 2009, but the ratio of the percentage of voting rights of the two countries in the IMF is 4.5. Currently 43 African countries represent 4.4% of the voting rights, while the G7 countries control 45%.

China has been the second largest economy in the world since 2010 and has been a member of the World Trade Organization since 2001. The positions adopted by China, as its economy grows and overtakes those of the other countries, will be very important to determine the future of the Bretton Woods Institutions. Instead of reforming them, China may prefer to seek the construction of an alternative model. It may prefer to develop regional structures such as an Asiatic Monetary Fund and a new Asiatic monetary unit, instead of becoming involved in the present monetary and financial system, which it considers dominated by the USA.

Even in 2006 Fan Gang, a member of the Committee of Monetary Policy of the Central Bank of China and Director of the National Institute of Economic Research said that: "The dollar is no longer a stable support for a global financial system and it is very unlikely that it will ever be again. It is therefore time to seek alternatives" (International Herald Tribune, 2006). More recently, in March 2009, at the height of the financial and economic crisis Zhou Xiaochuan, the governor of China's Central Bank, said that "the desirable goal of reforming the international monetary system should be to create an international reserve currency that is disconnected from individual nations" (Xiaochuan, 2009).

3.75 The New Ascent of Asia Led by China

During the 21st century, the centre of gravity of the world economy will very probably move to Asia, led by China. This shift, driven mostly by demography, is a return to the pre-colonial era, when East Asia was the main economic region in the world. There are indications that, even as early as the 9th century, trade between

Chinese, Indians, Arabs, Siamese, Javanese, and Japanese was greater than within Europe. At the beginning of the 18th century, the Jesuits recognized that China was a flourishing empire with a vastly greater internal market than Europe, and in 1776 Adam Smith wrote: "China is a much richer country than any part of Europe." It is estimated that, in 1750, Asia represented about 80% of the world GDP (Bairoch, 1997). Other estimates indicate much lower values of the order 30%, but still greater than Western Europe (Maddison, 2010). The main economies were India, which dominated the world market for cotton textiles, and China due to its high productivity in the industrial, agricultural, transport, and trade sectors.

Asian economic prevalence started to weaken with the strengthening of the European colonial expansion and the development of the industrial revolution. One of the most important driving forces in this process of decline was the destruction of local industries caused by competition from European goods, manufactured with more advanced technology, and by imposed trade tax systems biased in favour of the colonizing countries. Another important factor in the case of China was the highly profitable smuggling of opium from India by British merchants, in defiance of Chinese prohibition laws. In the spring of 1839, Chinese authorities in Canton confiscated and burned the opium, and in response the British occupied positions around the city, starting the first opium war of 1839–1842. Under a humiliating treaty for the Chinese, Hong Kong became a British colony.

After the war, British antagonism of Chinese officials in the trading ports continued to increase, leading to the second opium war (1856–1860), which also involved France. The war ended with the signing of the Convention of Beijing in 1860, which opened more ports to foreign trade, imposed on China the payment of heavy indemnities to Great Britain and France, and legalized the opium trade.

The end of the cycle is approaching, with a return to a predominantly Asiatic world economy. Western supremacy had its origin in the Age of Discovery through maritime exploration, the emergence of modern science in Europe, and later, the industrial revolution. The increasing worldwide access to science and technology and to the instruments and benefits of the industrial revolution is creating a new economic order at the global level.

Curiously, Oswald Spengler was probably correct when he predicted the future predominance of the East, although he misjudged when he claimed that the cause for the shift would be the collapse of Western civilization associated with the decline of modern science. On the contrary, science and technology support the processes of development and globalization that are generating new modes of equilibrium on the global scale. Future historians will probably look at the economic supremacy of the West from the 18th century up to the middle of the 21st century as an episodic phenomenon that made significant contributions to improving the quality of human life, creating a civilization with remarkable scientific and technological achievements but profound inequalities, opening the way to globalization, and generating very complex problems of sustainability.

China displayed a remarkable economic performance in the 32 years that followed its market reforms and especially between 2002 and 2008. From June 2002 to June 2008, China's GDP has more than tripled and its exports more than quadrupled according to the Chinese National Bureau of Statistics. The development of the state-controlled Chinese banks in recent years has been remarkable. Four Chinese banks are now among the ten biggest banks in the world by market value (The Economist, 2010), and they were much more resilient to the recent financial and economic crisis than Western banks. The state owns a majority stake in the banks and the Communist Party appoints the directors, with much smaller salaries than their counterparts in the West.

The main goal of the Chinese leaders has been to assure an annual GDP growth rate around 8%. They are deeply concerned that, below this level, there would be so much unemployment and so many discontented people that major convulsions would lead to an ungovernable country. The economic objective of growth and prosperity at any cost has replaced the ideological objective of communism, but the problems facing this development paradigm are immense. The population is likely to increase from the current 1.3 billion to 1.5 billion in 2020 and the land area that can accommodate, feed, and employ the growing population is only half the area available 50 years ago. The impact of the growing economic activity on the environment is becoming unsustainable. Furthermore, China has to secure long-term supplies of food, oil, natural gas, iron, copper and other natural resources needed to fuel its continued growth.

The Chinese Communist Party has adopted the essentials of the Confucian political philosophy, which defends the viability of a harmonious society where the governed and the governors respect and protect each other's role. The rulers have the obligation to support and promote development, while the people have the duty to obey. Chinese history is full of examples of peasant rebellions caused by the failures of the rulers to ensure them an acceptable livelihood, some prosperity, education, and defence against foreign invasions. The Chinese leaders are fully aware that if they do not fulfil the rising expectations of the people, they face the risk of dramatic upheavals and crises.

The Western political philosophy has different values. It assumes a permanent engagement with the confrontational nature of society fostering what it considers to be the essential freedoms of speech, assembly, political expression, and democracy. Interestingly, there is no significant difference between the West and China as regard developing science and technology. The USA still dominates the world's science output, but its share of publications in the period 1993–2008 dropped from 26 to 21%, while China's increased from 4.4 to 10.2% (RS, 2011). China is likely to overtake the USA as the world's dominant publisher of scientific research papers between 2013 and 2020, close to the time when it is also likely to become the world's largest economy.

China's authoritarian brand of state capitalism has provided rapid growth and stability and is becoming a role model for some developing countries. Chinese officials are now speaking about and defending a China model based on a one-party government, an eclectic and pragmatic approach to free markets, and state intervention in the economy with strong state enterprises. Inspired by China's success, Joshua Cooper Ramo (Ramo, 2004) argued for a Beijing Consensus to substitute the collapsed Washington Consensus, characterized by new approaches to promote development and establish a new balance of power. Some view the emergence of the China model as posing a strategic challenge to the West and particularly to the USA. The future relations between China and the USA constitute the most important long-term foreign policy issue in the USA, and probably also one of the most important in China. "Just as globalization is shrinking the world, China is shrinking the West," argues Stefan Halper (Halper, 2010). This is very probably inevitable.

In spite of all the recent successes, the Chinese rulers are continually facing tremendous challenges. One of these results from the fact that most of the economic growth over the past 30 years has been driven by a continuous supply of cheap migrant labour flowing from the rural areas into the cities. Official government figures indicate that about 153 million people left their rural homes to work in urban areas in the past 12 years (Financial Times, 2011). These waves of poor workers are not entitled to the same social infrastructure and security provided to the urban residents. Their integration into the cities is a daunting problem, far from being resolved. There is often civil unrest, as in the recent June 2011 clashes between migrants and police in the southern industrial region of Guangdong.

Although China is fast increasing its military budget, there is no sign that it wants to enter into an arms race. Nevertheless, members of the US national security community are concerned with China's growing power and defend the idea that the country has to maintain its military supremacy everywhere in the world, including the Western Pacific and the South China Sea. This objective is likely to be increasingly difficult to achieve. On the Chinese side, the intention is to project the asymmetry that results from its fast-growing economy to limit US power in its region and counteract the attempts of some fringes of US politics to isolate the country.

The most important foreign policy goal of the Chinese leaders is very probably to foster good international relations in order to secure robust economic growth, believed to be the main guarantee of internal social peace. In the USA, the main geopolitical strategy is probably to reassert America's role as an Asian power, peacekeeper, and partner to China's neighbours. The main thrust of this policy is to develop a deeper relationship with India and to explore and take advantage of its tensions with China. An important recent move in this direction has been the nuclear deal between Washington and New Delhi.

3.76 How Sustainable Is Growth in China, Other Emerging Economies, and the World?

Another issue, more difficult to solve in the long term, is the sustainability of the increasing consumption of natural resources required by the current growth in the emerging economies. Since 2000, they were responsible for about 85% of the increase in the world energy demand, with China responsible for a third of that value. China generates more than 80% of the energy it uses, but in the future will have to import enormous amounts of energy. It was a net oil exporter in the 1980s, became an oil importer in 1993, and in 2009 imported more than 50% of its crude oil consumption. On 19 July 2010, an official of the International Energy Agency stated that in 2009 China consumed 2 252 billion tonnes of oil equivalent, becoming the world's largest energy consumer, 4% above the USA (Le Monde, 2010a). Chinese government officials refuted the estimate, saying that it was too high, but without revealing any other figures, which shows how uncertain and sensitive these assessments are.

Recent estimates indicate that non-OECD countries will account for 93% of the increase in global energy demand between 2007 and 2030, driven largely by China and India (IEA, 2009). One of the first priorities of those two countries is to secure access to the oil-producing countries, especially in the Middle East, Latin America, and Africa. If the use of oil in China increases by a factor of 10 in the next 30 years, the consumption per capita will still be 30% less than the current value in the USA. Transport is one of the sectors that contribute most to oil consumption growth. The world total number of motor vehicles is expected to grow by 80%, reaching a value of 1 250 million in 2025 (IGAS, 2006). Estimates for China in 2008 indicate that there are 13 motor vehicles per 100 people, much lower than the 77 per 100 people in the USA, but the number is fast increasing. The same applies to India, where there are currently 1.2 motor vehicles per 100 people.

The demand for metals in the emerging economies has grown rapidly since the 1990s, contributing to sudden increases in prices. In China, the demand for copper grew on average by 14% per year in the 15 years up to 2005 (The Economist, 2006). In 2009, China was responsible for 39% of global copper consumption (Newsweek, 2011). The demand for iron also increased vigorously during the same period. In 2005, China was the world's leading importer of cement (55% of world production), iron (25%), nickel (25%), and aluminium (14%), and the world's second largest importer of crude oil after the USA (Le Monde, 2006a). In the three years up to 2010, China became a net importer of coal, to fire the increasing number of its thermal power stations. It is likely to overtake Japan in 2010 as the world's leading importer of coal, five years earlier than expected (FT, 2010).

Prices of most minerals and metals increased by factors of two to three in the first few years of the 21st century preceding the global 2008–2009 recession, due to increased world demand, particularly from the emerging economies. In many cases, mineral suppliers were unable to satisfy demand by expanding output through discovery, development, or expansion of mines. Arguably, this was the first instance in history of a wide-ranging scarcity of non-renewable natural resources. It is predictable that, as the recession recedes and world economic activity becomes more robust, global non-renewable natural resource scarcity will return and probably intensify. Eventually this increasing scarcity is very likely to change the way of life of the 1.2 billion people in the industrialized countries and to frustrate the expectations for a better quality of life of many billions of people in the developing countries.

It is important to distinguish between non-renewable natural resources that experienced temporary global scarcity during the pre-recession years of this century from those that experienced permanent global scarcity. Only in a few cases were supplies physically unable to keep pace with the ever-increasing demand in the 2000–2008 period. Estimates of the world's reserves of the various non-renewable natural resources are very difficult to make and involve large uncertainties. However, we will eventually reach their Hubert peak, since they are exhaustible. Consider the cases of copper and iron: recent estimates indicate that, at the current rates of consumption, the known copper reserves will last another 107 years, while iron reserves will last 151 years (The Economist, 2006). What we can predict with high probability is that, in 100 to 150 years from now, copper and iron will be very scarce metals, more difficult to exploit and hence much more expensive.

Some chemical elements that were once only laboratory curiosities are now essential for the development of new energy technologies for a low-carbon economy, including the production of wind turbines, solar energy collectors, and electric cars. They have been called energy critical elements and include most of the rare earth elements, gallium, germanium, helium, indium, platinum, rhenium, selenium, tellurium, cadmium, and lithium. Lithium, for instance, is used for a new generation of hybrid car batteries and tellurium for solar power cells. Some are extremely rare, such as tellurium, rhenium, and many rare earth elements. Others, such as cadmium and indium are already scarce. Among the energy critical elements, some are also used for the development of telecommunications and computer networks.

Industrialized countries and emerging economies are particularly concerned with the increasing demand for these strategic elements, well above traditional levels of supply, and about their likely future scarcity. Although these are well-founded preoccupations, it should be emphasized that the scarcity of metals, minerals, and more generally non-renewable natural resources will stimulate recycling, more efficient use, and the search for alternative products and replacement technologies.

3.77 Growing Environmental Degradation in China and India and the New Development Paradigm

The fast growth of the emerging economies is also having a major impact on the environment. The situation is particularly critical in China due to the increase in water and atmospheric pollution. The main source of energy in China is coal, which accounts for about 70% of total energy use. Annual coal consumption is currently higher than in the USA, Europe, and Japan put together. Coal combustion in thermal power plants with outdated technology produces large amounts of soot and SO_2, and these cause acid rains in 30% of the territory. Due to the high dependence on coal, China is currently the single largest emitter of carbon dioxide into the atmosphere.

Assuming the continuation of the current rate of energy consumption based on the intensive use of coal without carbon capture and sequestration technologies, China's emissions in 2030 will become larger than all current industrialized countries, including the USA. To reduce greenhouse gas emissions in China while maintaining its fast economic growth is a challenge of gigantic proportions. China is well aware of the need to address it. The country is already the world's leading renewable energy producer with an installed capacity of 152 GW. One of the objectives of the five-year plan that ends in 2010 was to reduce the energy intensity of GDP by

20%. Furthermore, in December 2009, before the Copenhagen Conference, China announced that it would decrease its GDP carbon intensity by 40–45% relative to 2005 intensity levels. More recently, on 20 July 2010, an official of the Chinese National Energy Administration informed that the country is planning to invest US $ 783 billion in renewable energies up to 2020 (Le Monde, 2010a). This is a very significant step since, according to the United Nations, the investment made by China in renewable energies in 2009 was 34.6 billion US dollars.

In 2006, the World Bank (WB, 2006) estimated that 20 out of the 30 cities worldwide with the worst air pollution are in China, due to the intensive use of coal and motorization. Linfen in China's inland Shanxi province, which provides about two thirds of the country's energy, is one the world's cities with worst air quality, caused mainly by the large number of surrounding coal-fired power plants. About 60% of the 340 Chinese cities that have air quality monitor programs had serious problems with atmospheric pollution in 2003, according to the State Environmental Protection Administration of China (SEPA, 2003).

Studies carried out in the West China Zaozhuang region with 281 million inhabitants indicate that atmospheric pollution was responsible for a death rate increase of 6% in 2000, and this is likely to increase to 13% in 2020. The human health costs of atmospheric pollution were estimated at 10% and 16% of the GDP in 2000 and 2020, respectively. If the human health costs associated with the current use of coal in thermal power stations were internalized, the price of coal would have been three times higher (Wang, 2006). Estimates of the Asia Development Bank presented in a December 2006 report indicate that air pollution in the large Asian cities is responsible for the premature death of about 500 000 people per year.

Pollution affects most of the largest rivers in China and about one quarter of the population does not have access to safe drinking water. To reduce the chronic water scarcity in the north and northeast, which is a serious risk for public health and an obstacle to economic development, the government is carrying out huge engineering works to transfer the waters of the Yangtze to the cities in the north, including Beijing and Tianjin.

In a surprisingly candid interview with Der Spiegel in 2005, Pan Yue, China's Deputy Minister of the Environment said:

> Because air and water are polluted, we are losing between 8 and 15 percent of our gross domestic product. And that doesn't include the costs for health. Then there is the human suffering: In Beijing alone, 70 to 80 percent of all deadly cancer cases are related to the environment. Lung cancer has emerged as the number one cause of death.

Answering another question, he said:

> China will have more than 150 million ecological migrants, or, if you like, environmental refugees.

For him, GDP growth cannot be the only yardstick by which to gauge the government's performance. Furthermore, he stated that a prospering economy does not automatically go hand in hand with political stability.

Recently, there have been clear signs of climate change in China that aggravate environmental problems and the sustainability of the use of some natural resources.

In the north, the Huang He river, which crosses the region where the Chinese civilization was born, and is one of the largest in the world, has stopped running into the sea owing to long-lasting droughts and the intensive upstream exploitation of its waters for irrigated agriculture. A warmer climate in the Sanjiangyuan region on the Tibetan plateau is causing the glaciers that feed the Yangtze, the Huang He, and the Mekong to melt, changing the river flow regime. On 31 August 2005, the Chinese government created the Sanjiangyuan National Nature Reserve with an area of 152 300 km^2, the second largest in the world. Its aim is to protect the Tibetan plateau ecosystem and its unique wildlife, to forest part of the area in order to regulate river flows, slow erosion and soil degradation, and promote sustainable development for the 200 000 people living in the reserve.

Rainfall is decreasing in the north and increasing in the south of China. These two tendencies are likely to be the result of climate change and will be aggravated in the future, according to climate change scenarios. More intense extreme weather and climate events are also expected in the future, implying a greater risk of droughts and floods.

Desertification is intensifying in the northern region of China. In the spring, the strong west winds from Inner Mongolia carry enormous amounts of sand over many hundreds of kilometres right up to Beijing. The desert is advancing eastwards at a fast pace, and there are now dunes only 70 km from Beijing. Deserts already cover about 23% of China, and large parts of these deserts result from deforestation and unsustainable livestock and agricultural practices over the last two to three decades. The Chinese government has recently launched a massive campaign for afforestation, reforestation, and grass planting, reducing over-grazing and over-cultivation, and promoting a sustainable use of water resources.

With 1 100 million inhabitants, India is the second most populous country. It is culturally closer to the West and has a very rich and diversified religious history. Furthermore, it has been a democracy since its constitution came into force on 26 January 1950. From that time up to 1980, India adopted a socialist-style planned economy that lead to a modest GDP annual growth rate of 3.5%. However, at present it is growing at around 9% a year.

Just like China, India has an increasing economic, political, and environmental influence in the world. It is probably 10 to 15 years behind China as regards average social and economic development. However, in the long term, its economy could become dominant, because the population is growing more rapidly and could exceed China's by the year 2025. China is a bigger, militarily stronger, and richer country as regards natural resources than India. It has a greater macroeconomic stability and higher rates of development in the areas of education and public infrastructures, but these differences may not last long. Many economists predict that India's rate of economic growth will overtake China's in the coming decade. In any case, India's economy will be a very important component of the world economy and have a strong impact on the global environment.

The relatively rapid increase of India's population is causing environmental degradation and adversely affecting the natural resources of the country. The most serious problems are air and water pollution, water scarcity, land degradation, deforestation, habitat destruction, and loss of biodiversity. In particular, the conversion of natural forests for palm oil plantations to satisfy the high internal demand is a threat to biodiversity and to sustainable ways of life. Water and air pollution is worsening, in part due to increased urbanization and migration from rural to urban areas. Recent estimates indicate that, over the next ten years, 300 million people will leave the agricultural areas in search of better living conditions in the urban areas. In the medium and long term, socio-economic development in India depends on how it addresses its enormous environmental challenges, just as in China. As the country's population and economy continue to grow, the need to find solutions will become increasingly urgent.

In 2006, India produced 1.36 metric tonnes of CO_2 emissions per capita, which is considerably lower than the corresponding value of 4.65 for China. Nevertheless, India is the only emerging economy to be rated as having a high vulnerability risk to climate change. This is due to the high population density, poor resource security, and strongly negative impacts on water resources, agricultural productivity, and coastal zones. Water scarcity is likely to increase in the drought-prone regions, particularly in Rajasthan. Changes in the monsoon regime will affect the livelihoods of many hundreds of millions of people. The rise in average temperature will reduce snowfall and accelerate the melting of glaciers in the Himalayas, increasing the risk of floods during the wet season and decreasing the river flow during the dry season. Furthermore, sea level rise in the low-lying coastal areas of India and Bangladesh will probably create millions of climate refugees.

A brief analysis of the current state of the world shows that the emerging economies, and in particular China and India, are going to contribute decisively to the way development and the quest for sustainability evolve in the 21st and ensuing centuries. The development models adopted by China, India, and the other emerging economies will largely determine the future of our global society.

The world population is rising and the global demand for food, energy, and natural resources have been growing at an accelerated pace in recent years. The environment is now part of the equation that we need to solve. As Thomas Friedman has said (Friedman, 2008), we have entered the Energy–Climate Era. It will be impossible for the emerging economies and in particular for China and India to keep on copying the American paradigm of consumption, building, and transportation. On the way to that utopian goal, natural resource scarcity and environmental degradation would trigger major global crises. The emerging economies and especially China, due to its huge population, must discover an alternative development paradigm. This new way is as important to these countries as it is to the whole world because of the increasing integration of national economies and the increasingly global nature of the environmental problems.

It is often said that we need a green revolution, but revolutions are both impossible to plan and unpredictable. It is unlikely that the developed countries and in particular the USA will lead the way to a new global development paradigm for the Energy–Climate Era. The West by itself, imagining that it could be isolated, may be closer to some form of sustainable development, but it only represents about 20% of humanity.

Technology transfer in the field of sustainable development from the West to the emerging economies and to the developing world will be very important. However, China and India are more likely to find new paradigms and to start practising them. While the developed countries are focused on fierce attempts to defend their relatively wealthy lifestyles, the developing countries must find new models of development that may fulfil the expectations of well-being for their huge populations. Probably it will be through a process of trial and error, battling against recurrent crises. But the main drive will be theirs.

Chapter 4
The Future

4.1 Science and the Future

Although deeply involved in the affairs of the present and near future — hours, days, weeks, months, or even a few years — people also ask themselves about the middle and long term future — a decade, decades, a century, or several centuries. Clearly, the former time scales are of vital importance, while the middle and long term future is irrelevant to the majority of the human population, especially those who live on the edge of survival. However, in the 21st century, more than in the past, there are many human activities with serious cumulative impacts on the Earth system and on the depletion of natural resources, and these will manifest themselves in a marked way in the middle to long term. Some of these impacts may even be irreversible on time scales that are long compared to the average duration of a human generation, on the order of thousands of years. Moreover, they are likely to aggravate the current inequalities in social and economic development, increase poverty and hunger on a global level, and raise the risk of critical and dangerous situations in the future.

This perception generates awareness that the medium and long term future is relevant, and above all that it is associated with a significant degree of uncertainty and risk. The study of possible futures, or prospective, is a discipline that tries to unravel the future through specific methodologies and an interdisciplinary analysis of the present and its current trends. It was initially created by the French philosopher and industrialist Gaston Berger, who coined the name. It is increasingly used in a variety of fields such as decision-making in regional and national planning and corporate strategic analysis. It can also be used to construct medium to long term future global scenarios that integrate social, economic, political, and environmental components. These scenarios range from ecological catastrophes to utopian futures, where all mankind enjoys wealth and quality of life, to futures dominated by the transformation of humans into post-human life forms.

Science, particularly the physical and natural sciences, has proven itself to have remarkable predictive power. We are now accustomed to the idea that certain aspects of the future may be predicted, and also that some will probably be positive and others negative. In this context, the ideal rational attitude would be to seek to moderate or avoid human behaviour and activities that lead to potentially harmful impacts and serious risks in the future. There are, however, three types of argument that work against this precautionary approach.

The first, most primitive, and most selfish argument is to consider that the distant future, even no more than one to three human generations away, is something that has very little to do with us directly, and as such should not condition our present behaviour. The dominant argument is that we will no longer belong to this world when pessimistic and probably unreliable projections may eventually become reality. So why not take full advantage now of all the opportunities offered to us by the current paradigm of consumerism, especially in the more industrialized countries, if we have the means to do so? The prevailing economic system is based on, depends on, and consequently stimulates this voracious consumerism. Over recent decades the consumer-oriented lifestyles of the well-to-do in the industrialized countries has evolved into a behaviour model that resonates ever more effectively with human nature. And the model is tending to become universal. Billions of people in developing countries strive to improve their quality of life and dream about reaching the simplest forms of those lifestyles.

A second, more mature and subtle type of argument is to recognise that the current model of development, coupled with a growing world population, is likely to involve serious risks and dangerous impacts on the environment and on natural resources, but to suggest that all these future problems can be solved through more or less complex technological solutions supported by science. This confidence stems from a belief in technological fixes. The fastest possible economic growth of all countries is considered to be an imperative whose sustainability will always be guaranteed by science and technology. Defended mostly by investors, economists, and engineers, it is a position based ultimately on the conviction that, since human ingenuity has no limits, it will always be possible to find sufficient natural resources or adequate substitutes to assure continuous economic growth and mitigate or solve all ensuing environmental problems through appropriate scientific and technological applications. This discourse is frequently referred to as Promethean, a symbolic analogy with the episode in Greek mythology in which Prometheus secretly stole fire from Zeus to give it to man, thus allowing him unlimited ability to progress.

Finally, we have a third type of argument, relevant regardless of the point of view adopted, which stems from the uncertainty about the future negative impacts of the current model of development. Even accepting the precautionary principle, it may be argued that it is not justifiable to adopt corrective measures addressing the causes of future risks, especially when they are expensive and may lead to a slowdown of economic growth, if the uncertainties associated with their occurrence are considered to be too great. Faced with a given risk, how should one define the lowest degree of uncertainty that justifies immediate action? The debate on the minimum level of uncertainty required to reach a conclusion that justifies action is a perennial question in the application of science. In some cases, the uncertainty can be quantified by means of probabilities, but in others it is not possible to evaluate it numerically.

The dominant paradigm of science has been the search for knowledge about nature, and especially for knowledge of its fundamental laws, by observation and experimentation, mainly in specialized and increasingly sophisticated laboratories. We know with great accuracy the four fundamental forces in nature, and this allows us to predict and interpret an immense variety of phenomena on Earth and throughout the Universe.

However, the paradigm is changing, because science today is called upon to analyse the behaviour of extremely complex systems, such as the terrestrial systems and the socio-natural systems that involve the interaction between natural systems and human societies. The study of global environmental issues is a good example of the new challenges facing science. It requires multidisciplinary and interdisciplinary analyses, covering very diverse spatial and temporal scales, from the local to the global, and from very short to very long periods of time. The behaviour of terrestrial and socio-natural systems is studied mostly through computer simulations using mathematical models, since it is obviously impossible to experiment with them. These models can be tested in various ways, for example, by requiring that they reproduce the observed past behaviour of the systems they simulate.

But this methodology nevertheless generates uncertainties concerning projections of their future behaviour. It is therefore very important for science to find out how to evaluate such uncertainties in quantitative ways, and to establish methodologies for managing and communicating them. This is a crucial step to ensure the reliable use of models that simulate the future behaviour of the terrestrial and socio-natural systems by decision-makers and by society at large.

4.2 What Is the Future of the Human Odyssey? Can We Trust the Long Term Projections Provided by Science?

A very succinct description of the adventure of humans on Earth from the origins of life until the present time has been presented here. The relationship between humanity and its planet, and in particular with its global subsystems, has been changing markedly, and rapidly. After emerging about 200 000 years ago, *Homo sapiens* eventually reached a stage where he was able to pursue a cultural evolution that led him to dominate the biosphere, often in a devastating way, and to provoke profound changes in the Earth system.

It is clearly impossible to predict the future of this human odyssey. However, many aspects of the future are predictable, because they result from a direct application of the fundamental laws of nature, especially the laws of physics. There are even certain large scale phenomena, way ahead in time, which, curiously, we are able to forecast with very little uncertainty, such as the evolution of the Earth and the Solar System on time scales on the order of billions of years. It is also possible to make projections about the future of the Earth on shorter time scales, on the order of thousands of years to hundreds of millions of years.

The short and medium term projections are based on the socio-economic scenarios up to the next century, and therefore, on the cumulative effect of the decisions and lifestyles of present-day and near-future generations. Collectively, through the very complex, diversified, and fragmented decision-making processes and government structures worldwide, we have the almost unconscious power to determine the extent to which humanity will interfere with the Earth system, increasing or diminishing the risks of critical situations regarding the environment and natural resources, and the extent to which humanity will deal with the risks of various forms of conflict, including organized crime, terrorism, and war. The majority of the human population is unaware of this power and of the concerns that it may raise. Indeed, only a very small minority are conscious of the potential risks ahead.

The current most representative example of this new type of situation for humanity is climate change. If the amount of greenhouse gas emissions into the atmosphere continues to grow uncontrollably, the resulting climate change will very probably cause serious food security problems and economic and environmental crises by the end of this century, with these perduring into the ensuing centuries. We still have the opportunity to avoid dangerous anthropogenic interference with the climate system, but it will require an ambitious post-Kyoto climate regime agreed by all countries. It will be necessary to reduce global annual emissions by 50 to 85% up to 2050, relative to levels in the 1990s. But it is becoming increasingly unlikely that this goal can be achieved, because it requires strong mitigation measures involving large investments in energy efficiency, carbon free technologies, and renewable energies which are politically unpopular in a period of economic uncertainty following the global financial and economic crisis of 2008–2009.

Many remarkable mitigation projects and initiatives are of course being undertaken by governments, businesses, non-governmental organizations, and private individuals, but they are simply not enough to address the daunting challenge that we are facing. If we do not act forcibly and jointly now, it will be impossible to avoid an increase in the average global temperature greater than 2°C relative to pre-industrial 18th century temperatures.

It may be argued that there are large uncertainties in the future climate change scenarios and in the impacts on the various socio-economic sectors and biophysical systems that they imply. Such uncertainties exist and must be dealt with, but as a matter of fact they are not sufficient to invalidate the main conclusion that unmitigated climate change will adversely affect food security, human health, a range of human activities, and major ecosystem services, increasing biodiversity loss, creating climate refugees, and enhancing or generating conflict throughout the world.

Climate science is very robust today. We are free to choose between science-based climate policies that avert dangerous interference with the climate system, affecting mainly the future generations, or choose a myopic policy that avoids the costs of strong mitigation now but leaves the higher costs of impacts and adaptation to subsequent generations.

The important point here is that science is now able to construct models that reliably simulate the Earth subsystems. Some of the conclusions derived from the short to medium term scenarios based on those models clash with the current development paradigm because they point to an increased risk of future crises if we persist on a business-as-usual trajectory. But there is no guarantee that the advice proffered by science will actually be followed. Science is influential in society when it helps develop technologies that are seen as promoting social development and economic growth, and improving health and the quality of life, but seemingly not so influential when it alerts to the dangers of transgressing some planetary boundaries on Earth-system processes that are likely to be reached through some human activities with significant financial and economic value.

4.3 A Few Trends and Challenges That Will Shape Our Short to Medium Term Future

Despite the major uncertainty associated with predictions about the behavioural aspects of human societies, there are some issues where it is probably safe to glimpse into the future. Population will be one of the most important factors determining the future of humanity. The uncertainty in the statement that the global population will increase up to 2050 is extremely small, because the greater part of that growth is already assured by the current younger generation, whose fertility is safely predictable. From 2050 onwards, the uncertainty is much greater, but the world population is likely to reach a peak and start decreasing before the end of the century.

As regards the economy, we are in a transition between a world dominated by the Western economies and a world centred on the emerging economies, including China, India, Russia, Brazil, Indonesia, Mexico, South Africa, and many others. The Chinese economy will rely as long as possible on lasting Western consumerism and greed to support its fast growth, but also increasingly on trade with other large developing countries and its growing internal market. The country has the potential to transform the world with new models of modernity that are quite different from those in the West. China is more like a civilization-state than a nation-state, characterized throughout its history by the forceful role of the state in society (Jacques, 2009). This imprint of the state on society is likely to influence the whole world in the future. However, the political aspects also exhibit dangerous weaknesses.

The Arab Spring of 2010–2011 is likely to promote democracy in other parts of the world. The Chinese government is well aware of this possibility and vigorously resists it. A clear sign of a growing movement for democratic political representation in China is the new wave of independent candidates in the elections for lawmakers at the county and township legislatures that started in May 2011. India and Brazil have the very important advantage that they are already democracies. The latter has a particularly robust growth potential. Among the BRIC countries, Brazil has a relatively low population of 200 million and a territory nearly three times the size of India. It had a GDP of 2 trillion US dollars, equivalent to Russia's, and a per capita income of 10 000 US $ in 2010, three times India's and nearly twice China's. It is easily self-sufficient as regards food and energy, and it is also the country with largest biodiversity on the planet. The other developing countries, particularly in Asia and later in Africa, will share an increasing percentage of the world GDP.

Meanwhile, most countries in the West will continue to battle with their large deficits and debts, stagnant and ageing populations, and structural unemployment. The percentage of the world GDP shared by the West will shrink inexorably. We will enter a post-Western world, but there will be much controversy over the exact definition of the transition, and hence also over the exact time it is likely to occur. In any case, the essence of Western culture and its artistic legacy will be co-opted and will prevail for the foreseeable future.

The global GDP will probably continue to rise over the next few decades, but its medium to long term behaviour is highly uncertain. An increasing number of people in the West are debating alternatives to the current paradigm of continuous economic growth. The idea is to find a way to achieve a global economy in a steady state, with sustainable non-growth. This discussion is still in its early stages, but the main idea is to define prosperity without growth and to see to what extent this program can be made sufficiently attractive (Jackson, 2009). At the present time, it is anathema for politicians seeking election to propose anything but economic growth. It seems likely that we may only reach some form of global steady-state economy as a reaction to very damaging ruptures and crises.

As regards quality of life, it is important to realize that the affluent people in the world, particularly in the more industrialized countries of the West, have never had it so good. They have easy access to excellent education at all levels and to all types of professional training, all imaginable comforts in their homes, availability of all forms of communication and information, magnificent cars, boats, and planes, very easy mobility all over the world, luxurious shops, hotels, and restaurants everywhere, high-technology gadgets for every type of indoor and outdoor activity, and easy access to culture and to the most diverse art markets and events all the world over. The well-to-do of the world are likely to continue enjoying this quality of life for a long time.

On the downside, they have been facing increasing security concerns, which have been addressed and solved to a large extent. Their main question is how to spend the money. The variety and splendour of luxury in a wide variety of products such as fashion clothes, jewels, watches, homes, health treatments, trips, cars, boats, planes, resorts, hotels, and restaurants is likely to keep on increasing. However, these splendid benefits are restricted to a very tiny minority of the world population. There is nothing new about this. It is just a continuation of past trends. The difference is only that the gulf between the very rich and the very poor is getting wider and deeper because of the diversification that the applications of science and technology have introduced into the manifestations of luxury. The significant inequalities of development between and within countries are very likely to persist and grow, since this is a characteristic outcome of the dominant economic paradigm. Extreme poverty and hunger are likely to continue and may worsen during this century.

However, it is important to emphasize that the same economic paradigm has also produced many positive results. These are particularly clear if we consider global averages of some of the development indicators. Over the last 60 years, billions of people have on average benefited from better living conditions, better health care, greater mobility, and much easier access to information and communication. In 2010, about 30% of the world population was connected to the internet, and in 2015 the percentage of people with access to the internet and with mobile phones is likely to reach 50% (The Millennium Project, 2010). This increasing global connectivity will profoundly change the way people live, study, trade, defend their individual and collective interests, and use political power. The global average for life expectancy increased from 45 years in the 1950s to 67 years in 2010. However, there are major differences between countries: in the period from 2005 to 2010, Japan had an average life expectancy of 82.6, while Swaziland was struggling at only 39.6 years (UNPD, 2007).

Capitalism is likely to remain predominant for a long time, since it is the most productive economic system available. Nevertheless, the shortcomings of capitalism will very probably become more clearly visible and well defined over the next few decades. A global poll conducted by the BBC World Service in 2009 found that dissatisfaction with capitalism is widespread, with an average of only 11% saying that it works well and that greater regulation is not needed (BBC, 2009a). In only two countries, the USA and Pakistan, did more than 20% feel that capitalism is working well as it stands. An average of 23% in the 27 countries surveyed felt that capitalism is fatally flawed and a new economic system needed.

As far as politics is concerned, there will be increasing divergence between various brands of authoritarian state capitalism and the Western type of modern democracy. Democracies are increasingly facing the problem of being unable to adopt policies that have long term gains but imply some inevitable short-term negative outcomes. Intergenerational solidarity is an essential component for the efficient functioning of a democracy. However, in Western countries, the Great Acceleration following World War II has generated the perception that social and economic development, social security, and continuous improvement in state supported health care and in the quality of life are somehow guaranteed forever. A hedonistic culture of immediate gratification has developed that tends to weaken democracy, and its future evolution is highly uncertain.

According to Freedom House's 2010 report (FH, 2010), there was an erosion of freedom in 40 countries, while it improved in only 16. The number of electoral democracies decreased by three to 116 countries. Perhaps more significant is the fact that the percentage of people voting in elections in the 15 largest democratic countries has decreased systematically since 1990 and is now less than 40% (The Millennium Project, 2010). According to the same report (FH, 2010), 46% of the world lives in 89 'free' countries, 20% in 'partly free' countries, and 34% in 47 'not free' countries.

The global information age will generate a new political power narrative and remap power relationships. Smart power strategies based on the extensive use of information technologies and advanced robots and drones will become increasingly important in relation to military strength. The capabilities to wage a cyber war will be relentless pursued. A recent example of a sabotage via cyberspace was the use of the Stuxnet computer worm, designed to disrupt the Iranian nuclear program.

The number of weak, dysfunctional, or failed states is likely to increase in the least developed countries and in the poorest regions of the world, increasing the security risks, and particularly terrorism. This is likely to become one of the long-lasting consequences of the growing inequalities of development in a globalized world, with greatest impact on industrialized countries, and one of the most difficult problems to solve in the future.

The other crucial concern for humanity's future is natural resources. Energy and water stand out as the most important. We may even say that we are already in a global crisis situation regarding energy and water and that these problems must be solved simultaneously and in a coordinated way, because these two resources are intimately related and vital for the well-being of humanity. Industrialization, economic growth, and population growth have been the driving forces for the large increase in their consumption worldwide. Energy and water sustainability is the most pressing issue regarding natural resources. There is encouraging news about energy efficiency and the increasing use of renewable energies. The price of fossil fuels, particularly oil, is likely to experience increasing volatility and periods of large increases and high prices over the next 50 years, leading to difficult economic situations.

Per capita availability of land will decline in the coming decades due to population growth. We can also expect more diversified diets in developing countries, occurring simultaneously with increasing food shortages, hunger, and malnutrition. As regards non-renewable natural resources, the current growth in consumption in the world is clearly unsustainable, particularly in view of the demand from the emerging economies. Prices of some resources are likely to increase considerably in the next few decades, leading to difficult economic problems. The rare earth elements used for clean energy development give a clear early example. Their prices increased on average by 300% between January and December 2010. The main reason for this increase is that China, which mines and refines about 97% of the global supply and has an estimated 36% of world reserves (CSIS, 2010), has introduced export quotas, considering them to be strategic elements. A recent report by the US Department of Energy (USDOE, 2010) recognizes the scarcity of rare earth elements and recommends the development of cooperative efforts with Japan and Europe to mitigate worldwide shortages.

Although it is impossible to predict how problems of this nature will evolve and when they are likely to become critical, we are clearly entering a new phase of depletion of non-renewable natural resources. The seriousness of future shortages will depend on how soon the Hubert peaks of each of the non-renewable natural resources, especially oil and gas, are reached, in particular as regards the world's population peak, and to what extent we are able to find adequate and affordable substitutes.

Regarding the environment we run the risk of very serious environmental change if we do not stay within defined boundaries for a range of vital Earth system processes. According to Rockstrom (Rockstrom, 2009), humanity has already transgressed the maximum acceptable boundary limit for the disruption of the nitrogen cycle, for biodiversity loss, and for anthropogenic climate change. Other processes are likely to be under very strong anthropogenic pressure, such as the phosphorus cycle, ocean acidification, global fresh water use, and change in land use.

It may seem strange that the short to medium term projections for our common future are mostly negative, in the sense that they involve increasing risks. They are not positive in spite of the fact that there is very good news in all the relevant domains of human activity. For instance, there is a fast increasing investment in energy efficiency and renewable energies all over the world. However, the magnitude of the challenge requires a much greater political commitment and economic effort to reach sustainability. The situation is similar in most other sectors. It will be essential to keep on reinforcing research in all areas from energy to food security, in order to find new solutions to the current problems.

A large part of what needs to be done is already well known. The challenge is to start addressing the problems as soon as possible in a vigorous and coherent way. Some countries are pursuing the quest for sustainable development much more forcefully than others. According to a recent United Nations report (Barbier, 2009), the Republic of Korea and China are ahead of other major economies with green investments as an important component of their economic and employment recovery strategy following the 2008–2009 crisis. As of August 2009, 79% of the total stimulus spent by the Republic of Korea was green stimulus, followed by China with 34% and France with 18% (UNEP, 2009). China is already the leading world producer of solar cells, solar water heaters, and wind turbines. Environmental security-related problems and concerns will become increasingly important and defining factors in the international political and military spheres. Preventing and responding to environmentally caused conflicts is likely to become an outstanding priority in the future.

The nature of conflict will continue to evolve. There is a clear reduction in the number of inter-state conflicts, but an increase in conflicts involving non-state actors. Violence is more fragmented worldwide and is very likely to grow because of the probable increase in social, economic, and environmental crises, and the increasing number of weak and dysfunctional states. Growing shortages will intensify the quest to secure access to natural resources. A few countries are already renting land overseas for agriculture and the demand for food-producing land will become severe. This kind of competition will be more visible in the future and is very likely to generate new forms of conflict. The territorial conflicts in the Middle East are unlikely to be solved in the foreseeable future. The event of a nuclear war in the 21st century or after 2100 is improbable, but still far from being impossible.

4.4 What Is Our Future?

The answer to this question depends largely upon ourselves, and the social, political, financial, and economic structures and organizations that we have constructed and to which we belong. Our collective cumulative actions are increasingly affecting Earth system processes and the availability of natural resources. Both are essential for the sustainability of life on Earth, and in the final analysis, controlled by the laws of nature, which cannot be changed by man. Science and technology do not have magical powers to solve all problems caused by dangerous anthropogenic interference with the Earth system, nor to regenerate or substitute all overexploited natural resources.

On the other hand science is irreplaceable in foreseeing the impacts of human activities, to evaluate the consequences of various options through a detailed and firmly grounded analysis, to assess the uncertainties involved, and finally to advise society and its political decision-makers. However, the decisions that shape our future at a local, national, regional, and global level are beyond the reach of science. They are of our own exclusive responsibility, both individually and collectively.

It is time to start a journey into the future, in order to do a brief tour of what science can tell us. The uncertainty will be smaller when dealing with issues and events that bear a close relationship with the fundamental laws of nature. This journey is going to take us away progressively from the present into very remote future times. It could be said that this exercise is irrelevant, because it takes us far beyond the spell of our lives and, moreover, because it is fraught with uncertainty and speculation. As regards the last point, the methodology and basic tools we use to reconstruct the past history of man, life on Earth, the Solar System, and the Universe are exactly the same as the ones we shall use to unveil some of the more probable facets of the future.

What allows us to discover the past is the systematic search for its vestiges and the application of the methodology of science to observe and interpret them. The future is just a continuation of the past. The better we know the past, the better we can project the future. There is continuity and an inherent coherence between the past and the future. We are merely active subjects and interested spectators of ephemeral and localized events in a Universe governed by laws we have deciphered but which transcend us. Moreover, the emergence of intelligent life on Earth was probably a very rare occurrence in the Universe, or at least in our galaxy. Perhaps by getting a better understanding of how we fit into the Universe, we will become less proud and better appreciate the little planet on which we have evolved from the most primitive unicellular forms of life, and hence have more respect for it.

In Greek mythology, Nemesis is the goddess who punishes those who have hubris, originally defined as the pleasure in humiliating their victims and maltreating others. Today hubris refers to the limitless pride, arrogance, and lack of humility that is frequently associated with subsequent suffering as indicated in the biblical proverb: "Pride goes before destruction, and arrogance before failure" (Bible, Proverbs Chap. 16, Verse 18). Humanity prides itself for increasingly exploring the Earth's subsystems, particularly the biosphere, for its own benefit, and for its ingenuity in solving increasingly complex engineering, technological, and environmental problems. But unlimited rapacity will inevitably lead to ruin.

Our cultural evolution unfolded very rapidly compared to the natural rhythms of change in the Earth's subsystems and the biological evolution of most other life forms on Earth. In the last 10 000 years, humanity has created wonderful civilizations that have deeply transformed our way of life. This evolution has accelerated in the last 500 years, mostly under the impetus of modern science and technology. The pace of change became faster with the industrial revolution, about 250 years ago, and the Great Acceleration since World War II. But five hundred years corresponds to only 0.25% of the approximately 200 000 years of the existence of *Homo sapiens*. It is a mere instant compared to the time elapsed since the beginning of the Cambrian era 550 million years ago, when a large number of diversified multicellular organisms first emerged.

At least as important, if not more so, is the fact that five hundred years is an extremely short period relative to the immensity of time that humanity has before it. How are we going to deal with all this time? Will humans live avidly in the fast lane, regardless of the future, or wisely, respecting sustainability and intergenerational solidarity? What are the frontiers at which utopia begins? There are many indicators signalling that the most attractive life styles recently created and adopted, and which are tending to become globalized, are not sustainable in the long term. Consumerism and greed are prevailing behavioural traits, particularly in top income brackets, which serve as role models for thousands of millions of people. How are we going to address these potential contradictions? Will we have a different *Homo sapiens* in the future, who is able to construct a globalized society in equilibrium with his natural environment?

4.5 Current Conditioning Factors and the Biological Evolution of *Homo sapiens*

Our remarkable achievements in the last 10 000 years were all accomplished by the same biological species that emerged in African savannahs, migrated to Europe during the last glacial period, probably about 60 000 years ago, and ended up dominating all Earth continents. These accomplishments resulted from a strictly cultural evolution without significant changes in the *Homo sapiens* genome. The culture, or 'software', has evolved tremendously, while the brain, or 'hardware', has been the same for at least 200 000 years.

Will our species evolve biologically, turning into some sort of superman capable of finding new paradigms for sustainable growth and progress? The most important modes of natural speciation all involve some degree of geographic separation between populations. In the case of *Homo sapiens*, these mechanisms are ineffective because we belong to a highly mobile global society where all human populations are in relatively close contact and maintain the ability to interbreed frequently. The genetic variation in our genotype is sufficiently abundant to serve as material for a natural selection process that could possibly increase our survival capacity under changing environmental conditions. However, unlike the small groups of *Homo erectus* that evolved into *Homo sapiens*, we now constitute a mass society where the mechanisms of natural selection are strongly suppressed. Our social culture allied with modern medicine makes natural human biological evolution almost impossible. For example, if medical research had not succeeded in the fight against AIDS and the pandemic had strongly intensified, the selection pressures would have increased the number of mutations, probably leading to a fixation of the most resistant among them. Evidence that a particular mutation has actually boosted reproduction is very hard to find in contemporary humans. There are only a few examples indicating that our species is evolving biologically in very subtle ways (Bacher, 2005).

The future biological evolution of *Homo sapiens* depends mostly on the external and internal factors affecting human societies. The external factors are largely determined by environmental changes and by the availability of natural resources on a regional and global scale, while the internal factors are determined by the social, political, financial, economic, cultural, scientific, and technological evolution, and by the intensity, characteristics, and extent of future conflicts. There is an increasing interdependence between the two types of factor. Transformations of society caused by serious crises related to energy, water, or other natural resources, or caused by continuous global population growth, unmitigated climate change, nuclear war, or generalized forms of violent conflicts, could disrupt social stability, significantly reduce social services and solidarity, and reduce the capacity for medical intervention, thereby stimulating natural selection mechanisms. The more stable and secure the future, the less likely it becomes that *Homo sapiens* will evolve biologically.

In the event that the more serious future socio-economic and environmental scenarios become a reality, we may witness a breakdown in our present mode of continuous 'software' development, that is, in our cultural evolution. Critical situations may break this continuity, reviving the currently almost nonexistent pressures of natural selection. The crucial point is whether or not we will be able to make our cultural development compatible with the conditions that are forced upon us, in the end, by nature and its underlying laws. We are held hostage by our cultural dynamism and to the immense capacity it has for making our ways of life so outstanding, even though these successes are far from reaching the whole human population.

To evolve biologically into another species will probably require a break in the current pattern. Long-term sustainable development can only be reached through adaptation by a process of continuous cultural evolution of *Homo sapiens*. The other way involves deeply upsetting and destructive crises that could intensify natural selection and cause speciation in the long-term. There are no short cuts for *Homo sapiens* to evolve naturally into a new species blessed with the exceptional abilities that would be required to avoid the contradictions, conflicts, and crises likely to result from the continuation of our current unsustainable development. Meanwhile, we will have to adapt to an environment that is permanently changing as a result of natural and anthropogenic causes.

4.6 The Third Way: Trans-Humanism and Synthetic Biology

There is a third way for the future of *Homo sapiens*, based on trans-humanism. The defenders of this movement support the use of science and technology to improve the mental, biological, and physical capacities of humans, transforming them into a post-human beings. They advocate that humans could control their own evolution through the application of emerging technologies, such as nanotechnology, biotechnology, information technology, artificial intelligence, and robotics. We would thus move into a post-Darwinian phase of human evolution.

The most daring versions of trans-humanism propose eugenics to modify the human body in ways that would alleviate some of its biological limitations, such as propensity for certain diseases, ageing, and memory loss, while enhancing its intellectual, physical, and psychological capacities. Eugenics was very popular in the early decades of the 20th century, but rejected all over the world after the Nazis used it in their programs of racial hygiene, human experimentation, and extermination of undesirable population groups. Today there is a revival of interest and a reassessment of its development potential (Lynn, 2001).

Post-humans could also benefit from a symbiosis between actual human capacities and the potential capabilities of artificial intelligence devices. The objective is to free us, although only partially, from the human condition, through technological transformations of our organism based on genetic engineering, life extension techniques, psycho-pharmacology, and the use, in particular through implantation, of physical systems that enhance various capacities, such as computers, sensors, interface systems, and robots (FM 2030, 1989; Ettinger, 1974).

The expression 'trans-humanism' was used for the first time by the biologist Julian Huxley (Huxley, 1957) to explain the idea of "man remaining man, but transcending himself by realizing new possibilities of and for his human nature". Currently, trans-humanism clearly assumes the transition to a post-human state. Pushed to its limit, the fusion of man and machine with the aim of drastically increasing our capacities, especially cognition, intelligence, and longevity, would transform *Homo sapiens* into *Machina sapiens*. Essentially, it is an endeavour with its roots in the Age of Enlightenment, founded on a belief in progress based on science and technology. It seeks to rationalize, explore, and promote deep transformations in our nature and ways of life, for our benefit and through technological innovations, up to the point of modifying the human being itself, physically and biologically.

We are facing a post-modern form of scientism that raises many questions of an ethical, social, and political nature. What are the ethical principles and moral values that would allow us to justify a more or less irreversible transformation of our physical and biological capacities? What are the risks involved in eugenics and genetic engineering applied to human beings? Who is responsible for the creation and proliferation of post-humans? All these questions are very difficult to answer, or even unanswerable, and they inevitably generate controversy. Trans-humanism involves the danger that it could create a dystopia similar to the one imagined by Aldous Huxley in 1932 (Huxley, 1932), even though its current defenders consider that it is not incompatible with democracy and the defence of human rights and civil liberties. However, the creation of post-humans is likely to change the social order and to generate new conflicts between post-humans and humans, possibly aggravating the already deep inequalities in the world.

Trans-humanist movements tend to consider the preservation of nature and of natural systems, particularly the integrity of the human organism, as secondary, compared with the benefits of the progress achieved by the application of technological

innovations to humans without significant boundaries or restrictions. Nature is looked upon as something that will progressively become a feature of the past, and
whose relationship with humanity will tend to weaken, eventually becoming ex-
pendable. The defenders of trans-humanism call its opponents bio-conservatives,
bio-Luddites, or neo-Luddites.

In spite of the reservations that have been expressed, the project of redesigning
humans is likely to advance. The main reason is that the technologies that are required to do so are becoming available and, as in previous times, what science and
technology allow us to do will be done no matter how dangerous and damaging the
consequences of any misuses might be. The argument that scientific knowledge and
technologies are neutral and that it will be possible to enforce responsible strategies
and legislation to apply them safely is likely to prevail. Furthermore, it is unquestionable that humans live in increasingly artificial conditions. This tendency is in
the mainstream of our cultural evolution. The defenders of trans-humanism propose
that we assume and develop that tendency to its limits, managing it for the benefit
of humanity, even if it leads to an entirely artificial human (Changeux, 2007). Gregory Stock, one of the most outspoken supporters of trans-humanism, states that:
"A thousand years from now the future humans will look back on our era as a challenging, difficult, and traumatic moment. They will likely see it as a primitive time
when people lived only seventy or eighty years, died of awful diseases, and conceived their children outside a laboratory by a random, unpredictable meeting of sperm
and egg" (Stock, 2003). Why should we not rework our biology and reproduction?

The fact is that we are already witnessing remarkable advances in synthetic biology. The discovery of restriction enzymes in the 1970s led to the development of
recombinant DNA technology that allowed very important medical breakthroughs,
such as the production of insulin for diabetics. With this new technology, it became possible to describe and analyse the genes of all the known living organisms
and, furthermore, to construct new gene arrangements, which opens the possibility
of creating new living organisms. Recently, in May 2010, Craig Venter, Hamilton
Smith, and their colleagues synthesized an entire bacterial genome and used it to
take over a cell whose DNA had been previously removed (Gibson, 2010). The
new bacterium cell was capable of continuous self-replication and made all the proteins necessary to sustain its life, following the specifications carried by the inserted
DNA. The long-term objective of the team is to produce useful new organisms, in
particular with the property of being optimized for biofuel production.

The next step along this line of research is to build an entire alga genome to make
algae cells especially efficient at photosynthesis, which would be used to produce
petrol and diesel fuel out of atmospheric CO_2. There are many other similar ongoing
projects and the likelihood that they will succeed has been helped by falling prices
for sequencing and making DNA, since the late 1990s. Synthetic biology is still at
its inception, but it clearly has the potential to create new living organisms that may
lead to useful crops, fuels, and drugs. This entire program involves increasing risks,
but they are very unlikely to stop it. Given sufficient time, synthetic biology will
naturally lead to a special form of trans-humanism.

4.7 Emerging and Converging Technologies and the Technological Singularity

It is becoming ever clearer that the emerging technologies are potentially capable of influencing future human evolution. The main set comprises nanotechnology, biotechnology, information technology, and cognitive science, identified by the acronym NBIC. Reports published by the US National Science Foundation (NSF, 2003) and by the European Union Parliament (EP, 2005) make a detailed analysis of the possibilities for improving man's capacities and the sustainability of human societies by using emerging and converging technologies. The latter correspond to different disciplines that tend to merge or become strongly linked towards a common objective, such as producing computers, developing information technologies and media to improve communication, entertainment, and alienation. Trans-humanism defends the convergence of emerging technologies as the most promising way to transform humans, improve their performances, and eventually succeed in creating post-human beings.

The emergent technologies involve several risks. Let us consider just one example of such risks. Molecular nanotechnology can be used to construct machines on a molecular scale, atom by atom, with specific functional properties that may be useful. However, if they have the ability to reproduce, these micro-robots could go out of control and cause significant damage, including the destruction of the ecosystems they inhabit. It is also plausible to admit that they may be produced with the deliberate destructive goal of ecofagia, that is, of reproducing fast and consuming the living matter in which they were placed. Eric Drexler (Drexler, 1986) was the first to warn of the danger of these agglomerations of mini-robots, referred to as 'grey goo'.

Throughout history, technological innovations have had a profound impact on our social, economic, and cultural life, thoroughly changing our lifestyles and the way we represent the world. These changes, according to Gerhard Lenski, Leslie White, and Lewis H. Morgan, among other authors, constitute the main driving force behind the evolution of societies and cultures, and in the final analysis, the evolution of civilizations. There are many examples of discoveries of great social and cultural importance that resulted from technological innovations, mostly based on modern science. Some of the more remarkable discoveries were the stone tools in prehistoric times, the different forms of representation, beginning with primitive paintings and sculptures in caves, the wheel, the way to make a variety of useful objects from gold, copper, bronze, iron, and other metals, writing systems, the printing press, the telescope, the microscope, the steam engine, the telephone, the internal combustion engine, the automobile, the airplane, the telegraph, radio, television, satellites, personal computers, mobile phones, and the worldwide web.

Looking at the past we perceive sharply accelerating technological change, which suggests that in the future innovations will be increasingly significant and less spaced out in time. Raymond Kurzweil takes the argument to its limit and predicts a technological singularity during this century, around 2045 (Kurzweil, 2005). One of the best known examples of the accelerated rhythm of technological change is Moore's Law (Moore, 1965), according to which the number of transistors that can be placed in an integrated circuit with minimum cost per component has doubled approximately every 24 months since 1965. Another example of exponential growth is the increase in the number of units of information stored on computer hard disks.

Kurzweil goes further and considers that that there is not just acceleration in the efficiency and capacity of a given technology, but also in the invention of new technologies that allow us to overcome barriers arising from the use and development of previous ones. Those that believe in a future technological singularity, or singularitarians, consider that it will be characterized by an immense number of innovations in a very short period of time, but they diverge as regards its nature and the risks involved. The most common characterization of the singularity is the creation of entities with an intelligence superior to human beings. These may be very powerful computers or large computer networks, interfaces between humans and computers, or mental transfer processes between robotic and human brains so efficient and powerful that their users turn into beings with superhuman intelligence.

Computers are getting so incredibly fast that there may come a moment when they are capable of something comparable and eventually superior to human intelligence. Brain–machine interfaces, which allow for activity in the brain to be sent to, or received from a computer are already helping paralysed patients communicate and control robotic arms, computers, and other devices. The enormous potential of artificial intelligence and the possibility of building machines with thinking abilities, capable of solving complex problems, endowed with various forms of conscience, and equalling and eventually surpassing human intelligence is considered to be a sure sign of an approaching singularity. The crucial point may be that these machines become able to continuously improve themselves using artificial intelligence. Another possibility is that biological transformations on humans, including genetic engineering, may lead to a greater than human intelligence.

The singularity arises because it is considered impossible to predict what these super-intelligent entities will do when they start behaving on their own. Some singularitarians believe that the singularity will announce the end of the human era and the beginning of an unimaginable process of rapid evolution of post-human beings. A Singularity University with the stated aim to "assemble, educate, and inspire a cadre of leaders who strive to understand and facilitate the development of exponentially advancing technologies and apply, focus, and guide these tools, to address humanity's grand challenges" was established in 2009 with the support of Google and NASA. The school is not an accredited university but is modelled as the International Space University, an international and interdisciplinary multicultural school. According to the Financial Times (FT, 2009), its proponents go as far as saying that the singularity "smarter-than-human computers will solve problems including energy scarcity, climate change, and hunger". Besides sounding scientifically and sociologically naive in 2011, it reveals that there is a tendency for humans on Earth to consider that they are confronted by superhuman problems, some of their own making, which can only be solved by post-humans. Nowadays, the singularity concept attracts people that form a community of believers in a post-human age.

Future prospects for the creation of post-humans probably remain repulsive to a large part of humanity today, and likely to involve risks to human life. This is yet another example where the development of science and technology brings opportunities to improve our capacities and achievements, but also generates new risks. According to past experience these risks will not prevent the development and application of whatever technological innovation becomes accessible and turns out to be lucrative. In any case the accelerated rate of technological innovation may be an erroneous conclusion, resulting from different kinds of memory perception for recent events relative to past events. Some authors consider that the rate of innovation is effectively decreasing, rather than increasing. According to Jonathan Huebner (Huebner, 2005), it reached its maximum at the end of 19th century, in the 1870s, and will attain relatively low values by the middle of this century.

While the frequency of technological innovations with a wide raging applicability throughout the world is increasing, the creation of super-intelligent entities is something that remains uncertain. Nevertheless, robots are increasingly powerful and are very likely to become part of our everyday lives. The South Korean government has the objective of getting a robot in every home by 2020. There are already active movements to give legal rights to robots, which will probably become very powerful once they begin to have humanlike capacities and especially consciousness. It will be a long and controversial process, at least in some countries, probably including some Western countries, likely to control the risks involved in the emergence of smarter-than-human intelligent machines.

4.8 Risks and Uncertainties of the Present and the Partial Predictability of the Long Term Future

We have briefly explored some of the presently imaginable limits of the contribution that science and technology can make to the improvement of human life. There are clearly risks and uncertainties in the evolution of societies, even in the medium term future of 50 to 100 years. Such an evolution takes place in a global natural environment that has its own dynamism, determined by the laws of nature. No matter how great the 'hubris' of humanity, we cannot escape this reality. We therefore find ourselves in a somewhat paradoxical situation. On the one hand, we sense risks and uncertainties in the short and mid-term future and we do not know how we are going to achieve sustainable development. On the other hand we have the capacity to predict the Earth's evolution reliably into the long term future. We have an enormous amount of time to adapt to this natural evolution of the environment, but very little time to make our own culture sustainable.

Knowledge of the fundamental laws of nature allows us to reconstruct the past and project the future of planet Earth, the Solar System, and the Universe. These exercises give us a clearer image of the virtues and limitations of the surprising human adventure. It is one of the ways to get to know ourselves better, and perhaps create the conditions for deciphering whether we can possibly sustain our cultural evolution or whether we are simply headed for collapse. It is also a way to find out whether the defence of that capacity for evolution is sufficient reason to overcome the dominance of randomness. Nature has no intentions or objectives, but we certainly can. In the history of life on Earth, and on the corresponding time scale, the culture of *Homo sapiens* may be like an ephemeral but bright flash of light, containing the degenerative seeds of ruin, or it may be a durable light, capable of achieving a sustainable harmony with nature and the cosmos.

Before discussing in Chap. 5 the various arguments that must today be considered to interpret and rationalize the risks and uncertainties of the near future, let us look further ahead to the long term evolution of planet Earth. This brief journey through time reveals the power of the forces that control the life and death of planet Earth, and the great vulnerability of our civilization. However, the greatest challenge to mankind is mankind itself, and it centres on the problem of finding a sustainable form of development in the coming decades and centuries.

4.9 How Will Mankind Adapt to the Next Glacial Period and the Ensuing Glacial Cycles?

The current interglacial, which corresponds to the Holocene, will eventually end in a glacial period. We have some ability to predict the lifespan of this relatively warm period, since we understand the climate system to a certain extent and can identify the principal causes of climate variability. The main mechanism for the oscillation between interglacial and glacial periods is the variation in the total solar irradiation (also known as insolation) at the Earth's surface caused by small changes in the Earth's orbital and rotation parameters. The last interglacial period, called Eemian and also Riss–Wurm in the Alps, lasted between 15 000 and 25 000 years and reached its maximum temperature about 125 000 years ago. At that time, according to paleoclimatic studies, the temperature in the polar regions was about 3–5°C higher than it is now, the climate was more humid, and the average sea level was probably about 4–6 meters higher than the current level, due to the lower volume of the ice caps.

In the present interglacial period, the amplitude of the variations in insolation resulting from astronomical forcing is exceptionally small, smaller than in the Eemian. In the next few tens of thousands of years, the eccentricity of the Earth's orbit is expected to reach very low values, causing it to be almost circular. This relatively rare event reduces the amplitude of the insolation changes, and this will tend to increase the duration of the interglacial. Furthermore, anthropogenic climate change is also likely to make it longer. Such conclusions are obtained through climate system models that simulate its behaviour over the next few centuries and millennia and incorporate the radiative forcing caused by the emission of greenhouse gases. Simulations obtained by André Berger and M.F. Loutre (Berger, 2002), with various levels of CO_2 emissions, produce an exceptionally long interglacial ending only in 50 000 years, followed by a glacial period that will reach its coldest temperatures in about 100 000 years. According to these calculations, the climate system will take about 50 000 years to adapt to anthropogenic climate forcing caused by the intensive use of fossil fuels since the beginning of the industrial revolution, about 250 years ago. During this time, the Greenland and Antarctic ice sheets are likely to thaw completely if the concentration of atmospheric CO_2 exceeds levels of 1 000 ppmv, causing an increase in sea level of about 70 meters (Alley, 2005). More recent estimates indicate that the critical threshold where the Greenland ice sheet becomes unstable is a CO_2 concentration somewhere between 400 and 560 ppmv (Stone, 2010). The Antarctic ice sheets are more stable.

The cumulative effect of some of our activities has the potential to produce global changes which will be irreversible over time intervals of tens of thousands of years. How are we going to manage this surprising power of interference with the planet Earth, whose consequences are likely to be adverse for the greater part of the world's population? It is an open question, which we will only begin to answer in the next few decades. The argument that anthropogenic climate change is, in the end, positive because it probably delays the next glacial period is unfounded since there is no way to compare the potential advantages with the negative impacts of climate change on many human activities, human health, food security, and biodiversity loss.

Whatever happens, the Earth will eventually dive into a glacial period. According to current climate change models, it only begins in about 50 000 years, but there is considerable uncertainty in this estimate. Its characteristics and effects will be very similar to those of the last glacial periods, which we know rather well, especially in the case of the last one. The ice sheets start to advance to lower latitudes because in the summer the reduced solar irradiation becomes insufficient to melt all the accumulated snow fall from the previous winter, especially in the very sensitive high latitudes of the northern hemisphere around 65°N. The increasing snow cover reflects more solar radiation, acting as a positive feedback on the climate system. Accumulating amounts of snow, year after year, increase the extent of glaciers, and create new ones. Eventually, they will combine to form gigantic ice caps 2 to 3 km thick. In the last glacial period, the northern ice cap covered Scandinavia, the northern half of Great Britain, a large part of Central Europe and Russia, most of Canada and parts of the northern United States. At lower latitudes, in the high mountainous areas, glaciers are expected to advance again and to merge forming large ice sheets.

It is impossible to imagine how human society will be organized in 50 000 years time. We may even wonder whether humans will still exist. I think we will, because we are very resilient. This resilience stems from our remarkable abilities, which we will surely strive to maintain, and from the pride and confidence that we have in ourselves. However, it is very likely that during that time period we will have to struggle with and overcome very serious and devastating crises, some of our own making. In the next glacial period, the territories in the higher latitudes of the northern hemisphere, where we now find the Scandinavian countries, a large part of Russia, and parts of Canada and the USA, will be buried under huge ice sheets, forcing populations to migrate. This profound climate change will not be limited to the polar regions. Paleoclimatology reveals that, in glacial periods, the ice caps are surrounded by extensive cold and arid or desert areas. The global climate is drier and the winds are stronger, due to the higher latitudinal temperature gradient, and carry large amounts of dust resulting from soil degradation and desertification. The amount of vegetation covering the planet diminishes noticeably, savannahs prolife-rate, and tropical rainforests become hidden in small widely dispersed niches. This is undoubtedly a catastrophic scenario for a civilization like ours at the present time, living at the limits of the available natural resources.

We can imagine that humanity would try to mitigate a glacial period through geophysical engineering, but it is very unlikely that it would be able to stop it. Such a program would imply finding gigantic amounts of energy to warm the planet without causing even more serious collateral impacts. In any case, it will probably not be a question of survival for *Homo sapiens*, whose capacity for adaptation is outstanding. The next glacial period will be the third in his history. Many more will follow with the same average periodicity of about 100 000 years. Each one will deeply change the Earth's biosphere and geography. New fjords will appear, similar to the ones left by the last glacial period in Scandinavia and Southern Chile, along with new lakes caused by the movement of glaciers, similar to the Great Lakes in North America and Lake Baikal in Asia.

It is well known that the alternation between interglacial and glacial periods fa-vours the process of speciation. In the next glacial period, for the first time, specia-tion will be affected by the pressures that human activities exert on the habitats and populations of many species, especially in a situation of great stress for the bios-phere and particularly for humans. How will it be possible for species to co-evolve biologically in the presence of man during the next glacial and interglacial periods? Loss of biodiversity is very likely to increase, driven by the synergy between an-thropogenic pressures and the impacts from climate changes caused by the glacial cycles. How will agriculture adapt to the cold and dry climate of a glacial period? Will the ecosystems still be able to provide the essential services required by hu-manity, especially as regards food security? Will it be possible to reach some form of sustainable development under the very harsh conditions of glaciations? These are just some of the many daunting adaptation challenges for future glacial periods. *Homo sapiens* has already experienced the approach and establishment of a glacial period about 110 000 years ago, but at that time the human population was only around 1.1 million, living in scattered regions of Africa at relatively low latitudes.

The next one will surely be a severe test for the continuity of our cultural evo-lution. One may think of a scenario where the living conditions are so difficult that fragments of the human population become isolated, creating favourable conditions for the speciation of *Homo sapiens*. There are large uncertainties regarding the fu-ture of mankind on time scales of hundreds of thousands of years, but coherent scenarios of future evolutions can be formulated (Ward, 2001).

4.10 The Consequences of Continental Drift Over the Next 300 Million Years

As we proceed in time, the uncertainty about the future of *Homo sapiens* continues to grow ever more rapidly. Nevertheless, it is possible to outline the essential features of the long term evolution of the Earth's subsystems, in particular the biosphere, with relatively small uncertainty.

We now move from a time scale of hundreds of thousands of years to one of tens or hundreds of millions of years. The study of the Earth's climate evolution in the latter time frame reveals that it is influenced by phenomena related to plate tectonics. The continental plates drift slowly, either in convergent or divergent relative movements that have cyclic characteristics. Currently, the main continents are far apart, but in the past, about 250 million years ago, they were joined together to form the Pangaea supercontinent. After a few tens of millions of years, Europe drifted away from North America and Africa from South America, opening the Atlantic Ocean. About 120 million years ago, India separated from Africa and collided with Eurasia, while Australia separated from Antarctica. It is very likely that in the future the various continents will again converge to form a supercontinent. The one that preceded Pangaea occurred about 600 million years ago and is called Pannotia.

The mechanisms of plate tectonics and continental drift are relatively well understood. The lithosphere is broken up into tectonic plates that fluctuate and move in a highly visco-elastic material called the asthenosphere. The movements are driven by the convective motions of hot material in the Earth's mantle. When the behaviour of the mantle and its convective motions are simulated with computer models, one finds a tendency for the lithosphere to form a supercontinent and then to separate again into the major continents. This process appears to be quasi-periodic with a period of about 250 to 300 million years. The result agrees with the formation time of the two last supercontinents, and there are records of large land masses up to more than 3 billion years ago.

By projecting the future movements of the current major continents with plate tectonic computer models, one finds that, within the next 50 million years, a collision will occur between the Eurasian plate and the African plate, causing the disappearance of the Mediterranean Sea and the formation of a long chain of mountains from South West Europe to the Persian Gulf. The region that was once the cradle for some of the first great civilizations will become unrecognisable. All the monuments that remind us of this brilliant past are likely to disappear forever. After about 250 million years, the Atlantic Ocean and the Indian Ocean will close up due to collisions between North America and Africa, and between Patagonia and South East Asia (Ward, 2002). A new supercontinent will be formed, surrounded by a much more extensive Pacific Ocean. The rearrangement of the tectonic plates, the new patterns of ocean currents, and the new chains of mountains will produce major changes in climate at the global and regional levels.

Let us look more closely at this evolution. Over the next 10 million years, due to continental drift, the current ice age, characterized by permanent ice caps in the

Arctic and Antarctic and by alternating glacial and interglacial periods, will come to an end, giving rise to a warmer global climate with average global temperatures that could eventually reach as high as 21°C, as in the Eocene optimum about 50 million years ago. The slow convergence and fusion of the major continents and the formation of just one very large ocean will have important consequences for ocean currents, the climate, and the biosphere. In the Super-Pacific Ocean, the much smaller latitudinal temperature gradient will lead to the collapse of the thermohaline circulation, driven by the sinking of cold and highly saline sea water. In the absence of vertical currents, the ocean will become more stratified and the amount of dissolved oxygen will decrease substantially, except close to the surface, and this will drastically reduce its capacity to support marine life. There are records of such anoxic events in past great extinctions of species, like the one that happened at the end of the Permian period.

The collapse of surface water sinking and deepwater upwelling also tends to build up concentrations of CO_2, methane, and hydrogen sulphide in sea water. The release of the latter gas into the atmosphere will destroy plant and animal life in the coastal zones. Meanwhile, CO_2 will accumulate in the atmosphere, since it dissolves less in the warmer ocean, increasing the average global temperature. The huge expanse of the supercontinent will originate extreme climates, adverse for many species. In the interior regions, because of the great distance from the sea, and the blocking of weather fronts by coastal mountain ranges, there will be a strong tendency to form large deserts with extreme temperature amplitudes. In coastal regions, due to the global ocean anoxia, biodiversity will be much reduced, and primitive micro-organisms, such as the stromatolites, may become dominant, as in Precambrian times.

The biosphere and climate scenarios for the next 250 million years are largely irrelevant for humanity today. Although the probability that such scenarios will actually occur is high, it is impossible to foresee what will happen to *Homo sapiens* or to the other species that may have evolved from him by that time.

4.11 Solar Global Warming and the Extinction of Plants and Animals

On a longer time frame of hundreds to thousands of millions of years, there is another driver of change that will certainly shape life's destiny on Earth. The luminosity of the Sun has been growing slowly since its formation about 4.6×10^9 years ago. Since that time it has increased by about 30%, and until the end of its journey as a main sequence star, before turning into a red giant, it will increase by a factor of approximately 2.2 (Sackmann, 1993). The reason for this phenomenon has already been mentioned in Chap. 2 and is quite simple. Essentially it results from the application of the fundamental laws of thermodynamics to describe the equation of state of the gas found in the central region of the Sun, where the fusion of hydrogen into helium takes place. Although the increase in the Sun's luminosity is extremely slow, it becomes significant on time scales of hundreds to thousands of millions of years. Its effect on the Earth's systems and processes will be far-reaching.

The systematic increase in the Sun's luminosity over billions of years will produce a solar global warming with devastating effects for life on Earth, through the slow perturbation of the regulating effect of CO_2 on climate. This form of global warming is currently unnoticeable at the moment, because the increase in solar luminosity is extremely small and hence irrelevant on the time scale of hominidae evolution. However, it will slowly shift the habitable zone of the Solar System away from the Sun to include Mars. On Earth, it will increase the average global temperature, leading to an intensification of the water cycle, and this in turn will lead to an increase in the global average rainfall and erosion processes. These increases will intensify the weathering of silicate rocks, removing large amounts of CO_2 from the atmosphere. The equilibrium of the carbonate–silicate cycle will break down and the concentration of atmospheric CO_2 will decrease significantly. Eventually the concentration will fall below the levels required for photosynthesis by plants. The extinction of plants and animals will be the first step toward the extinction of all life on Earth.

James E. Lovelock and M. Whitfield (Lovelock, 1982) were the first to estimate the longevity of plants, on the assumption that their survival becomes impossible with atmospheric CO_2 concentrations below 150 ppmv. They concluded that they would become extinct within the next 100 million years. However, some plants are more resistant to low CO_2 concentrations than others. They are known as C4, because they use a chemical compound in photosynthesis with four carbon atoms instead of three, like all the other plants, which are known as C3. C4 plants, which include maize, sugar cane, sorghum, and switch grass, arrived after C3 plants, in the Cenozoic era. They have a competitive advantage over C3 plants, in that they adapt better to drought, high temperatures, intense solar radiation, and atmospheres with low concentrations of N_2 and CO_2. Currently, they represent about 5% of the terrestrial biomass, but they fix about 18% of the carbon captured in photosynthesis. The minimum CO_2 concentration that enables photosynthesis falls from 150 ppmv for C3 plants to about 10 ppmv for C4 plants.

More recent models of carbon cycle evolution in a climate system subject to forcing by solar global warming indicate that plants and animals will only become extinct between 800 and 1 200 million years from now (Franck, 2006). The eukaryotic unicellular organisms, because of their greater resistance to higher temperatures, will survive longer, from 1 300 to 1 500 million years. Finally, the prokaryotic organisms, the first to appear on Earth and those that can stand higher temperatures, around 100°C, will disappear later, about 1 600 million years from now. There are of course major uncertainties regarding these quantitative estimates, but the main features of the forthcoming evolution of life on Earth are quite well established.

Curiously, the various types of living organisms are likely to disappear in the reverse order to which they appeared on Earth. The more complex multicellular organisms, including the large diversity of plants and animals that arose in the Cambrian explosion around 530 million years ago and all those that evolved from them, will probably be extinct within 800–1 200 million years. This means that they would have been on the planet for around 1 500 million years. We are about one third of the way through this cycle, and this is precisely the time it took for intelligent life to emerge, in the form of *Homo sapiens*. Science has enabled him to foresee, rationalize, and react to the probable future evolution of his terrestrial environment. The permanence and development of intelligent life on Earth is certainly not unlimited. How long it will last remains an open question.

The extinction of plants interrupts oxygen production, leading to a rapid decrease in its atmospheric concentration due to various oxidation processes. Without oxygen, animals will die and the ozone layer will tend to disappear, allowing the penetration of ultraviolet solar radiation and making life at the planet's surface much more hazardous and precarious. A few fungi and algae will survive, but above all bacteria will once again dominate the biosphere, as they did in the first few billion years of Earth's history.

During the huge time intervals that we are considering, life will certainly evolve, seeking to adapt to new environmental conditions by natural selection. Life's skill in finding new solutions to adapt to global solar warming and to the fall in atmospheric CO_2 concentrations will probably be remarkable and surprising, just as in the past when it had to face other major challenges. However, in the long run survival will inevitably become a lost cause. Today's plants and animals cannot tolerate average temperatures above 45°C. At such temperatures, mitochondria, the organelles present in most eukaryotic cells, stop working. Yet their activity is essential, because they transform the organic molecules that serve as food for the cell, such as glucose, into energy in the form of ATP, the nucleotide that transports chemical energy within cells. It will be extremely difficult for eukaryotic cells to evolve in such a way as to overcome this problem. Even if they did, their victory would be short-lived because the average global temperature of the atmosphere at the surface will keep on increasing.

It will reach values around 50°C within 1 300 million years (Franck, 2006). In the equatorial regions, the temperatures will be even higher, forcing the remaining living organisms to migrate to the polar regions and to develop underground and nocturnal lifestyles. When the average global temperature reaches values around 70°C, only thermophilic bacteria similar to those living in present-day hot springs will be able to survive on the surface. The last complex forms of life will probably live in the deepest regions of the oceans. We are now irreversibly far away from the lush tropical rain forests that contain most of the Earth's biodiversity today.

4.12 The Oceans Evaporate and the Atmosphere Becomes Like the Atmosphere of Venus

Meanwhile, the oceans will become unstable due to the increase in the temperature of the atmosphere. Currently, almost all molecules of evaporated sea water remain in the troposphere and circulate in the water cycle. There is, however, a very small amount which, propelled by the ascending air currents, penetrates the stratosphere where the ultraviolet radiation dissociates the molecules into hydrogen and oxygen. Because hydrogen is lighter, part of it manages to free itself from the terrestrial gravitational field and escapes into outer space. This phenomenon is visible around the Earth through the detection of the ultraviolet radiation, corresponding to the spectral line known as Lyman-α, emitted by excited hydrogen atoms. This radiation was observed for the first time using a highly ingenious telescope, the far ultraviolet camera/spectrograph invented by George Carruthers and left by the astronauts of the Apollo 16 mission on the Moon's surface in 1972. It is not possible to observe this effect from the Earth due to the protective effect of the stratospheric ozone layer. The loss of oceanic water is currently insignificant, partly because the vertical temperature profile reaches a minimum in the tropopause, of the order of $-60°C$, depending on the latitude and season, and the air at these relatively low temperatures cannot contain much water vapour. Thus the water molecules are trapped in the troposphere and only a very few reach the stratosphere. The present loss of oceanic water corresponds to a shell just 1 mm high every million years. In the future, as the Sun's luminosity increases, this loss will accelerate.

There is still one other crucial aspect to consider. When the atmosphere becomes warmer, it will be able to hold more water vapour, which is a greenhouse gas. Therefore the accumulation in the atmosphere of water evaporated from the oceans will intensify the greenhouse effect, leading to a rise in the average global temperature of the atmosphere. This positive feedback will generate a runaway greenhouse effect, similar to what happened in the primordial atmosphere of Venus, where the higher temperature relative to Earth maintained water in the gaseous state without forming oceans. With the increase in the global temperature, the average height of the troposphere will grow from the current value of 10 km to about 100 km, and the convective air currents reaching these heights will give rise to much more violent storms. The weather will become brutal and dangerous.

About 3 500 million years from now, the luminosity of the Sun will be about 40% greater, and the Earth atmosphere will resemble that of Venus. The oceans will disappear, but their floors will be covered by deep layers of halite, barite, and gypsum, like a huge salt flat. Earth will no longer be the Blue Planet, but will take on a brown colour with shades of red, just like Mars today. When the surface temperature reaches $374°C$, water will no longer be able to remain in the liquid state at any pressure. Life based on this solvent will cease to exist. Meanwhile the planet will suffer other profound transformations. Plate tectonics and continental drift, which, as we have seen, play an essential role in the carbon cycle, will slowly grind to a halt due to exhaustion of radioactive decay in the Earth's interior, and possibly also the absence of oceans. Mountains will cease to form, and erosion will reduce the height of those still existing until the planet becomes almost flat. The beautiful mountains with diversified landscapes and ecosystems that we so much enjoy will disappear forever.

4.13 The Final Phases of the Evolution of the Sun

Planet Earth, once dynamic, complex, and full of life, will turn into an almost inert, deserted, and scorching world, inhospitable, uneventful, and without much to tell about. The new centre of activity will be the Sun, whose rate of change will fast increase. From here on the uncertainty about what is going to happen on the time scale of hundreds and thousands of millions of years is small. This confidence results from our very detailed knowledge of stellar evolution. It is now possible to validate the theoretical models of solar evolution by comparing with stars in the Milky Way that are similar to the Sun and have already reached the various stages of their evolutionary cycles.

About 10 to 11 billion years after the Sun's formation, or within 5.4 to 6.4 billion years, the hydrogen fusion reactions that convert nuclear energy into radiative and thermal energy in its central region will finally come to an end because there is no more hydrogen left (Sackmann, 1993). However, such reactions will continue in a spherical shell that surrounds the central core, now composed almost entirely of helium. The migration of energy production to the more peripheral regions of the stellar volume will cause the Sun to undergo a huge expansion that will transform it into a red giant. The term giant is appropriate, because its radius will become up to 250 times larger than it is today. Mercury and its orbit will then lie inside the Sun. The same may happen with Venus. The Sun's surface temperature will go from 5 780 K down to 2 600 K, giving it a red colour. In fact, it will look like Betelgeuse, a red giant that is the second brightest star in the Orion constellation.

In the final stage of its life as a red giant, the Sun will attain a maximum luminosity up to 2 700 times the current value. This gigantic energy generation will cause large amounts of matter to be ejected into the surrounding space, reducing the mass of the Sun by about 30%. All that remains of the terrestrial and Martian atmospheres will be expelled by the extremely strong solar wind. Due to the Sun's mass loss, the radii of planetary orbits will increase and the Earth will migrate, about 7.6 billion years from now, to a wider orbit where it may avoid being engulfed by the Sun. However, recent calculations (Schroder, 2008) indicate that the Earth will also induce a tidal bulge on the Sun's surface that will eventually slow the Earth down and drag it into the Sun's interior. Meanwhile, in the Sun's central region, extraordinary phenomena will be taking place.

Surprisingly the star will not implode, although there is no thermal energy generation in its central core, now formed exclusively of helium. What is the nature of the pressure that supports the enormous weight of all the shells wrapped around the core? To find the answer, one must travel along the paths of quantum mechanics. In a simplified way, the pressure results from the fact that the central core, formed by a mixture of helium nuclei and electrons, will constitute a degenerate electron gas when the density becomes higher than 10^6 kg/m^3. In this type of gas, the electrons occupy all available quantum states of lowest energy. In accordance with the Pauli exclusion principle, each electron must be in a different quantum state. Because electrons cannot all go into the lowest energy state, they occupy a relatively large volume, and this requires very large amounts of energy to compress it. That is the reason why a degenerate gas of electrons is practically incompressible and hence will be able to prevent the collapse of the star.

As more helium produced in the outer shells is added to the core, its temperature and pressure will keep rising. When the temperature reaches about 10^8 K, the fusion of helium into carbon by the triple alpha process will start in an explosive way, releasing enormous amounts of energy. This mechanism, called the helium flash, is responsible for all the carbon production in the Universe. Carbon formed in successive generations of Milky Way stars was dispersed throughout interstellar space. In particular it enriched the molecular cloud whose gravitational collapse gave birth to the Solar System. All the carbon that supports life on Earth today was formed in triple alpha fusion processes throughout the Milky Way.

When it begins to fuse helium into carbon, the Sun regains some stability, but it will be short-lived. After about 100 million years, the central core will run out of helium and the star will enter a phase of great instability, provoked by the very large sensitivity of the helium fusion reactions to temperature, in which the luminosity and radius will vary cyclically. Finally, after a life of about 12 billion years, the fusion reactions will become ever more peripheral and the Sun will eject its surface layers, launching hydrogen, helium, carbon, oxygen, and nitrogen into space, along with smaller amounts of other elements. A planetary nebula will form, consisting of an expanding spherical shell of gas, similar to many observed from Earth today.

All that remains of the Sun in the central region of the nebula will be a very special star called a white dwarf, made up mainly of carbon and oxygen nuclei surrounded by a degenerate gas of electrons, with a mass about half the initial mass of the Sun. The radius will be about 100 times smaller than it is today, so the white dwarf will have dimensions similar to those of Earth, and extremely high densities, on the order of 10^9 kg/m^3. A cubic centimetre of white dwarf weighs about 1 000 kg. The luminosity will be close to one tenth of the Sun's present luminosity, but the surface temperature will reach 100 000 K. There are no nuclear reactions in these stars, and the radiated energy comes from the heat generated in the core during previous stellar evolution phases.

Slowly, the white dwarf Sun will begin to cool and dim continuously until it becomes a black dwarf. It will be a long process. Models indicate that it will take hundreds of billions of years. There are probably no black dwarfs yet, because the Universe is only 13.7 thousand million years old. As for Earth, its destiny is probably to plunge and disappear into the Sun's fiery surface, before the latter reaches its white dwarf phase. More unlikely, it may continue to orbit the Sun, but at twice the distance it has today, cooling down to a temperature of about $-200°$C.

4.14 How Long Will *Homo sapiens* or His Descendants Live?

How long will mankind be able to share the Earth's evolution? The answer depends largely on the kind of relationship that *Homo sapiens* is able to establish with terrestrial systems, especially with the biosphere and its biodiversity, in the short, medium, and long term. At the present time, humans exploit the ecosystem services and the natural resources in an unsustainable way. The rate of species extinction is increasing, and this tendency seems likely to continue into the future in spite of much dedicated effort by remarkable people and organizations, who have so far been unable to reverse the global trend. We run the risk that natural ecosystem services will no longer be able to support the growing requirements of our development paradigm.

From this point on, uncertainties become large and we can only make conjectures. It may be that *Homo sapiens* or his biological successors will find a sustainable equilibrium with the Earth's systems that will allow them to survive a succession of glacial periods, the end of the ice age, and the beginning of a warm age and even the formation of a supercontinent. Beyond that will come the challenge of the solar global warming and the ensuing reduction in the atmospheric CO_2 concentration, caused by the Sun's increasing luminosity.

Would it be possible to keep the Earth inhabitable beyond 500 to 800 million years? Various geoengineering and planetary engineering solutions can be imagined. Solar radiation management is a kind of geoengineering that seeks to reflect or scatter part of the solar radiation reaching the Earth, and it is already under consideration to counteract current anthropogenic climate change (RS, 2009). One proposal is to place mirrors, diffraction gratings, or lenses in space to reduce in a controlled way the amount of solar radiation reaching us. A particularly convenient position in space is the Lagrange L1 point between the Sun and the Earth, at a distance of 1.5 million km from the latter. A small object placed at L1 would theoretically remain stationary relative to the Earth and Sun, due to the balance between their gravitational forces. By reflecting solar radiation, a mirror at L1 would create a zone of shadow on Earth, protecting it from the increasing solar radiation. Initially this proposal was made to combat the effects of anthropogenic global warming (Gorindasamy, 2000). A more risky response to solar global warming would be to perturb the Earth's orbit around the Sun into one with a greater radius (Korykansky, 2001). However, this kind of planetary engineering would be essential to avoid the likely possibility of the Earth being engulfed when the red giant Sun reaches its maximum radius.

A different type of strategy to ensure the survival of humans, our natural biological descendents, or engineered post-humans is to emigrate from Earth. The closest alternative is Mars, where the total solar irradiation is less than half the value on Earth. Mars will probably be colonized by man long before living conditions on Earth deteriorate significantly and irreversibly. It is important to realize that Mars is a particularly inhospitable place for present-day humans to live in. The atmosphere is extremely rarefied and therefore the pressure very low. The surface pressure on Mars is equal to the pressure in the terrestrial stratosphere at an altitude of about 35 km. The composition of the atmosphere is 95% CO_2 and 3% N_2, along with mere traces of oxygen and water vapour. Furthermore, there is no ozone layer to protect against solar ultraviolet radiation. Frozen water exists only beneath the Martian surface, although one billion years ago it was probably relatively abundant. The planet was more habitable in the past than it is today, but it is unclear whether simple forms of living organisms ever existed there.

A journey to Mars would take several months and the astronauts would have to survive long periods in weightless conditions, subject to the permanent bombardment of radiation and particles from outer space. It is now well known that the absence of gravity for long periods has harmful effects on human health, such as the loss of bone mass. As the planetary habitable zone of the Solar System moves away from the Sun, to distances of 10 to 50 AU, it will become necessary to colonize ever more remote planets and satellites. To achieve living conditions for humans in artificial environments without the possibility of ever returning to our natural home habitat will be an awesome challenge. Migration to far less hospitable regions of the Solar System is likely to be a tremendous odyssey, threatened by a permanent risk of fatal failure.

Emigration from Earth and colonization of the Solar System will necessarily be a highly complex process, controlled by elites that have privileged access to the human and material resources required for such a Herculean project. It is extremely unlikely that all humans would be able to emigrate, and this implies a difficult process of selection, likely to be fiercely combated, especially in critical situations where the survival of humans on Earth is at stake.

One may also imagine a search for new dwelling places on extra-solar planets similar to Earth. However, these could only be reached after a long interstellar voyage. The Solar System is located in a peripheral zone of the Milky Way, between two spiral arms, and at a distance of about 25 000 light years from the centre. It is a relatively starless area, which means that we would have to travel large distances before encountering a planetary system with hospitable planets. The nearest star, Alpha Centauri, is situated at a distance of 4.39 light-years. In spite of these shortcomings, we live in what is considered to be the habitable zone of the Milky Way, an annular region with a distance to the centre between 22 900 and 29 300 light-years, and composed of stars that formed between 8 and 4 billion years ago, which gives sufficient time for the development of complex life forms (Lineweaver, 2004). Closer to the centre, the density of stars is larger, but so too is the risk of high energy radiation and bombarding particles emitted by various types of high mass star and explosive stellar phenomena, such as novas and supernovas. In the more peripheral regions of the Galaxy, the abundance of heavy elements, such as iron and elements with higher atomic number, is lower. But such depletion is unfavourable for the formation of terrestrial planets, where the emergence of life is likely to be more probable.

The journey to an extra-solar planetary system is far beyond our current technological capabilities and likely to be extremely risky. The enormous distances that must be travelled require very powerful and reliable means of spaceship propulsion and means of survival aboard over long periods of time. Since interstellar travel at speeds much higher than a few percent of the speed of light is improbable with foreseeable technologies, a journey to stars with planetary systems likely to be interesting would take decades or centuries. To solve this problem, we can imagine generation starships with sufficient people on board to ensure that the descendents of the original astronauts would reach their destination.

Another possibility is to use suspended animation techniques that would allow astronauts to have their life processes slowed down for several years and be safely reanimated when required. Suspended animation is at present very far from being sufficiently developed to be used on people. Experimentation is currently being carried out on animals with cryonics and chemically induced hypothermia.

Another problem with long journeys through interstellar space is the risk of being hit by high energy galactic cosmic rays, made up of protons, alpha particles, nuclei of heavier elements, and electrons. Their energy can reach 10^{21} eV, which is much higher than the highest energies of about 10^{13} eV achieved in the highest energy particle accelerators currently available. On Earth we are naturally protected from cosmic rays by the atmosphere. They collide with the nuclei of atmospheric components, producing a shower of harmless lower energy particles, instead of hitting our bodies directly. The problem is more difficult to solve on starships, which have to be hardened to resist cosmic rays. This type of protection increases the mass of the starship and the amount of energy required to propel it to the chosen destination.

Human biological limitations in the context of space colonization may be overcome by transhumanism, replacing *Homo sapiens* by *Machina sapiens*. If we obtain a complete understanding of the operating mechanisms and organizational structures of the human brain, we might attempt to reconstruct it, but replacing organic components by molecular nanostructures. A non-biological brain with much reduced dimensions could be associated with a robot, which could achieve autonomy and sustainability, while retaining a form of intelligence and consciousness similar to our own. This type of android, heir to the human civilization, would be better adapted to interplanetary and interstellar journeys and have a greater chance of survival in extreme interstellar space environments and in other planetary systems. Another imaginable pathway to create androids would be to use synthetic biology to create intelligent living beings with superhuman capabilities. These may conceivably end up populating the whole Milky Way, although we are now in the realm of speculation. In this perspective, trans-humanism may indeed have a future.

4.15 The Long Final Path of the Universe

Let us return to what we can foresee with much less uncertainty. What will be the future evolution of the Milky Way and the Universe? Our galaxy is part of a Local Group of galaxies, with a diameter of about ten million light-years, which includes Andromeda, Triangulum, and at least 35 others of smaller size, mostly gravitationally bound to the three big ones. The first recorded observation of Andromeda was made by the Persian astronomer Abd al-Rahman al-Sufi in 964 AD. He described it as a small cloud. Visible in the constellation of the same name, it is actually a spiral galaxy with comparable dimensions to the Milky Way (Karachentesev, 2006). The two galaxies are about 2.5 million light-years apart, but on a collision course, approaching one another at a rate of 100 to 140 km/s, and they are likely to collide some 3 000 million years from now, to form an elliptical galaxy, which has been called Milkomeda (Cox, 2007).

At that time, it is unlikely that there will be any complex life forms on Earth to witness this cyclopean collision. Possibly bacteria will be the only remaining living organisms. The constellations in the night sky will be much changed, but the collision is not likely to be disastrous for the Solar System, because the density of stars in the colliding galaxies is low, and the process will be slow, taking about one billion years. Nevertheless the whole Solar System may be pulled away to the outer regions of the elliptical galaxy or even ejected into intergalactic space.

During the galactic collision, the rate of star formation will increase significantly, and it may become possible to witness a supernova event each year, whereas at the present time there is on average only one every century in our galaxy. Furthermore the massive black hole that will form at the centre of the new elliptical galaxy may gobble up a sufficient mass of gas to transform it into a quasar. In that case, the voracious black hole will become a luminous object that outshines the entire surrounding galaxy. The centre of Milkomeda, now a quasar, would become the brightest object in the Earth's daytime sky.

Collisions between spiral galaxies to form elliptical galaxies are a frequent event in the Universe. These mergers end up reducing the rate of star formation which happens mostly in the arms of spiral galaxies. Eventually the formation of new stars will cease due to the increasing scarcity of molecular hydrogen clouds. Meanwhile the existing stars will reach the end of their lives, becoming white dwarfs, neutron stars, or black holes. The greater part of the matter that makes up the atoms and molecules, called hadronic matter, will be in a degenerate state. We know that matter and energy are two forms of the same concept, that matter can be transformed into energy and vice versa, and also that the total amount of energy in the Universe is conserved. It is estimated that hadronic matter represents only 4% of the total energy of the Universe, whereas dark matter, whose nature remains unknown, represents 22%. The remaining 74% is dark energy, responsible for the accelerating expansion of the Universe (Spergel, 2003). As time goes by, the distance between galaxies is increasing and the Universe will contain ever more free space, sparsely populated with degenerate stars and black holes.

It is well known that an isolated system, formed by a large number of particles held together by the gravitational force will disintegrate over a long period of time, slowly ejecting the particles into the surrounding space. Computer models indicate that the planets of the Solar System will eventually dissociate themselves from the Sun and evaporate into interstellar space. The same type of phenomenon will happen with galaxies over a very long period of time, on the order of 10^{19} years. Degenerate remains of stars, once luminous, and black holes of various masses will slowly free themselves from the gravitational forces exerted by the galaxies they once belonged to, and scatter throughout the immense free space around them. Various types of physical phenomena will still occur, but at ever slower rates. Progressively, the Universe will become more inhospitable, empty, cold, dark, and inert. In the far distant future, time will dissolve and its meaning will be lost forever.

Chapter 5
Human Development and Environmental Discourses

5.1 Human Development and the Dilemma of Continuous Economic Growth

After visiting the far distant future, it is time to return to the reality of the current problems regarding human development and the environment from local to global scales. The human development concept used here encompasses all its aspects but is mostly centred in the educational, social, political, and economic components. There is a strong, although far from universal, convergence of opinion to the effect that there are troublesome difficulties at the present time concerning global economic growth and its impact on the environment. This fairly generalised perception is an important step in defining the problems we are facing, from which it should be possible to construct appropriate strategies to address them. A point to bear in mind from the outset is that the challenges of human development and the environment are closely related and hence inseparable.

The essential characteristic of the social and economic development models adopted by contemporary societies is their foundation on robust and continuous economic growth based on industrialization. Economic growth is assumed to be a necessary condition to assure a better quality of life, social development, and lasting prosperity. Countries unable to achieve significant economic growth have a much slower rate of social development and tend to become socially unstable and ungovernable.

Improvement in the quality of life, especially in developed countries which already have high standards of living, is widely identified with an increase in the capacity to consume, access to ever more sophisticated goods and services, greater availability of free time, and a continuously increasing mobility. There are profound psychological motivations for consumption. The impulse to acquire and to possess is an essential behavioural trait of our heritage as a biological species, spontaneous, immediate, irrepressible, and insatiable. Consumption favours the exercise of our basic psychological tendencies to construct an identity and to establish differences in social status and power relative to others, while promoting self-esteem.

Novelty is a very important factor in consumption because it carries important information about status. Thus the profit motive stimulates a permanent search by the economy for more innovative and better products. The basic drive to consume constitutes the fundamental basis for the economic growth paradigm. Political history since the industrial revolution shows that the most diverse ideologies, from liberalism to Marxism, have advocated economic growth through industrialisation, albeit along different paths and with different emphasis.

The paradigm of continuous economic growth is fiercely competitive and unequivocally favours the countries that are more successful in its implementation. It generates a global dynamics of human development, involving different forms of integration for its social, political, and economic components, and this dynamics is increasingly dominated by the emerging economies, such as the BRIC countries. Furthermore it contributes decisively to the process of globalization, while at the same time the forces it creates in the global economy are at the core of the current problems regarding human development and the environment.

There are many signs that continued worldwide acceptance of the economic growth paradigm involves an increasing risk of environmental degradation and instability. Furthermore, it has been argued that the 2008–2009 crisis proved that the paradigm is also financially and economically unstable. In fact the growth imperative was in part responsible for relaxing financial regulations and promoting the development of the complex financial derivatives that became toxic. The speculative expansion of credit for the housing and commodities markets was deliberately used as a mechanism to stimulate economic growth.

On the other hand, if the economy does not grow, unemployment increases, output falls, the ability to service public debt diminishes, the economy enters into a recession, there is social instability, and people lose their quality of life and security. Apparently, continuous economic growth is the only mechanism available to prevent collapse. We are forced to conclude that both economic growth and degrowth are unsustainable. So what is the solution to this dilemma? Before addressing this question, let us consider the various types of environmental discourse.

5.2 The Origins of the Environmental Movements

Environmental concerns were only recently identified and expressed clearly, more precisely in the second half of the 19th century. Nevertheless, since the middle of the 18th century, some natural philosophers and statesmen showed interest in the preservation of nature, adopting views that were to become the precursors of the environmental movements. One of the most striking figures in this group was José Bonifácio de Andrade e Silva (1763–1838), a naturalist, poet, and statesman, and the patriarch of Brazilian independence. He was born in Santos, Brazil, and went to Coimbra University where he became professor, then Secretary of the Lisbon Academy of Sciences, and was well known throughout Europe for his scientific work, especially in mineralogy.

Influenced by the Enlightenment and by other European naturalists, he developed in his works a harsh criticism of the predatory use of natural resources, with a special emphasis on deforestation. When he returned to Brazil in 1819, he defended and tried to implement in his country an integrated and sustainable form of development, especially in agriculture, forestry, and fishing, through better practices in the use of natural resources. His goal was to achieve a balance between human activities and nature, and he clearly identified some of the adverse consequences of breaking such an equilibrium (Pádua, 2004).

The roots of environmentalism in the USA are associated with George Perkins Marsh (1801–1882), David Henry Thoreau (1817–1862), John Muir (1838–1914), and Gifford Pinchot (1865–1946), and resulted from a preoccupation with the impacts of intensified natural resource use, particularly deforestation and damming of rivers, and from the will to protect the magnificent beauty of nature, especially in the west of the country. There were two key strands in the early environmental movement. The protectionists, such as George Marsh, David Thoreau, and John Muir were primarily concerned with the protection of nature in its pristine state, while the conservationists such as Gifford Pinchot were mainly interested in the sustainable management of natural resources.

John Muir was born in Scotland and his family immigrated to a farm near Portage, Wisconsin, when he was eleven years old. He enrolled at the University of Wisconsin in Madison, where he learned some botany and geology, but did not graduate. At the age of 26 he left school and went to Canada, wandering in the wilderness near Lake Huron, collecting plants. Following a serious accident, where he nearly lost his sight, he was determined to "be true to myself" and follow his dream of exploration and the study of plants (Wolfe, 1945). He then walked for about 1 600 km from Indiana to Florida, through the wildest, leafiest, and least trodden ways he could find and went on to explore the Yosemite in California. There, in 1871, he met the naturalist Ralph Waldo Emerson, who was delighted to finally meet the prophet-naturalist he dreamed about and offered him a teaching position at Harvard. John Muir declined and later wrote: "I never thought of giving up God's big show for a mere profship" (Tallmadge, 1997). His studies and activism were crucial for Congress to establish the Yosemite and Sequoia National Parks in 1890. Amazed by their pristine nature, almost free from the transformations and impacts of human activity, he dedicated his life to preserving it.

The conservationist movement founded by Gifford Pinchot had distinct concerns. Its first objective was to ensure that natural resources such as forests, water, energy, and minerals were used in a controlled way so that they could support a growing economy, and never exploited in an irresponsible or unsustainable way. There is a fundamental ideological difference between these two types of pioneering ideas for the environmentalist movements. While the protectionists wanted nature to be set aside and protected for its own sake, the conservationists focused on the relation between man and nature, seeking forms of economic growth that were compatible with conservation of the environment.

In Europe, the origins of the environmental movements were centred on a reaction to industrialization, urbanization, and air and water pollution. The first clear example of confrontation between an organized protectionist movement and the defenders of social change and economic growth at the local level took place in 1876, when the Thirlmere Defence Association tried to block the building of the Thirlmere reservoir in a valley of England's Lake District to supply water to Manchester (Ritvo, 2003). The point of view of those advocating the building of the dam prevailed in parliament, but environmental protection in the Thirlmere reservoir and surrounding countryside and throughout the whole of the Lake District became a very important objective that has been granted ever-growing support right up to the present day.

This sequence of an initial conflict, followed by a resolution in favour of economic growth associated with an increased awareness of environmental problems has been very typical and frequent, especially in the developed countries. In most developing countries the situation is different, as the predominance of problems regarding food security and social and economic issues leads to a lower sensitivity to the environment, and a reduced capacity to prevent its degradation.

5.3 The Environmental Discourses

What kinds of environmental discourse address the current situation and its projection into the future at the local, national, regional, and global levels? An environmental discourse is a coherent and shared way of apprehending the world, its environmental problems and the possible solutions to these problems (Dryzek, 2005). Each type of discourse enables those who subscribe to it to interpret and process the relevant information and to construct coherent accounts of the state of the environment and future scenarios at various scales, so that they can form substantiated opinions and plans of action. They are universal in the sense that we are all affected by the environment and by its problems. We all have the responsibility to obtain reliable information, study, analyse, debate, formulate opinions, and adhere to a coherent point of view regarding those problems. To deny or ignore the current environmental problems is also a discourse, but of an antithetical nature.

Michel Foucault (Foucault, 1980) has clearly shown that political discourses embody power since they condition the way we perceive, act, and defend what we consider to be our interests. With environmental discourses, the situation is analogous. The power of an environmental discourse is often revealed by the capacity it has for attracting followers and by its efficiency in achieving given environmental objectives and making these objectives compatible with others of a social, economic, political, or institutional nature. The evolution of environmental policy at the various scales is strongly influenced by the substance and dynamism of the most powerful environmental discourses.

Success in the application of environmental policies depends largely on the institutional capacity to implement them, the quality of the environmental laws and regulations, the degree of compliance, the level of environmental education and training, and the awareness and willingness of the public sector, businesses, and citizens in dealing with and contributing to the solution of environmental issues.

It is possible to identify a wide variety of environmental discourses throughout human society, regarding both content and form of expression. They range from the sophisticated intellectual circles of the more industrialized countries to the hundreds of millions of people that live in the ever-expanding slums of large urban areas, especially in the developing countries, and the last remaining indigenous populations in remote and nearly pristine areas, who still maintain primordial forms of relationship between human life and the natural environment. All these people, in fact all of us, have an environmental discourse, because we depend on the natural resources provided by Earth subsystems: air, water, soils, oceans, terrestrial and marine ecosystems, fossil fuels, and ores. We all have a more or less structured and cultured opinion about the subject. Across the broad range of social groups, there are common concerns shared by the greater part of the world's population, but they manifest themselves in very different ways. To promote information, communication, dialogue, and sharing of experiences between these manifold perceptions is a very important contribution to the creation of a well-founded and diverse global conscience, with sufficient strength to face current and future environmental challenges.

A person's professional activity often has an influence on the choice of environmental discourse and these correlations tend to become stronger with the industrialization of the society in which they are imbedded. If, for example, the environmental discourses of biologists and economists are compared, one is likely to find significant differences in emphasis on subjects such as the negative consequences of biodiversity loss, the vulnerability of ecosystems, and the sustainability of their services, or the capacity that market mechanisms have to solve the scarcity of a given natural resource, with the help of science and technology.

In a given environmental discourse it is possible to distinguish an interpretive and an active component. The former refers to the analysis and interpretation of the current state of the environment, along with the trends and projections of future evolution, while the latter refers to action programs that should be implemented in society on the basis of the interpretive component. Coherence requires that the active component of discourses be in accordance with the behaviours and attitudes of its practitioners toward the environment. However, there is often a significant gap between interpretation, commitment, and action. This inconsistency partly results from the deep-rooted and often subconscious association between the current concept of prosperity and the unquestionable paradigm of economic growth.

Environmental ethics is the discipline that studies the moral relationship of human beings to, and also the value and moral status of, the environment and its nonhuman contents (SEP, 2008). In this field it is very important to distinguish between instrumental and intrinsic values. One of its main challenges is to find rational arguments to assign intrinsic value to the environment and to its nonhuman contents, for instance to a wild flower with no instrumental value. Many traditional Western ethical perspectives are anthropocentric in that they only assign intrinsic value to humans, or a much greater intrinsic value to humans than to any nonhuman beings.

The human-centred tradition was already clearly present in Greek civilization. Aristotle expressed it by saying that "nature has made all things for the sake of man" (Politics, Book 1, Chap. 8), which implies that the value of nonhuman things is merely instrumental. The essential connection between man and the environment is also reflected in the fact that an environmental discourse is always associated with a certain way of apprehending and assessing the questions of human development throughout the world. Thus an environmental discourse is simultaneously a discourse about human development.

5.4 The Limits Discourse and the Difficulty to Act

The assessment of the state of the environment at the various spatial levels, along with current trends and future projections, varies significantly between two extremes. At one end there is the limits discourse, also known as the survivalist discourse, and at the other the Promethean discourse. The first was initially formulated in a detailed way by the *Club of Rome* in *The Limits of Growth*, published in 1972 (Meadows, 1972). The projections of this report indicate that, if the contemporary growth tendencies of production, consumption, environmental degradation, and world population were to prevail, disastrous crises would become inevitable within the coming 100 years, that is, up to around 2070.

These results, obtained through computer models, were at the time interpreted as an update of the discredited Malthusian catastrophe, and heavily criticised. On the one hand they were considered to be relatively obvious, because approximately exponential growth in consumption and world population cannot be indefinitely sustainable in a finite system. On the other hand the timing of the presumed crises was regarded as erroneous, because the role of price mechanisms in the control of consumption and the role of science, technology, and innovation in the calculation of the theoretical expiry limits for the various natural resources was judged to be underestimated.

Nevertheless the book had an enormous influence and was the first to clearly characterize the limits discourse. The authors also presented an alternative model of a global economy in a stationary state, with stabilised consumption and world population, but they did not fully explore the means to achieve that goal. Meanwhile the limits discourse evolved and became much more sophisticated and credible (Meadows, 1992; Brown, 2003; Ehrlich, 2004; Meadows, 2004; Constanza, 2007). A 2002 report from the US National Academy of Sciences (Wackernagel, 2002) concludes that the total human demand on ecosystem services provided by the biosphere has already exceeded its carrying capacity.

More recently, a comparison of historical data for the 1970–2000 period on industrial and food production, pollution, and environmental degradation with the scenarios presented in *The Limits to Growth* accords with the key features of the business-as-usual standard run of the report, which results in the collapse of the global system around the middle of the 21st century (Turner, 2008). This result can be interpreted as indicating that we should move away from the current growth pattern of business-as-usual and seek a steady state economy.

The key idea in the limits discourse is that sooner or later the present pattern of economic growth will inevitably lead to critical situations and an eventual collapse, because it relies on unsustainable levels of consumption of natural resources and interference with the Earth subsystems, especially the biosphere. The specific nature, incidence, and spatial extent of these crises is uncertain, although they are very likely to cause high levels of hunger, destitution, morbidity, suffering, and misery, and a significant global decline in the quality of life. They do not constitute a serious threat to the survival of humans on Earth, but they may generate violent conflicts that seriously increase the mortality rate.

The active component of the survivalist discourse was initially characterized by arguments proposing radical action, such as intervention and controls established by governments and by the scientific elite to coercively abandon the paradigm of continuous economic growth. According to its advocates, only in this way would it be possible to avert the so-called tragedy of the commons that consists in the eventual collapse of limited common resources due to conflicts between individual and collective interests (Hardin, 1977; Hardin 1993).

According to Robert Hellbroner, the only hope to cure humanity's profligate ways and avoid the serious critical situations that will arise when the voracity of consumerism far exceeds the carrying capacity of the Earth systems would be monastic forms of government, capable of combining "religious orientation with a military discipline" (Hellbroner, 1991). More moderate attempts to find an active survivalist discourse rely on the democratic process and on the capacity of people and their non-governmental organizations to overcome the business-as-usual growth path (Brown, 1992).

The globalized nature of the limits discourse combined with the profound inequalities of social and economic development and per capita use of natural resources make it difficult to construct an active discourse that is coherent at the local, national, and global levels. The main difficulty is to formulate a program of moderate action directed specifically at uprooting the paradigm of business-as-usual continuous economic growth. The active survivalist discourse has a very important environmental component and justification, but it should also address the economic and political questions of development and quality of life in a very unequal world that is becoming increasingly polarized economically (Dumont, 1973). If it does not try to meet this challenge, the survivalist discourse will be unable to provide a credible active component, despite its increasing relevance.

Anthropogenic global warming exemplifies this problem. According to some authors (Rockstrom, 2009), the safe planet boundary for atmospheric CO_2 concentration is 350 ppmv and this limit has already been largely exceeded. However, it is very difficult to achieve world agreement for a significant reduction in the global emissions of greenhouse gases because of the large disparities of historical and present-day emission responsibilities between the developed and developing countries. Without addressing this question in an effective way, it is very unlikely that we will be able to solve the problem of climate change.

5.5 Success of the Promethean Discourse
and Its Dependence on Energy

The limits discourse was fiercely opposed by the defenders of the current paradigm of economic growth after the publication of *The Limits of Growth* (Meadows, 1972). The opposing discourse, frequently referred to as Promethean or Cornucopian, is based on a practically unlimited confidence in the ability, determination, and inventiveness of humans to solve all the problems created by the unrelenting pursuit of economic growth, including those of an environmental nature.

In fact the major concern of virtually all governments is to promote economic growth. It is considered essential for development and the key to promoting political and social liberalization throughout the world (Friedman, 2005). It tends to increase wealth, income, profits, and employment, and also to produce a greater number of houses, cars, domestic appliances, energy and communications equipment, number of miles travelled per person, gadgets, toys and so on. The political and economic discourse is presently based on the virtues of growth, with moderate concerns of an environmental nature, mostly in the industrialized countries. News reported by the media that covers the economy is always made with the underlying assumption that growth is good. Environmental problems are recognized and addressed, but they are usually considered to be disconnected from growth, under the conviction that they are secondary and that it will always be possible to solve them in a way compatible with continued growth.

The arguments against the limits discourse were developed mainly by economists, notably Wilfred Beckerman (Beckerman, 1974; Beckerman, 1995) and Julian Simon (Simon, 1981; Simon, 1984; Simon, 1996). These authors claim that energy, natural resources, and commodities in general will continue to be available and have accessible prices for consumption in such a way that economic growth can be supported indefinitely. Questioned as to how far into the future, Simon replied (Simon, 1984): "We expect this benign trend to continue until at least our Sun ceases to shine in perhaps seven billion years, and until exhaustion of the elemental inputs for fission (and perhaps for fusion)." This is a bold statement that disagrees with current models of Solar System evolution and the long term duration of life on Earth.

In 2001, Bjorn Lomborg, a political scientist and former professor of statistics in the Department of Political Science at Aarhus University in Denmark, published a very successful book entitled *The Sceptical Environmentalist* (Lomborg, 2001), thereby becoming one of the most visible leaders of the Promethean discourse. Like Simon, he opposed the survivalist discourse and most of the theses of the environmentalist movement, defending the idea that natural resources, energy, and food are becoming more abundant and accessible, while the world environment has been improving. He also claimed that the quality of life indicators show generally positive trends worldwide.

The book was received ecstatically by the majority of the media with stronger links to the economy, industry, and government, in many developed countries, and widely used to alleviate the pressure applied by environmentalists. *The Economist*, in its issue of 6 September 2001, considered that: "This is one of the most valuable books about public policy — not merely on environmental policy — to have been written for the intelligent general reader in the past ten years. *The Sceptical Environmentalist* is a triumph." Some US media figures claimed that its publication marked a critical turning point that would relegate into oblivion the limits discourse of Paul Ehrlich and Lester Brown, and initiate a new age of eco-optimism.

However, the book did not stand up to critical analysis by several scientists, some of whom had publications cited in it. Its scientific credibility was placed in doubt, and *Scientific American* dedicated one of its issues (January 2002) to debunking some of Lomborg's arguments and conclusions. The author was criticised for misleading use of statistical methods, selective presentation of evidence, and inadequate treatment of the uncertainties associated with complex systems. Faced with several complaints, the Danish Committee on Scientific Dishonesty ruled in January 2003 with a mixed message stating that the book did not reach the standards of good scientific practice, but that the author could be excused because of lack of expertise in the fields in question.

However, in December of the same year, the Danish government revoked that ruling, and in April 2004, Lomborg was designated by *Time* magazine as one of the 100 most influential people in the world. Lomborg, now a professor at the Copenhagen Business School, has long opposed international curbs on greenhouse gas emissions to combat climate change, saying that they would be too expensive and not cost-effective. Recently, however, he reversed his opinion in a new book (Lomborg, 2010), arguing that global warming is a challenge that humanity must confront and proposing the investment of 100 billion US $ a year in mitigation.

The Promethean discourse looks more like a human development discourse, while the limits discourse is more oriented towards environmental issues. Furthermore, the fundamental entities and assumptions of the limits and Promethean discourses are quite different. The first is based on the concept of ecosystem, the principle of sustainability of ecosystem services, and the finiteness of natural resources, both renewable and non-renewable. The second tends to ignore the concepts of ecosystem, nature, and natural resources and focuses on the abstract physical concept of matter, which is assumed to be transformable into whatever product we may need, given enough energy to do it. It defends the idea that it is possible to produce all the required resources for continuous economic growth by means of transformation processes applied to brute matter.

Thus energy and access to energy play an essential role in the Promethean discourse. The discourse is based on the belief that human ingenuity and perseverance in science, technology, and innovation will find the primary energy sources needed to replace scarce natural resources by suitable substitutes and to depollute the environment. Energy is thus the key to business-as-usual continuous economic growth. Private greenhouses claim that, with access to enough energy, it will always be possible to obtain sufficient water by desalinization, to transform arid or hyper-arid lands into cropland, and to produce all the required scarce elements by the transmutation of the more abundant ones through nuclear reactions in appropriate accelerators, for instance, producing copper and nickel from iron.

All this is in fact possible with our scientific and technological knowhow, but to do it on a large scale would require huge amounts of investment and energy. Pollution in the air, water, soils, and oceans is regarded as just matter in undesirable forms and places. Hence, theoretically, it can be transformed or removed by the skilful application of appropriate technologies. This program, although feasible, would be very expensive and therefore inaccessible to the vast majority of countries in the world. But confidence in science and technology, provided mainly by the success of the industrial and energy revolutions, has persuaded people to believe in the Promethean discourse.

In contrast to the survivalist discourse, the Promethean discourse has an active component with a long history of success (Friedman, 2005). Its practical application flourished throughout the industrial revolution, especially since the end of World War II, with the growing domination of capitalism and globalization. The main function of the political and institutional systems in the democracies of the developed countries has been to facilitate the conditions for economic growth, on the assumption that it will always lead to an improvement in the prosperity and quality of life of their citizens.

Although the Promethean discourse is widely used throughout the world, it is in the USA that it has been most visible and active. The decision of the George W. Bush administration not to sign the Kyoto Protocol on climate change mitigation, on the grounds that this would slow down US economic activity, is a good example of the prevalence of economic concerns over environmental ones. The underlying assumption is that it is preferable to adapt, and maybe develop geoengineering to combat climate change, rather than to mitigate, since this response involves the risk of negatively affecting the economy. There are many other examples where, at an international level, especially in the environmental organizations within the United Nations system, the USA has preferred to privilege economic interests over environmental ones.

Between the more extreme expressions of the limits and Promethean discourses, there is a whole spectrum of more moderate discourses. The current trend is that the limits discourse is mainly regarded as an environmental discourse with an increasingly influential interpretive component, and a much weaker action component that tends to be radical. On the other hand the Promethean discourse is mainly regarded as a prosperity discourse with an interpretive component considered often to be unrealistic and an action component that enjoys a great success. Its success in developed countries has often led to increased environmental degradation in the developing world. In part, the improvement of environmental conditions in the more industrialized countries results from exporting some of the more polluting industries to developing countries where environmental legislation and pollution controls are weaker.

5.6 Radical Environmental Discourses: Deep Ecology and Social Ecology

The interpretive component of the limits discourse is probably the main inspirational source for the radical environmental discourses. These question or reject the presuppositions, values, beliefs, and power structures of industrialized societies, and the way in which environmental problems are addressed by them. They are radical and imaginative discourses seeking new ways of life that are less aggressive towards the environment. Because of their creative and innovative character, they propose a very wide range of analyses and solutions. This diversity tends to generate lively debates and controversies.

Regarding politics, their views are graded from complete rejection of the social, economic, and political systems, to a collaborative and participative attitude toward the institutions and political activities of the democracies in industrialised countries. There are two main trends with regard to the active component of the radical discourses. One aims at raising people's awareness and modifying the way they think and behave in relation to the environment. The other prefers a direct intervention on institutions and on political and economic activities. These two tendencies are not mutually exclusive: some movements strive to combine behavioural change with politics.

Deep ecology is a movement that practices one of the more radical environmental discourses. The name and initial formulation was established by Arne Naess, a Norwegian philosopher born in 1912 and strongly influenced by Spinoza, Buddha, and Gandhi. The fundamental principles of deep ecology (Naess, 1989) are based on self-realization, through the identification with a larger organic unity embracing all nature's ecosystems, and on the idea of biocentric equality. According to this concept, there is no species that can be considered more intrinsically valuable or in any sense higher than another, humans being no exception to this rule.

This is indeed a point of view that is radically opposed to the anthropocentrism that results from acknowledging the domination of all species by humans. Deep ecology vigorously defends the idea that the preservation and expansion of broad areas of wilderness, in pristine condition or only slightly modified by human presence, is vital to ensuring ecological integrity, the preservation of biodiversity, and the possibility of biological evolution. Reducing the human population is considered essential, but there are no plans of action to solve current environmental problems, such as pollution in most megacities and the environmental degradation in many areas of the world. The deep ecology discourse does not address the contemporary problems of human development and how to solve them in a way that might be compatible with the preservation of the environment. Some of its more extreme forms adopt manifestly utopian and misanthropic positions (Manes, 1990).

The main philosophical opponent of deep ecology in US radical environmental circles is social ecology, a movement established by Murray Bookchin (Bookchin, 1982). Contrary to deep ecology, it emphasizes the social dimension of human life and claims that the main problem in human societies is not the relation with nature, but their strong hierarchical structure considered to be an unnatural phenomenon without counterpart in the nonhuman living world. It points out that the relations that we perceive as competitive in nature are in fact cooperative, always involving some benefit for the competing actors. Human societies do not conform to that pattern according to the eco-anarchist social ecology movement.

Bookchin, a long time socialist and ecologist, interprets the current ecological crisis as a consequence of modern global capitalism. Social ecology proposes an anarchist solution of radical municipalism, where small self-sufficient local communities live in harmony with each other and with the environment. This may be regarded as a utopian proposal, when confronted with the size, complexity, and interdependence of urban areas all over the world. However, when compared with deep ecology, it has the virtue of indicating a social and political strategy.

5.7 The Environmentalism of the Poor

Deep ecology and social ecology are radical environmental discourses that have emerged in the developed countries and that have the majority of their followers in the USA, Canada, Europe, Australia, and New Zealand. To have a balanced view one should also consider the environmental discourses arising in the developing countries, in an ever more vehement and frequent manner, and which have been called the 'environmentalism of the poor' (Guha, 1997). In a large number of developing countries, local populations suffer from, and some protest against, the degradation of the environment and their livelihood conditions.

The main problems result from deforestation, overexploitation of water resources, unsustainable forms of agriculture, soil erosion, desertification, implantation of industries that produce high air, water, and soil pollution, and the relocation of populations caused by the installation of large scale projects, like dams, extensive agricultural and livestock farms, oil exploration and extraction, and mining. The environmental arguments used by the affected populations are relatively simple and spontaneous. They are based on empirical knowledge derived from a long past experience of equilibrium with the environment. The agents in such conflicts do not usually see themselves as environmentalists, but they are aware that the degradation of the environment is threatening their livelihoods.

Well known examples are the fight for the preservation of Amazonia by rubber tappers led by Chico Mendes, and the struggles of the Ogoni, the Ijaw, and other ethnic groups in the Niger Delta against the environmental damage caused by oil extraction conducted by Shell. These protests combining economic and environmental concerns, frequently against large multinational corporations, tend to be readily associated with anti-globalisation movements. When they manage to call the attention of international non-governmental environmental organizations and the media, the resulting negative publicity for the corporations can lead to a solution or to the mitigation of the problems. However, in most cases, the protests of rural populations concerning the degradation of their environment are seldom heard and rapidly forgotten.

Deep ecology and social ecology are very far from being able to provide solutions to these problems. Furthermore, some of the characteristic objectives of deep ecology may go against the interests of the populations in developing countries. The preservation and expansion of huge areas of wilderness may seriously endanger the livelihoods of the indigenous populations that live in and around these areas. A well known example occurred in Kenya at the beginning of the 1990s, when the Masai, Turkana, and Ndorobo tribes were forcibly displaced from their lands to build a game reserve for the protection of elephants (Haynes, 1999).

More recently, such trade-offs have started to be addressed by emerging markets for ecosystem services, such as biodiversity protection, which can help mitigate the development impacts of conservation. These payment systems for biodiversity can take several forms, such as biodiversity offsets, habitat credit trading, biobanking, and a growing number of conservation standards or certificates. The main idea is the preservation and creation of protected areas in the biodiversity hotspots of developing countries through public and/or private partnerships and investments from developed countries, including compensation mechanisms for the indigenous populations.

Another promising mechanism is to promote responsible eco-tourism. Nature in high biodiversity wilderness areas, especially in the tropical regions, is a wonderful and ever rarer spectacle that attracts the curiosity and admiration of a growing number of people. To visit these sanctuaries of wild natural beauty, which still exist mostly in Africa, Asia, and the Americas, is an opportunity to rediscover the ecological foundations of the primordial human communities and feel the magnitude of our cultural evolution since that time.

5.8 Eco-Feminism

In contrast with the deep ecology movement, eco-feminism considers that the root of the environmental problems does not lie in anthropocentrism, but rather in androcentrism. Françoise d'Eaubonne, born in Paris in 1920, proposed the concept of eco-feminism in her book *Le Feminisme ou la Mort*, published in 1974 (d'Eaubonne, 1974). The movement considers that there is a profound relationship between the prevalence of a society of a patriarchal nature, in which women are frequently oppressed, and environmental degradation and the destruction of nature. The eco-feminists defend the idea that the search for new and radically different cultural sensibilities could lead to the replacement of the relationships of domination, control, and transformation of nature by others more harmonious and ecological. Val Plumwood (Plumwood, 1995) goes a step further and defends that, while democracy is under the control of liberalism, it will not be possible for it to serve as a basis for the development of a truly ecological way of life for human societies. The current forms of liberalism are considered to be socially unjust and a threat to the environment.

Eco-feminism has a more universal following than the deep ecology and social ecology movements, as it is present in both developed and developing countries. In the latter, one of the most active voices is Vandana Shiva, who has a degree in theoretical physics and has dedicated her life to the defence of eco-feminism and the anti-globalization movements. She became well known in the Chipko movement of the 1970s for the protection of forests in the north of India, where people embraced trees so that they could not be cut down.

Shiva defends the idea that the current imperative of economic growth, associated with the overwhelming influence of science and technology in human development that started with the industrial revolution, tends to destroy the diversity of human life and its sacred aspects. She opposes scientific reductionism and proposes as an alternative the search for holistic forms of knowledge and the valorisation of traditional ways of life. She points out that in many developing countries women have a very important role in environmental protection. However, their contribution is overlooked and diminished by the implantation of new agricultural and industrial technologies in these countries, which frequently have negative impacts on the environment and on the quality of life of the populations.

She considers that biodiversity is intimately linked with cultural diversity and has campaigned against biopiracy, the illegal appropriation of biological material, from microorganisms to plants and animals, by a technologically advanced country or corporation without fair compensation to the peoples or nations where they were found. Regarding agriculture she criticises the 'green revolution' because of its negative impact on ecology, agriculture, politics, and social relations, particularly in the state of Punjab in India, and defends an organic model of agriculture. She is the founder of the Navdanya movement that trained more than 500 000 seed keepers and organic farmers in India. In ecology, she proposes solutions inspired by the wisdom and traditional knowledge contained in the Vedas (Shiva, 2002).

It could be argued that the generalized implementation of her proposals in developing countries, particularly in India, instead of the green revolution methodologies, would decrease grain production and increase the risk of malnutrition and hunger. It would also increase imports from the more industrialized countries, where production has significant subsidies. From this point of view, bearing in mind the continuous growth of the world's population, this approach would considerably decrease food security, and become rapidly unsustainable.

However, others would argue that organic farming is a solution to global rural poverty, although its implementation would require profound transformations in the prevailing international economic and trade systems. Identifying and implementing new forms of sustainable agriculture that are able to feed the world is clearly one of the greatest challenges facing humankind.

5.9 Environmental Discourses and the Use of Genetically Modified Organisms

Some radical environmental discourses share a romantic rejection of the Enlightenment principles when they emphasize that science and technology are the primary cause of the degradation of the environment and the continuous destruction of nature. However, the dominant tendencies in contemporary environmental radicalism accept the value of science and technology and search for a conjugation of human development with ecology, based on the protection of the environment, but not a return to the ideals of an outmoded romanticism (Hay, 2002).

It is significant that the radical environmental discourses have distinct expressions and values in the more industrialized countries and in the developing countries. The basis for the deep ecology discourse is the absolute valorisation of both the natural environment surrounding us and of our human nature, integrating its physical and spiritual expressions. The goal is to bring together and harmonize these two natures by changing sensibilities and values. This is an ideal and abstract individualistic objective that implies a discontinuity with the past but has no definite proposals on how to deal with the current problematic relationship between human development and the environment. In the developing countries, radical environmental discourses tend to be more pragmatic. Some propose the return to the environmental wisdom and practices of the ancestral civilizations, and they look with suspicion upon the contributions of science and technology to improve their livelihoods. However, without such contributions to enhance agricultural productivity, levels of malnutrition and hunger would be much higher.

The spectacular increase in grain production achieved by the so called green revolution, initiated in the 1940s, was only possible through the selection of high-yield varieties, mechanized agriculture, modern large scale irrigation systems, distribution of hybridized seeds, and the intensive use of synthetic fertilizers, pesticides, and herbicides. These requirements have negative collateral effects, such as the loss of biodiversity among species of agricultural value and among wild species, a greater dependence on fossil fuels needed for mechanization and manufacture of agrochemicals, flooding of fields due to water table rise caused by irrigation, soil erosion and degradation, increased salinity, perturbation of the natural nitrogen and phosphorus cycles, and pollution caused by fertilisers, pesticides, and herbicides.

Nevertheless global food production has grown remarkably with the green revolution, through scientific and technological advances, and through investments that depend strongly on the globalization process and on its continuity. More recently, since the early 1990s, genetically engineered, transgenic, or genetically modified organisms (GMO) have been used to produce food in many countries, but especially in the USA, Argentina, Canada, and China. GMOs are obtained through a biotechnology that allows the introduction of foreign genes into the genome. Genetically modified plants, such as rice, corn, canola, banana, tomato, cotton, and soybeans, have the advantage of being resistant to pesticides, herbicides, pests, or viruses.

The use of biotech crops and foods started a fierce and often polemic debate regarding possible health and environmental hazards. The main concern is that it may introduce new allergens, generate higher levels of toxins, and spread resistance to antibiotic. Furthermore, the use of GMOs in food production introduces gene pollution into the environment, which in the medium and long term may have irreversible negative impacts.

Environmental movements emphasize these risks and oppose the use of GMOs. However, up to now, there has been no undisputable evidence that the use of GMOs presents health risks greater than those associated with other breeding processes. GMOs may have medium to long term harmful effects which are probably impossible to identify at present, since they have been in use for less than 20 years. The governmental institutions of the countries that allowed the use of GMOs considered that the resulting advantages outweighed the benefits of a strict application of the precautionary principle as regards its potential medium and long term effects.

Another important aspect of GMO food production is that the intellectual property rights regime makes the improved seeds unaffordable to poor farmers in developing countries. Moreover, research in plant breeding is mostly oriented toward the requirements of farmers in developed countries rather than to those of the poor countries which involve a much higher number of people. It is important to emphasize that increasing food productivity does not necessarily solve the problem of malnutrition and hunger. If the poor are unable to buy food because of their very low income, increased production would not allay their destitution.

The introduction of new agricultural technologies without addressing the problem of a biased social and economic system that favours the rich against the poor in accessing technological benefits will not improve global food security. Redistribution of economic power, especially as regards access to land and purchasing power are essential to solve the current problems of malnutrition and hunger.

5.10 Eco-Theology

There is yet another type of radical environmental discourse, namely, eco-theology, which focuses specifically on the relationship between the current ecological crisis and the various religions and forms of spirituality. The basic idea is that the root of our arrogance, aggression, and disregard for nature that currently coexists with human development is a shortcoming of spiritual or religious nature, and so too must be the appropriate response.

The front line of the debate was initially established by Lynn White (White, 1967), who argues that the origin of the present environmental situation, critical in many respects, can be traced back to Judeo-Christian religious principles, which consider God above nature, with man created in his own image. God and man are placed outside nature, although both are God's creation. According to White: "Christianity is the most anthropocentric religion the world has seen" (White, 1967). God commands man to subdue the Earth and rule over nature, as stated in the Bible:

"God blessed them and said to them: fill the earth and subdue it. Rule over the fish of the sea and the birds of the air and over every living creature that moves on the ground" (Genesis, 1:28).

By eradicating animism, Christianity made it possible to exploit nature while being indifferent to the guardian spirits of trees, springs, streams, rivers, hills, and mountains. The new dualism between man and nature became stronger in the Christian Middle Ages and led to a long tradition of unlimited exploitation of nature. St Francis of Assisi is a remarkable but lonely exception, because of his profound love for animals, and was declared patron saint of the environment by Pope John Paul II in 1980.

The oriental religions, Hinduism, Buddhism, Taoism, and Shintoism, have no dualism between man and nature, in contrast with the Judeo-Christian ones. Divinity is not placed above nature, but nature and its very diverse manifestations have a divine value that must itself be venerated. Even though this closer relationship between man and nature strengthens respect for nature and therefore also for the environment, critical ecological situations are nevertheless relatively frequent nowadays in countries where the oriental religions are dominant. The beneficial influence of such religions has been unable to counteract the increasing risk of environmental degradation.

Eco-theology promotes a form of constructive theology for the well-being of humanity in harmony with nature and the cosmos. There is also a connection between this movement and the active component of the survivalist discourses. Some survivalists (Ophuls, 1977; Heilbroner, 1991) claim that the solution to the current environmental crisis requires a strong and enduring change of behaviour, which must be deeply motivated, as deeply as spirituality and religiousness. A mass conversion to new lifestyles would be needed, similar to those that in the past led to the growth and territorial expansion of the major religions, a conversion based on the conviction that the present paradigm of unlimited economic growth will become unsustainable due to the increasing irreversibility of its negative environmental impacts and the inequalities it generates. It would not be a movement of a religious nature, but a social phenomenon with a strong collective expression, only comparable with conversions moved by faith.

Clearly, for this transformation to be achieved at the personal level and reach its objectives, simultaneous structural changes would also be required at a macroeconomic level. It would be necessary to change the system that rewards and reinforces the more materialistic and egotistical forms of behaviour, change the social, economic, and employment structures that tend to isolate people and exhaust the free time people need in order to be well informed, to study, to understand, to act, and to create, and develop various types of active organizations in society at large.

In the current global economic system, most conflicts between the imperatives of the market, especially those resulting from the need to reassure the confidence of investors, and the defence of environmental values end up being resolved in favour of the former. Thus, for a conversion of the kind proposed by some of the more radical environmental discourses to be successful, it would be necessary to change those structures that now provide a relatively stable foundation for the established

economic order. This program is generally considered to be utopian and highly unlikely to be adopted in the near future. Still, it is also likely to be less utopian in
the middle and long term, as the unsustainability of the current global financial and
economic model of development becomes increasingly apparent.

5.11 The Green Parties

The environmental movements have been especially active at a political level since
the beginning of the 1980s. Since that time green parties have been represented in
the parliaments of various Western countries. The first were the Belgian green parties — Agalev, Flemish, founded in the 1970s by the Jesuit Luc Versteyler, born
out of a combination of progressive Catholicism and environmentalism; and Ecolo,
Francophone, founded later, in 1980. In the 1981 elections, these two parties won
seats in the Chamber of Representatives and the Senate. There is a wide variety
of green parties, although mostly adopt the four pillars of the German green party
as fundamental values: ecology, social justice, participatory democracy, and peace
and non-violence. Most of those that have significant parliamentary representation
are found in the democracies of Europe, North America, and Oceania. There are
few green parties in the developing countries and most of them have been organized with the help of their partners in the developed countries. Some are considered
clandestine organizations, such as the Green Party of Saudi Arabia.

One of the most influential is the German Green Party, founded in 1980, because
of its success as a social movement, its ability to attract votes in the elections, and
its record of political parliamentary intervention. It is divided into two tendencies:
the 'Realos', or realists, who believe in the need and capacity for effective action
in party politics, and the 'Fundis', or fundamentalists, who privilege the characteristics of a social movement over those of a political party, and denounce what they
consider to be the irrationality of the current political system. This division was
attenuated in 1998 when the Realos, led by Joschka Fisher, joined a coalition that
governed Germany until 2005. Curiously, during that period, Joschka Fisher was the
member of government who received greatest public support in the opinion polls.
One of the most striking achievements of the German Green Party at that time was
the 2000 decision to phase out the use of nuclear energy in Germany. A common
feature of environmental movements and in particular of green parties, throughout
their history, is a systematic opposition to nuclear weapons and nuclear power reactors.

The representation of the green parties in parliamentary politics remains very limited, almost always less than 10%, both in votes and in seats. They are very far
from acquiring a sizeable majority and leading parliament. This situation can be
interpreted as a signal that Western democracies welcome the parliamentary representation of the environmental movements, but are unwilling to give them more power because they believe that it would threaten economic growth, viewed as a *sine
qua non* condition for the improvement of well-being and quality of life. For the

other parties, especially those situated more to the left, the green parties represent a potential threat, due to their ability to attract votes.

Perhaps the most important contribution of the green parties is the implementation of environmental policies, usually in a weak form, by the parties in power, as a means of facing the 'green challenge' in the context of parliamentary and local elections. The presence of green parties in parliament and especially in coalition governments, as has already happened in Belgium, Germany, Finland, France, Ireland, Italy, Latvia, and Sweden, has served to influence and reorient the political debate towards a greater awareness of environmental problems, and consequently to a stronger commitment to solve them.

Recently, in September 2010, the German Green Party received a record support of 22% of the electorate in the polls, just two percentage points behind the Social Democrats (Der Spiegel, 2010a). However it remains to be seen whether this is a stable tendency or results mainly from the government decision to extend the lifespan of Germany's nuclear reactors by up to 14 years. In 2011, following the Fukushima nuclear disaster of March, the German government announced a plan to end the use of nuclear power by 2020, replacing it by renewable energy sources.

5.12 Merits and Limits of the Environmental Movements

Many supporters of the more radical environmental movements consider that engagement in parliamentary party politics is ineffectual, dispensable, or secondary compared with other more direct forms of action with greater impact on society and on the media. They prefer to organize activities at the level of local communities, including education and dissemination programs, demonstrations, sit-ins, boycotts, sabotage, and other actions that catch the attention of the largest possible number of people.

There is an enormous variety of non-governmental organizations in the area of the environment. Some act forcefully and with maximum visibility against activities they consider to have strongly negative environmental impacts or that lead to the destruction of nature. Greenpeace specialises in the fight against pollution of the oceans, the loss of biodiversity, whaling, deep sea bottom trawling, global warming, nuclear energy, the destruction of the rainforests, and the use of genetically modified organisms. Other organizations are more radical in their objectives and methods, such as Sea Shepherd, which seeks to protect the oceans and marine life, Earth First, and the Earth Liberation Front.

All have practiced 'ecotage', a word derived from the 'eco-' prefix and 'sabotage'. It consists in direct actions of civil disobedience and sabotage to defend the environment, like lying down in front of bulldozers about to destroy particularly valuable environmental assets, staying in the top of trees of primary forests that are about to be felled, sabotaging earth-moving vehicles, destroying roads opened for deforestation and burning SUVs in car showrooms. These environmental organizations are extremely careful to avoid human aggression, direct or collateral, and there are no examples of physical harm to people. Nevertheless, in some cases, economic losses have been inflicted. Because of such incidents, the US FBI classified the Earth Liberation Front as a terrorist organization, and imprisoned some of its members. However, the FBI has been less diligent in seeking out those responsible for acts of violence against radical environmental activists (Dryzek, 2005).

Over the past 30 years, the environmental movements have made important contributions to the development of a coherent critical and integrated analysis of the environmental, social, political, and economic situation at the local, national, and global scales. They have disseminated and strengthened an awareness of the major environmental problems worldwide. These achievements constitute one of the most important ideological advances to face the challenges of the deteriorating relationship between mankind and the environment, both today and in the future. In spite of their diversity the environmental movements converge when they attribute the ecological crisis to the current paradigm of a global economy that does not incorporate the protection of the environment. However, they are far from having the capacity to change that paradigm, at least in the short term future.

Regardless of the 2008–2009 financial and economic crisis, economic liberalism is ever more firmly implanted and expanding throughout the world. Environmental movements do not offer a competitive and coherent social, political, and economic alternative. Nevertheless, they propose various solutions with the common objective of developing active social and political structures to defend the environment, based on a participatory democracy that is much stronger at the regional and local levels.

5.13 The Origins of the Sustainable Development Discourse

What are the viable solutions leading to a transition to a sustainable global civilization? To answer this question, one must first analyse the origin and development of the conceptual and institutional aspects concerning the environment at the global level.

The United Nations Conference on the Human Environment held in Stockholm in June 1972 marked the beginning of international political awareness of global environmental problems. This was only possible through the recognition of the very strong relationship between human development and the environment, and the deep social and economic inequalities between developed and developing countries. Initially, the governments of the latter group considered the environmental concerns of the developed countries to be a luxury of the rich, compared with the urgent problems of hunger, poverty, disease, and lack of water, sanitation, and electricity in their own countries. In her address to the representatives of the 113 countries convened, Indira Gandhi, the Indian prime minister and the only national leader to participate in the Stockholm Conference, apart from the prime minister of Sweden, said: "Poverty is the worst pollution". One of the most significant results of the conference was the establishment of the United Nations Environment Programme.

Its headquarters were established in Nairobi, and its first executive director was Maurice Strong, a Canadian who chaired the Stockholm Conference.

Strong introduced the term 'eco-development' as a contribution to reconcile the desires of human development with the protection of the environment. Stockholm succeeded in placing environmental problems on the international agenda. Significantly, only two years later, the idea of building a 'sustainable society' emerged at an ecumenical conference on Science and Technology for Human Development, organized by the World Council of Churches (WCC, 1974). The most important aspect of these new ideas was to address simultaneously and coherently the need for equity among people and nations, the need for the sustainability of access to natural resources, and the need for the democratic participation in decision-making processes. Shortly afterwards, in 1980, the concept of sustainable development emerged in a document published by the International Union for Conservation of Nature and Natural Resources, defined as "the integration of conservation and development to ensure that modifications to the planet do indeed secure the survival and well-being of all people"(IUCN, 1980).

In 1983, the United Nations General Assembly created the World Commission on the Environment and Development, chaired by the Norwegian prime minister Gro Harlem Brundtland. The commission was asked to formulate "a global agenda for change". The first term of reference for the Commission's work was "to propose long-term environmental strategies for achieving sustainable development by the year 2000 and beyond". It is amazing how far away we are in 2011 from achieving that goal. Probably, even further away than in 1983. The commission's report, published in 1987, recognized that "many critical survival issues are related to uneven development, poverty and population growth. They all place unprecedented pressures on the planet's lands, waters, forests and other natural resources, not least in the developing countries" (WCED, 1987).

The report focused on the importance of the principle of intergenerational equity and defined sustainable development as "development which meets the needs of the present without compromising the ability of future generations to meet their own needs". This definition did not satisfy everyone and other definitions arose. It gradually became clear that sustainable development is not a concept of a strictly scientific nature which can be defined without ambiguities, because opinions differ on what precisely should count among the human needs that should be considered for the application of the principle of intergenerational equity. These can be categorized into the social, economic, and environmental realms, but the relative importance of the different components is a matter of opinion.

Sustainable development is a discourse used by distinct concerns and interests, sometimes contradictory. Those more concerned with the environment stress above all the need to ensure nature preservation, ecosystem services, and the sustainable use of natural resources. Some survivalists tend to adopt the sustainable development discourse and claim that its active component implies a non-growth economy. Those more concerned with the profound inequalities of human development and the needs of the poor consider that the priority for sustainable development must be to combat hunger, poverty, disease, lack of water, basic sanitation, commercialized energy, and institutional capacity for education and professional training, especially in the developing countries. Finally, those closely linked to business and to the economic sphere tend to emphasize the idea that sustainable development means primarily economic growth. This growth is considered essential to eradicate poverty and hunger, to sustain the growing world population and eventually to stabilize it.

5.14 The Challenge of Sustainable Development

The diversity of discourses that claim to implement the concept of sustainable development is not necessarily a disadvantage. Sustainable development is nowadays a meeting point for the debate about the state of the world and how to respond to the social, economic, environmental, and institutional challenges we are facing. An increasing number of people believe in and work toward sustainable development, although they may not know all the challenges and problems that must be overcome to reach it at the world level. They feel that it is the discourse that generates the greatest consensus, while being somehow inevitable on the global scale. For many it is an acceptable utopia that must be pursued in spite of all the odds. Sustainable development is nowadays a politically correct discourse that many politicians all over the world adopt and claim to apply.

The sustainable development discourse has its own frontiers that distinguish it from the limits and Promethean discourses. It recognises that there are ecological limits to growth, but contrary to the limits discourse, claims that economic growth can continue probably indefinitely if the right policies and measures are adopted. It emphasizes that the limits for the use of energy, water, land, and natural resources will manifest themselves in the form of rising costs and diminishing returns, and not in the form of sudden losses with critical consequences (WCED, 1987). The discourse assumes that it will be possible to satisfy the sum of all the growing needs of humankind, without specified limits and without changing the current fundamental social and economic models, through an intelligent and integrated management of human and natural systems.

This is a very ambitious proposition, especially if it is intended to be applicable in the distant future. The discourse endorses a vision where economic growth, global equity, population stabilization, peace, and environmental protection can all coexist. Critics point out that there is no demonstration of how to put such a program into practice and therefore that it is utopian. Some go as far as saying that there is an intractable contradiction in the juxtaposition of the words 'development', interpreted as implying continued economic growth, and 'sustainable', meaning the uninterrupted availability of natural resources in a protected environment. In any case, it should be remembered that sustainable development is a discourse, and not a concept that can be unambiguously defined.

The sustainable development discourse is distinguishable from the Promethean discourse because it recognizes the need to integrate the social, economic, and environmental components, rather than limiting itself to relying on human ingenuity and endeavour, supported by science and technology, to solve the problems of human development. The latter looks at nature as raw matter that can be transformed into whatever resources we may need, while the former acknowledges the complexity of the problems and searches for equilibrium between the requirements of social development, economic growth, protection of the environment, and nature preservation.

The greatest challenge of sustainable development is to bring the quality of life in the developing countries up to standards comparable to those currently enjoyed by the majority of people in the developed countries, while preserving environmental integrity. Let us try to evaluate the scale of this challenge. The average per capita income in the 30 richest countries is about seven times the average income in the other 134 countries, excluding those where no reliable information is available (Friedman, 2005). To bring all those 134 countries to the same standard of living as the last country on the list of the 30 richest ones in the next 50 years, admitting an annual increase of 1.3% in the total population of those countries, would require an increase in the world economic output by a factor greater than four. This calculation does not include the economic growth that would result from the rise in living standards of the richest countries during the same period of time. The consequences of quadrupling the world economic output in 50 years as regards natural resource availability and environmental degradation are staggering.

When making such estimates, various questions arise. Is it possible to significantly reduce the inequalities in the world with the current paradigm of economic growth? Will it be possible to fully globalize the lifestyles in industrialized countries? How could we find alternative development paths and can they become attractive to a considerable part of humanity? What are the primary energy sources that will support the transition to a much more equitable world? How could we ensure the sustainability of the transition as regards natural resource availability, especially water, soils, ecosystem services, and biodiversity? How could we ensure that the transition does not seriously aggravate global changes, in particular climate change? These are some of the major challenges that face us on the path to sustainable development. One of the main virtues of this discourse is that it presents a global and integrated framework to address them through coordinated collective efforts.

In spite of all its shortcomings, the sustainable development discourse is currently the most balanced and promising. Its innovative re-conceptualisation of the environmental problems through integration with social and economic concerns stimulates creativity for new solutions acceptable to the different interest groups involved. Nevertheless, the discourse is mainly interpretive, and has great difficulty in transforming itself into effective action plans at the local, national, and global levels.

5.15 Hope in Rio de Janeiro, Uneasiness in Johannesburg, and Resignation in Rio+20

After the publication of the Brundtland report in 1987, the sustainable development discourse acquired increasing visibility and notability, especially in international governmental and non-governmental organizations dedicated to environmental issues. This process reached a high point at the United Nations Conference on Environment and Development, often called the Earth Summit, held in Rio de Janeiro in June 1992, with the participation of 182 governments and 108 heads of state or government.

The Rio Earth Summit enjoyed a climate of hope and confidence in the future and proposed a significant change of course that would put the community of nations on the path to sustainable development. Even so there was a considerable tension between the G77 group of southern countries and the group of more industrialized countries. The south insisted on restructuring global economic relations so that it could obtain debt relief, increased official development aid, higher commodity prices, increased technology transfer and access to markets in the north. At the time, the industrialized countries had the financial and economic power that enabled them to negotiate from a much stronger position, and their main concern was with global environmental problems.

Significantly, the Stockholm Declaration was more constructive when it emphasized environmental protection and international cooperation. In the Rio Declaration on Environment and Development, there is more emphasis on development and national sovereignty, which reflects the growing north–south confrontation regarding inequalities of development and how to overcome it. Nevertheless the Rio Declaration included two new and very important principles: the precautionary and the polluter-pays principles. Another breakthrough was the approval of the Agenda 21 intended to be the framework of action for sustainable development.

In addition, the Rio Earth Summit produced the United Nations Framework Convention on Climate Change, the Convention on Biological Diversity, and a non-legally binding document on forestry known as Forest Principles. Immediately after the summit, in December 1992, the United Nations General Assembly established the Commission on Sustainable Development, with the objective of implementing Agenda 21 throughout the world. The sustainable development discourse was disseminated globally and became well implanted in some countries, creating the minimum conditions needed to start acting.

However, over the next 10 years, new sources of resistance were encountered and little was done for its effective worldwide implementation. As expected there was much more reservation and pessimism at the World Summit on Sustainable Development held in Johannesburg, from 26 August to 4 September 2002, than in the Rio Earth Summit. President George H.W. Bush was present and participated actively in Rio, but his son George W. Bush boycotted the event, rendering it partially inoperative. This attitude had the political support of conservatives in the USA. A letter written to President Bush by various US corporate-sponsored think tanks, including the American Enterprise Institute and the Competitive Enterprise Institute, praised his firm decision in spite of all the efforts that were made for him to be present at the summit. The industrialized countries, led by the USA, saw Johannesburg primarily as an opportunity to safeguard global trade rather than a chance to agree on an effective plan of action to combat poverty and stop the continuous deterioration of the natural environment.

The Johannesburg Declaration reaffirms the commitment to agreements made at the Stockholm and Rio Summits but sets weaker goals than those agreed upon at those meetings. It also lacks the provisions for substantial enforcement, making it very difficult to evaluate future progress. The most important innovation from Johannesburg was the launching of about 300 voluntary partnerships of non-state parties among themselves or with governments, involving about US $ 235 million of pledged resources. This development is welcome and reflects the increasing role of NGOs and business in international environmental issues. In the end, some said that sustainable development had lost the edge it had at Rio and become more of a rhetorical discourse, although it remains the politically accepted discourse internationally. The summit left many people worried about the sustainability of our collective future. The hope of Rio de Janeiro had given way to a greater unease about the coming decades.

The next United Nations Conference on Sustainable Development, known as Rio+20, will take place in Rio de Janeiro in 2012. The emphasis continues to be on sustainable development. Its overall objective is "to secure renewed political commitment for sustainable development, assessing the progress to date and the remaining gaps in the implementation of the outcomes of the major summits on sustainable development and addressing new and emerging challenges" (UNGA, 2009). Rio+20 will take place in a very different world to the Stockholm, Rio, and Johannesburg meetings. We are now facing multiple and interrelated crises regarding development and the environment: financial and economic uncertainty, increasing food prices and decreasing food security, water scarcity, volatile energy prices (particularly for oil), continued lack of safe drinking water and sanitation for hundreds of millions of people, high unemployment in many countries, particularly in the most industrialised countries, unsustainable consumption patterns, climate change, increased desertification, accelerating ecosystem degradation, biodiversity loss, and many more.

Moreover, the balance of economic and financial power in the world has changed significantly since Johannesburg. The emerging economies are now the main engines for economic growth and employment, while the more industrialized countries are still battling the consequences of the financial and economic crisis and struggling with high structural unemployment and persistent deficits and debts. Rio+20 will focus on "a green economy in the context of sustainable development and poverty eradication and the institutional framework for sustainable development" (UNGA, 2009). The emphasis on the green economy echoes the appeals for a Global Green New Deal (UNEP, 2009) to face the challenges of the credit crisis, poverty, environmental degradation, climate change, and high oil prices.

Furthermore, it is significant that the countries with the highest green funding percentage of the national GDP in the world are two Asiatic countries, currently with high economic growth: South Korea, a developed country, with 6.99%, and China, a developing country with an emerging economy, with 5.24% (UNEP, 2009). In view of the new world landscape regarding the economy and environment, at Rio+20, Western countries are likely to lose part of the leading role they had in previous Earth Summits. Rio+20 may be the time to renew our hope for a global sustainable development, but it may also be the opportunity to acknowledge our incapacity to move significantly in that direction during the last 20 years. Maybe this recognition could assist in making a more precise identification of the obstacles that must be overcome.

5.16 Sustainable Development and Global Governance: Present Situation and Future Prospects

The sustainable development discourse is relatively weak in the USA, both at the federal and state government levels, and also in civil society. It was only during the government of W.J. Clinton, in 1993, that a President's Council on Sustainable Development was established, largely due to the support of Vice-President Al Gore. In Europe, there is generally greater knowledge and receptivity with regard to this discourse. The European Community established an idealistic environmental program through its Environmental Action Plans beginning in 1973, following the recommendations of the Stockholm Earth Summit. From the beginning of the 1990s sustainable development gradually became a normative principle for EU environmental policy.

The principle of sustainable development was included in the text of the EU Amsterdam Treaty, approved in 1997. In 2001 the EU Council, meeting in Gothenburg, formulated a sustainable development strategy for the EU for the first time. Moreover, sustainable development was established as a key objective in the Lisbon Treaty of 2007, although its compatibility with other EU plans, especially the Lisbon Strategy, the development plan of the EU for the economy between 2000 and 2010, was problematic. In spite of all these achievements, the implementation of the EU environmental policies has always been relegated to a secondary priority during economic downturns.

The environmental component of sustainability in different countries has been compared using an index developed by the Universities of Yale and Columbia, the World Economic Forum, and the European Commission Joint Research Centre (ESI, 2005). According to this methodology, the five countries with highest Environmental Sustainability Index in 2005 were Finland, Norway, Uruguay, Sweden, and Iceland, a clear majority of North European countries. In 2006 the Environmental Sustainability Index was superseded by the Environmental Performance Index developed by the same institutions, which has a stronger emphasis on outcome-oriented indicators. In 2010, the five countries with higher Environmental Performance Index were Iceland, Switzerland, Costa Rica, Sweden, and Norway. Costa Rica is a remarkable exception among developing countries, and it may become a model for this group of economies. At the bottom of the list, the countries with lowest Environmental Performance Index were Togo, Angola, Mauritania, the Central African Republic, and Sierra Leone, all in Africa.

There is no well established comparative index of sustainable development that includes the social, economic, and environmental components. This is quite comprehensible since it is very difficult to establish universal criteria to compare the sustainable development performance of different countries with the current dualism of development in the world. If all countries in the world share the same objective of continuous economic growth, how should we measure sustainable development in that area? Does sustainable development impose any limits on economic growth? What are those limits? To what extent are the more industrialized countries good models for sustainable development? If they are, and if sustainable development is in fact an acceptable guideline for the future of humans on Earth, the desirable outcome could be for the developing countries to achieve the average level of consumption and production and the quality of life currently enjoyed in the industrialized countries.

Nevertheless this program is very likely to be incompatible with the sustainability of natural resource use, the stability of ecosystem services, the protection of the environment, and the preservation of biodiversity. Thus it would not be a program likely to achieve a higher degree of global sustainable development. The way out of this impasse would be to identify and establish the constraints that sustainable development should impose on economic growth, primarily in the more developed countries.

Since the Rio Earth Summit of 1992, businesses have been increasingly concerned, and many have been mobilized to contribute to sustainable development. This is a very important and positive development. The World Business Council for Sustainable Development established in 1995 is a global association of some 200 companies entirely dedicated to the promotion of the role of business in achieving sustainable development. Its aim is to share knowledge, experience, and best practice, to advocate business positions on sustainable development, and to establish partnerships with governmental and non-governmental organizations.

Recently, it published a courageous report entitled *Vision 2050: The New Agenda for Business*, which indicates the pathway to reach 2050 with enough food, clean water, sanitation, safe housing, mobility, education, and health for the 9 billion people that will probably be living at that time within the limits of what the Earth can supply and renew. The critical pathways of change identified in the report are (WBCSD, 2010):

[...] addressing the development needs of billions of people, enabling education and economic empowerment, particularly of women, and developing radically more eco-efficient solutions, lifestyles and behaviour, incorporation of the costs of externalities, starting with carbon, ecosystem services and water, doubling agricultural output without increasing the amount of land or water used, halting deforestation and increasing yields from planted forests, halving carbon emissions worldwide (based on 2005 levels) by 2050 through a shift to low-carbon energy systems and highly improved demand-side energy efficiency, providing universal access to low-carbon mobility and delivering a four-to-tenfold improvement in the use of resources and materials.

It is significant that this revolutionary program is proposed by a business association. Its implementation requires a very high degree of political commitment throughout the world and an effective coordination between the policies of all governments. How can we achieve that level of commitment and coordination in a highly fragmented world riddled by deep inequalities of development, corruption, and bad governance? Vision 2050 demonstrates that there are indeed pathways to sustainable development in the next 40 years, but it is difficult to see how they can be reached without the establishment of strong institutional structures for international governance. The move to sustainable development in an increasingly complex and diverse world requires a broad political agreement, an enormous coordination effort, and the capacity to enforce the agreed plans of action.

This brings us to the questions of international governance. As early as 2000, the Global Ministerial Environment Forum, organized by the UNEP, stated in its final declaration that (UNEP, 2000):

> The 2002 conference (Johannesburg Earth Summit) should review the requirements for a greatly strengthened institutional structure for international environmental governance based on an assessment of future needs for an institutional architecture that has the capacity to effectively address the wide-ranging environmental threats in a globalizing world.

However, no significant progress toward that goal has been achieved since then. As frequently happens with environmental problems, there are alarming discrepancies between commitments and action. All too often the institutions of the UN system have insufficient funding or authority to efficiently and effectively tackle the issues under their responsibility. UNEP remains one of the smaller UN programs with very little authority over other sections of the UN.

The creation of a World Environmental Organization standing side by side with the World Trade Organization and the World Health Organization would be a very important step toward addressing the global challenges of sustainable development in an institutionally balanced way. It will be impossible to integrate the environment into the mainstream of decision-making without a global inter-governmental institution with the powers to take legal action, to enforce laws, and to use a binding dispute settlement system.

A more ambitious goal is to strengthen the institutions of global governance, giving them the power to enforce compliance. If we are to follow the pathway of strongly institutionalized global governance, we should start by strengthening a world community interested in the process that would then proceed to establish a global constitution with clearly defined rights, duties, and objectives of global sustainable development. It is essential that people recognize the benefits and the legitimacy of strong global governance institutions and that states agree on their goals and procedures.

The United Nations is well aware that it should play an important role in this process. Note for instance that the issue proposed by the United Nations General Assembly President for the 2010 General Assembly general debate was *Reaffirming the central role of the United Nations in global governance.* Also recently, on 9 August 2010, the United Nations Secretary-General Ban Ki-moon announced the establishment of a High-Level Panel on Sustainability formed by 21 members, to be co-chaired by Finland's President Tarja Halonen and South African President Jacob Zuma.

The panel is expected to come up with practical solutions to address the institutional and financial arrangements needed to promote a low-carbon economy, enhance the resilience to climate change impacts, and tackle the interconnected challenges of poverty, hunger, water, and energy security. The final report of the panel is to be delivered at the end of 2011 in order to serve as input to the Rio+20 Earth Summit. The main objective of these initiatives is to increase the commitment of Member States to the pursuit of sustainable development. Let us hope that they will be successful in the Rio+20 Conference.

Support for sustainable development is very far from unanimous. Many people, some with high visibility, are firmly against it, particularly among US conservatives. Their argument is that sustainable development corresponds to the implementation by the government of the universal principles of Agenda 21, which implies a form of global governance that affects all aspects of human life (Lamb, 1996). This interference is considered to be a severe limitation to human freedom and to the free market economy, and should therefore be repudiated.

Pronunciations against science have recently become more frequent in the USA, especially among the conservatives. This tendency is part of the anti-science movement, analysed in Chap. 2, and is likely to become stronger and more widespread in the future. For the conservatives, science is very probably flawed, or should be ignored when its application leads to the establishment of norms that regulate the market, restrictions on certain forms of economic activity, increased government intrusion in people's lives, or indeed any challenge to the sustainability of the paradigm of continuous economic growth. In particular, it should be ignored when it collides with religious dogma.

In such a movement, science is not considered to be neutral since science should always mean progress: only devious liberal and socialist minds can try to break that relationship. Rush Limbaugh a very popular radio host in the USA has stated that "science has become a home for displaced socialists and communists" and called climate change science "the biggest scam in the history of the world" (Nature, 2010). These anti-science voices are unable to influence the majority of the American people, but they can gradually reduce the competitiveness of the USA if they become more powerful.

5.17 Environmental Economics, Ecological Economics, and Ecological Modernization

The relation between the environment and the economy is at the core of the current challenges of sustainable development. Not surprisingly, there is a great wealth of theoretical analysis and practical proposals on how to make the sustainable use of resources and the protection of the environment compatible with economic growth. Environmental economics follows neoclassical economics in its aim to shift the economic system towards an efficient allocation of natural resources. The objective is to bring the question of increasingly scarce natural resources into mainstream economic analysis and practice. When resources are not allocated efficiently, in part because of the environmental incompatibility of their use, this is interpreted as a market failure that can always be corrected. One of the major concerns of environmental economics is the valuation of externalities so that they can be internalized. There are specific methodologies for doing so, such as hedonic pricing, the travel cost method, and the contingent valuation method (Perman, 1999; Markandaya, 2002).

Ecological economics is a trans-disciplinary field of study that goes beyond the usual scope of analysis of conventional neoclassical economics by addressing the interdependence between economics and natural ecosystems. The core idea is that the economy of the human society is a subsystem of an overall economic–ecological ecosystem that co-evolves with the natural world. There is a recognition that the environment poses natural constraints on the provision of natural resources and on the absorption of the wastes of production and consumption.

Some concepts of the natural and physical sciences are frequently used in ecological economics, such as throughput, carrying capacity, and entropy (Georgescu-Roegen, 1971). In accordance with the second law of thermodynamics, all physical processes that occur in the economic system convert low-entropy energy and materials into high-entropy wastes. For instance, when transforming a high entropy copper ore into a low entropy sheet of copper, the decrease is more than compensated by a larger entropy increase associated with the mining process. In terms of entropy, the cost of any economic or biological activity is always higher than the product (Georgescu-Roegen, 1971). The question is whether or not the increase in the open system of the human economy relative to the surrounding closed system of the global environment on Earth leads to entropic constraints on economic growth. The emphasis on natural and physical constraints to economic growth in ecological economics places it closer to the limits discourse. On the other hand, environmental economics has more affinity with the Promethean discourse.

Most industrialized countries already have a significant experience of environmental problem-solving within the framework of their economic policies. These experiences give varying emphasis to regulatory administrative mechanisms based on the advice of experts, to a more participative democratic process, or to greater reliance on market forces. Nature conservation has improved significantly and the negative impact of some economic activities on the environment at the local and national level has generally decreased. Finland, Germany, Japan, the Netherlands, Norway, and Sweden are among the countries with the most successful environmental policy performance in the 1980s and 1990s (Dryzek, 2005).

One of the reasons for these successes was the practice of ecological modernization, a process proposed at the beginning of the 1980s in Germany by the sociologists Joseph Huber (Huber, 1982; Huber, 1985) and Martin Janicke (Janicke, 1985).

The main idea is that environmental degradation can be stopped in the industrialized countries within the current political and economic system through a restructuring of the processes associated with production and consumption. Ecological modernization is based on the conviction that it is possible to decouple economic growth from environmental degradation. Furthermore, it considers that economic growth and the protection of the environment can mutually reinforce one another, and that only this synergy can guarantee economic sustainability. Environmental problems can harm the basis of production, which implies that environmental regulation is not necessarily in conflict with economic growth. The success of this strategy is synthesised in the well known slogan 'pollution prevention pays' (Hajer, 1995), which presupposes that business has sufficient money to adhere to ecological modernization and is willing to wait for the beneficial returns from its investment rather than expecting quick profits.

Science and technology are the main driving forces of reform in the ecological modernization of industry. The most important social agents in this restructuring process are scientists and engineers, engaged in the search for innovative solutions, and businessmen, motivated by the advantages of a greater environmental sustainability and a higher competitiveness resulting from their public green image. A promising sign is that there is a growing desire in the market for products and processes that are environmentally friendly. Some companies take advantage of this trend through greenwashing, a form of deception in which their production processes and products are misleadingly promoted as protecting the environment.

For ecological modernization to be successful, it will be necessary to reach compromises involving the whole of society, but especially business and the political decision-makers. It is also important to develop a strategic vision of the environmental problems in the medium to long term, implement mechanisms that support the sustainability of the ecosystem services, understand in detail the processes of production and transport of pollutants, and systematically apply the precautionary principle. Ecological modernization relies on pollution prevention, waste reduction, product life-cycle assessment, and the analysis of material and energy flows. It promotes 'cradle to cradle' as opposed to 'cradle to grave' forms of manufacturing (Braungart, 2002). Instead of repairing the adverse effects of pollutants at the end of the production and consumption processes, the objective is to try to identify the processes that cause pollution in order to modify or replace them. Another important goal is to evaluate the various forms of environmental degradation in economic and financial terms so that the response measures can be quantified and optimized. Only in this way will it be possible to internalise the costs of the negative environmental externalities of industrial activities, production processes, and patterns of consumption.

The success of ecological modernization depends largely on a political and economic system able to adopt environmental policies and regulations that prevail over liberal business practices privileging competitiveness and profit above all else. To achieve this goal it is necessary to promote negotiations and consensual compromises between the public and private stakeholders. The followers of the Promethean discourse consider ecological modernization a waste of wealth and economic growing potential on excessive, onerous, and expensive regulations. Their proposal is to let the market function freely, and to rely on science and technology to solve the environmental and natural resource scarcity problems on a case-by-case basis when they become pressing.

Compared with the sustainable development discourse, ecological modernization is much more focused since it has the specific aim of decreasing environmental degradation through technology-driven innovations in production and consumption. Its practical application has been successful in a few developed countries with a consensual and interventionist policy style, particularly Germany and the Netherlands, where it was first developed. In countries with a more adversarial policy-making style and with strong liberal economic competition, as in the English-speaking developed countries, particularly the USA and Australia (Jahn, 1998; Fisher, 2001), it has been much less successful.

Ecological modernization is a reforming discourse that promotes confidence in the future of the developed countries as regards the environmental sustainability of their economic growth. It is a discourse of reassurance for the citizens of the more prosperous and industrialized countries. The applicability of ecological modernization in the developing countries is very limited because of the relatively low industrialization. Moreover, in these countries, the problem of environmental degradation has a relatively low priority, and production and consumption are at much lower levels than in the more industrialized countries.

The analysis of economic and environmental indicators reveals that it has been possible to advance considerably toward decoupling economic growth from environmental degradation at the local and national levels in several developed countries, particularly in Europe, such as Finland, Germany, Netherlands, Norway, and Sweden. However, the situation becomes much more complex when considering the contribution to environmental degradation at the global level. All countries, and in particular the developed countries, contribute directly to global environmental degradation and pollution and also indirectly through the products they import from outside countries.

The distinction between the two components becomes clearer if we consider the environmental Kuznets curves (Kuznets, 1955). These curves represent the environmental indicators in a given country as a function of the income per capita. They are supposed to have an inverted U-shape, which results from a greater demand for improved environmental quality when the GDP per capita in the country goes above a certain threshold. Although some indicators at the country level, such as SO_2 emissions, follow a Kuznets curve, others such as CO_2 emissions do not.

The more industrialized countries are effectively 'outsourcing' their greenhouse gas emissions to the developing countries. About one third of the CO_2 emissions associated with the goods and services consumed by the more industrialized countries are emitted outside their borders, and mostly in the developing countries. Europe is responsible for about four tonnes of CO_2 emissions per capita in the form of products containing embedded carbon imported from abroad, while in the USA the corresponding value is two and a half tonnes per capita (Davis, 2010). Moreover, tighter controls on CO_2 emissions in Europe and in other developed countries tend to drive factories that produce aluminium, steel, iron, and cement to relocate abroad, where they will increase CO_2 emissions, a process that is known as carbon leakage.

Another example of environmental degradation outsourcing by industrialized countries is provided by Japan, where the effects of its large ecological footprint are felt mostly outside the country, in particular with the massive timber importation from the Southeast Asian tropical rainforests and the very intensive fishing activity on the global scale.

Other examples are the relocation of polluting industries to less developed countries that export most of their products to the industrialized countries. If the economy continues to depend on the importation of relatively large amounts of energy and natural resources, the negative environmental externalities persist and are just relocated. It is therefore necessary to analyse the decoupling between economic growth and environmental degradation in the more industrialized countries not only at the local and national level, but also at the global level, taking into account the external impacts of the internal production and consumption patterns.

As for the developing countries, the behaviour of the environmental and economic indicators shows that, in most of them, there is no decoupling of economic growth from environmental degradation at the local and national level, and that it is much more difficult to apply ecological modernization. Nonetheless, it is a very important theoretical framework to reduce environmental degradation and pollution in developed countries, and it may be possible to apply it to all countries in the future.

5.18 Current and Future Global Energy Scenarios and the Patterns of Growth

Energy is the main basis for the sustainability of the current paradigm of economic growth. Without abundant and relatively inexpensive primary energy sources, it is impossible to ensure economic growth in the more industrialized countries, in the emerging economies, or in the other developing countries. Energy plays a central role in determining the future development trends in the short, medium, and long term. It is therefore very important to analyse the global energy situation, to determine the trends of its recent evolution, and to discuss the future projections of supply and demand within the framework of the different environmental and development discourses.

We are manifestly in a transition phase from an energy age dominated by the supremacy of fossil fuels to another age still permeated by many uncertainties. There are two main reasons that should encourage us to reduce our dependency on fossil fuels. One is that they are not environmentally compatible because of the CO_2 emissions. The other reason is that, with the current rate of consumption, we are probably less than a century away from the fossil fuel Hubert peak. As already mentioned in Chap. 3, we do not now the exact position in time of the Hubert peak for conventional oil, but most experts think that we are very close to it. Nevertheless, fossil fuels continue to be very competitive in the energy market, especially coal, whose reserves are relatively abundant. It is very likely that both the conventional forms of fossil fuels — coal, petroleum, and natural gas — and the unconventional forms, such as oil sands, oil shale, shale gas, and other types of deposits will be exploited until all reserves are completely exhausted.

What solutions are available for this transition to a non-fossil fuel age? The time factor is very important in this analysis. If we consider the short and medium term, up to 50 years, it is essential to improve energy efficiency, to develop and deploy the modern renewable energies and to use clean coal technologies, particularly carbon capture and storage. For the next 20 to 30 years, most probably, we will have to continue to rely heavily on fossil fuels, and increasingly on coal. The modern renewable energies (small hydro, modern biomass, wind, solar, geothermal, and biofuels) are growing fast but at present they account for only 2.7% of the global final energy consumption. Energy from nuclear fission is an important primary energy source, especially in the emerging economies, but its contribution to the global market will probably remain relatively small.

In the medium to long term, after 50 and 100 years, the problems become different. If the current rate of global consumption of energy is maintained, fossil fuels, particularly oil and natural gas, will have to be replaced on a massive scale by other primary forms of energy. The new renewable energies are likely to become increasingly important, sharing more than 50% of the global final energy consumption, but this goal will require huge investments worldwide in deployment, grid infrastructures, research, and innovation. Research and development may lead to more attractive nuclear fission reactors. Moreover it is hoped that we will be able to exploit nuclear fusion energy within 50 years, although as previously shown the uncertainties are very great.

From the environmental point of view, insistence on the use of fossil fuels up to their exhaustion is likely to bring about severe problems. Let us imagine that all the known reserves of fossil fuels are burned to produce energy and that all the resulting CO_2 emissions are released into the atmosphere. In this scenario the atmospheric concentration of CO_2 is likely to reach values of the order of 4 500 ppmv. The average global temperature will then increase by more than 9°C, and the average sea level will rise by more than ten metres above the present level (Hasselmann, 2003). The transition to the new equilibrium state of the Earth climate system would be long, taking many hundreds of years, but it would be irreversible. This is a brutal scenario, with dire consequences that would profoundly transform human society and the geography of the world. The scenario becomes less harsh, but still very serious, if we fully deploy capture and storage of the CO_2 emissions, especially in coal-fired power plants.

An analysis of the behaviour of energy demand and price trends in recent years reveals the challenges that we are facing at the beginning of the 21st century. In the five years from 2001 to 2005, the world economy grew significantly, with an average annual rate of about 4.4%, measured at purchasing power parity exchange rates, while in the previous five years it had been only 3.5% (Davies, 2007). The emerging economies, in particular China, were largely responsible for the acceleration in growth. In the OECD countries, the average annual growth rate was lower, at 2.5%.

What was the energy consumption growth associated with this global economic growth? Before answering this question, it is important to emphasize that the global energy intensity has been decreasing, being approximately 33% lower in 2006 than in 1970 (IPCC, 2007). During the last 25 years the decline in energy intensity has been about three times faster in the OECD countries than in non-OECD countries (IEA, 2008). The energy consumption grew faster relative to GDP growth from 2001 to 2005 than in the previous five year period. Energy consumption accelerated from an average annual growth of 1.2% in the period of 1996 to 2000 to 3% in the period of 2001 to 2006 (Davies, 2007). China accounted for almost half of the global energy growth in the last five year period. In the same period, the average energy consumption growth diminished in the OECD countries, which shows the large and increasing weight of the developing countries, particularly the emerging economies, in world energy demand.

The large increase in energy consumption from 2001 to 2006 occurred when the cost of energy, especially oil, was also fast increasing. From 1996 to 2000, the cost of fossil fuels remained relatively stable, with an average increase of around 8%. However, from 2001 to 2005, they increased significantly. Nominal prices of oil more than doubled, natural gas rose by around 75%, and a weighted average of coal prices by 46%, relative to the previous five-year period (Davies, 2007). Market mechanisms worked well to promote robust world economic growth in an environment of high energy prices. However, the resilience lasted only until 2008, when the price of the oil barrel reached US $ 148 in July. This oil shock contributed significantly to transforming the financial crisis in major oil importing countries into a full-blown recession (Roubini, 2010).

Data from the US Energy Information Administration indicates that in the period 2004–2010 global conventional oil production has oscillated in a band between about 72 and 74 million barrels per day, reaching a plateau that some oil experts have interpreted as an announcement of a peak in conventional oil (EIA, 2011). On a yearly basis, conventional oil production had a peak in 2005 with an average daily production of 73.719 million barrels (EIA, 2010). In 2008 the average daily production in a month reached an overall maximum of 74.666 million barrels in July (EIA, 2010), but due to the very significant increase in oil prices, both demand and production crashed.

It is relatively well established that growth in global conventional oil production depends mostly on an increase in Saudi Arabia's production, and also that it will probably be very difficult to go above a global output of 77 million barrels a day. We are likely to be near the conventional oil peak, which implies future higher price volatility and increased supply uncertainty. The growing demand for oil has been met by the availability of biofuels and non-conventional oil. The all-liquid supply, including conventional oil, reached an average of 87.5 million barrels per day in July 2010 (EIA, 2010).

Regarding the carbon intensity of energy consumption (the amount of CO_2 emissions per unit of primary energy consumed), recent statistics are upsetting. The average global carbon intensity decreased systematically after the 1970s, remained constant in the period of 1996 to 2000, and then increased due to the growing dependency on coal, led by China. This recent trend accelerated the global CO_2 emissions into the atmosphere, which up to 2008 were close to the highest emission scenarios considered by the IPCC in 2001 (IPCC, 2001). Between 1970 and 2004, greenhouse gas emissions from the energy sector increased by 145%, and in 2004, CO_2 represented 77% of global anthropogenic emissions (IPCC, 2007).

Things changed in 2009. In that year, for the first time since 1998, CO_2 emissions from fossil fuels decreased by 1.1% relative to 2008, from 31.55 to 31.13 billion tonnes of CO_2 (BP, 2010), because of the decrease in fuel consumption and industrial output resulting from the financial and economic crisis. However, the change in emissions was profoundly differentiated throughout the world. In the USA, emissions in 2009 fell by 6.5%, while in China they increased by 9%. Emissions from emerging economies, which now account for about half of the world's emissions, grew by more than 5%, while in the OECD countries they fell by 6%.

Regarding the future, the IPCC (IPCC, 2007) business-as-usual scenarios, meaning with no specific mitigation measures, project an increase in greenhouse gas emissions between 25 to 90% between the years 2000 and 2030. In these scenarios, fossil fuels maintain their dominant role among primary energy sources, and CO_2 emissions increase by 45% to 110% in the same period. More recent estimates that take into account the recent financial and economic crisis project an increase in annual global greenhouse gas emissions of 45% from 2005 to 2030, reaching 65.6 $GtCO_2e$ (McKinsey, 2010). This business-as-usual scenario, made after the crisis, estimates an overall reduction of emissions by 3.6 $GtCO_2e$ by 2020 and 4.3 $GtCO_2e$ by 2030, relative to the business-as-usual pre-crisis scenario.

Again one finds that the projected changes in emissions are strongly differentiated throughout the world. The more industrialized countries experience the largest decline in the after-the-crisis scenario, estimated at 11% in the USA and 6% in Europe. Large reductions are also expected in many developing countries in Africa and Latin America, and also India. However, China and the rest of developing Asia are expected to increase their emissions relative to the business-as-usual pre-crisis scenario, by 1% and 2% respectively.

The McKinsey study (McKinsey, 2010) identifies a mitigation potential of 38 $GtCO_2e$ (corresponding to 58% of the annual emissions) through technical measures costing below 80 euros per tonne of CO_2e, relative to the business-as-usual emission scenario of 65.6 $GtCO_2e$ in 2030. With an additional 8 $GtCO_2e$ mitigation potential of more expensive technical measures in all regions of the world and in all sectors, the atmospheric concentration of greenhouse gases would peak at about 480 ppmv of CO_2e, a value likely to lead to a global average temperature increase above pre-industrial levels at equilibrium between 2.0 and 2.4°C. This would mean that dangerous anthropogenic interference with the climate system would very probably be avoided. Thus it is still possible to avoid such interference, but the probability of the required measures being implemented is very low.

The International Energy Agency (IEA, 2010) has also developed a scenario where the greenhouse gas concentration peaks at 450 ppmv of CO_2e, and shown that it is possible to achieve that goal through appropriate mitigation measures. The total cost of these measures relative to the business-as-usual scenario is estimated at 10.5 trillion US $, to be spent until 2030. This may seem an enormous and prohibitive amount of money, but it is not. To have a measure for comparison, note that on 15 October 2010 the total outstanding public debt in the USA was 13.6 trillion US $, of which approximately 66% is debt held by the public. Regarding citizens, their mortgages and other debts amount to around 13 trillion US $, which is almost 120% of their annual disposable income (The Economist, 2010). There is a very large discrepancy between the international consensus about what should be done to avoid dangerous anthropogenic climate change and what is actually done to avoid it. At the moment the world remains squarely on a trajectory of rising atmospheric concentrations of greenhouse gases, and there are no clear signs that it seriously intends to change its course.

The shift in economic power from the more industrialized countries to the developing countries is very clearly reflected in the energy sector. At the beginning of this century, the rich countries contributed about two-thirds to the world GDP, allowing for purchasing power parity. They now contribute just over half, and in 2020 they are likely to drop to 40%. China began to dominate the change in the structure of world energy consumption at the turn of the century. In the period from 2001 to 2006, it accounted for 46% of world energy growth, of which 73% was coal. China's share of world energy consumption increased to 16% from 9% in 1991. Among the fossil fuels, coal is now the fastest growing fuel. Meanwhile, oil is losing its global share and gas has stabilised.

As regards energy trade, the balance of energy markets and energy production is shifting geographically. The imbalance between imports and exports is becoming stronger, aggravating the costs of transportation and security risks. About 64% of the world's oil, 26% of natural gas, and 17% of coal are traded internationally. Security of oil supply in OECD countries has been decreasing since its production peaked in 1997, due to the decline in North Sea production. Oil remains a very powerful instrument of potential conflict between the oil importing industrialized countries and the oil exporting developing countries. These risks coupled with the looming prospect of reaching the conventional oil peak could create potentially dangerous situations.

A recent study by the Future Analysis Department of the German Bundeswehr Transformation Centre that was leaked to Der Spiegel (Der Spiegel, 2010) recognizes the threat of imminent peak oil, when supply gradually starts to decline and prices tend to increase significantly. The study indicates that there is some probability that we are very near peak oil and that major consequences will be felt within the next 15 to 30 years. The document shows how preoccupied the German government is, and also how unwilling it is to share its concerns openly. There are indications of a similar situation in the UK where documents from the British Department of Energy and Climate about a potential oil supply crisis and its foreseeable consequences on society are kept secret. In the USA Glen Sweetnam of the Department of Energy recently acknowledged that there is a chance of a decline in world liquid fuel production as of 2011 (Le Monde, 2010). Higher investments would have to be made to continue to satisfy oil demand, and this implies an increase in its price.

In spite of these risks, worldwide government budgets for energy research, development, and demonstration have been decreasing from about US $ 20 billion in 1980 to about half that value in 2006 (IEA, 2008a). Private sector spending has also been gradually declining. These trends appear to be reversing, leading to a renewal of private and public investment in energy research and development. The many encouraging efforts to improve the efficiency of energy conversion from renewables and to develop new renewable energy sources need to be adequately supported. They cover the whole range of modern renewable energies: modern biomass, wind, solar, ocean, geothermal, and biofuels.

One of the most promising is to optimize photosynthesis, the outstanding natural conversion process that originally made the development of higher life forms possible. The average efficiency of plant photosynthesis is only about 5%, but there are organisms, such as the green sulphur bacteria *Chlorobaculum tepidum*, that can convert around 10% of incident sunlight into chemical energy. These live in the deep, dark layers of oceans and lakes and have special photosynthetic antennae called chlorosomes that are very efficient solar power converters. The idea is to understand how these chlorosomes function and use them as a model to develop more efficient energy conversion systems.

Another approach is to improve the machinery of photosynthesis at the molecular level, for instance by modifying a protein called rubisco that plays a fundamental role in the process, or by applying synthetic biology to improve the efficiency of photosynthesis in algae. A promising line of research is to develop photobioreactors to produce oil from microalgae and other organisms. Cultivated algae can also be used to reduce greenhouse gas emissions by making them grow off the carbon- and nitrogen-rich exhaust from traditional power plants.

An example of physics research that may lead to far-reaching rewards in the future is the quest to increase the efficiency of photovoltaic solar energy converters. The current best commercially available photovoltaic devices are semiconductor photodiodes that reach energy conversion efficiencies of only 20–25%. It is possible to reach much better values. The efficiency of a photovoltaic system is limited by the Carnot efficiency, derived from the second law of thermodynamics. In the case of a photovoltaic device, the hot source is the Sun's surface at a temperature of about 6000 K and the cold source is the Earth environment with a temperature around 288 K. Thus it is theoretically possible to reach efficiencies of 95%. Conversion efficiencies of about 85% are very likely to be achieved through various methods, such as the use of photonic devices, optimised absorption materials, and heat engines. Multiple-junction photonic devices involving very advanced fabrication technologies have already attained a maximum energy conversion efficiency of 42.8% in the laboratory.

If we are able to solve the problem of energy storage associated with intermittence, solar power has the potential to provide more than 1000 times the current global energy consumption. However, solar power satisfied only 0.02% of the global energy demand in 2008. The main problem is the lack of competitiveness with fossil fuels and the low efficiencies of the solar power devices. The two problems are likely to be solved in the next 5 to 10 years if the necessary investments are forthcoming. A solar energy revolution may lie in the not too distant future. Two technological roadmaps (IEA, 2010; IEA, 2010a) released by the International Energy Agency on May 2010 state that solar photovoltaics and concentrated solar power could account for between 20 and 25% of electricity production worldwide by 2050.

Before ending this analysis, it is important to emphasize that water and energy are deeply and increasingly interdependent. Huge quantities of water are needed to generate energy and huge quantities of energy are required to obtain clean water. The availability of each resource is increasingly dependent on the availability of the other, and this forces us to manage them in a strongly integrated way. Countries affected by water scarcity can desalinate seawater or brackish water from deep aquifers, recycle wastewater, or transport water over long distances from water-rich regions, but all these solutions require vast energy supplies.

On the other hand energy generation requires increasing quantities of water. In the transport sector the replacement of petrol and diesel by biofuels and by plug-in electrical vehicles requires the consumption of much larger quantities of water. Furthermore, thermal power plants functioning with coal, oil, natural gas, or uranium consume large amounts of water in their cooling systems. Energy and water resources must therefore be managed in a coherent and integrated way. Under extreme conditions water is more essential to human life than commercialized energy systems. Without water, life as we know it is simply not possible.

5.19 Can We Have Prosperity and Well-Being in a Non-Growth Economy?

This is an intriguing question when all countries in the world, except possibly for a very few exceptions, make economic growth their foremost objective. Two years after the peak of the 2008–2009 financial and economic crisis, the world is economically on two tracks. In 2010–2011, most developing countries, especially emerging economies like China and India, are growing fast, while some of the industrialized countries have weaker growth, others are not growing, and still others, like some EU countries, are actually suffering recessions. For countries that are not enjoying robust economic growth, the most important question is how to grow or how to grow more.

Economic growth has been the dominant myth of human society since the end of World War II. Since that time the global economy has grown by a factor of more than five. We have created a situation where growth is necessary to prevent collapse. Although some of the most important contributors to neoclassical economic theory, such as John Stuart Mill and John Maynard Keynes, foresaw a time in which growth would have to stop, we still do not have a macro-economic model for achieving a steady-state economy. To reach it, we need a new framework that incorporates the medium and long term dependence of the economy on ecological variables, such as natural resources and biodiversity. Furthermore this new model must lead to extreme poverty eradication, reduce inequalities, protect employment, and ensure distributional equity of natural resource revenues. All of these are objectives with regard to which the current model has failed.

At present about a fifth of the world's population earns 2% of the global income. De-growth or economic contraction is just an expression created by radical critics of economic growth theory who promote the search for alternative models (Fournier, 2008). Non-growth would be a more suitable expression since the main objective is to reach a steady-state economy that incorporates ecological variables and social equity. What we need then is a non-growth economic theory that can lead to practical implementation. It may even be utopian to think that we can reach a steady-state economy in a controlled, rational way without causing repeated financial, economic, and environmental crises. It is nevertheless worth exploring the ideas that may lead to a non-growth economy.

The proponents of non-growth economies wish to create integrated, materially responsible, and self-sufficient societies among both the developed and developing countries. Thus non-growth is not a way to prevent the developing world from resolving its current problems. The point is that developing countries should disentangle themselves from the orthodox path of development that has been followed by the industrialized countries, and seek alternative pathways. It is important to emphasize that the straight application of Western forms of development to the poorer countries has led to a decrease in self-sufficiency, an increase in corruption, and much greater inequalities (Sachs, 1992), except for a few success stories.

On the other hand, the creation of non-growth economies would have much better chances of success if the more industrialized countries led the way by adopting some form of economic stabilization. We are clearly very far from this situation. The two-track world economy is a welcome process of economic convergence between the more industrialized countries and the developing countries, especially the emerging economies. Still, the developed world is uncomfortable with the current situation and is forcefully trying to increase its rates of economic growth in order to avoid another recession and stop the widening disparity with the growth rates of the emerging economies.

The current paradigm of economic growth is very strongly linked with the predominant notion of prosperity. Different visions of prosperity would lead to other paradigms and in particular to a non-growth economy (SDC, 2009). Each culture has its own concept of prosperity, which co-evolves with it. Furthermore, there is a plurality of notions of prosperity which are operative in any society, particularly in the most industrialized ones.

The origin of the dominant present-day view of prosperity can be traced back to the Reformation and to the global expansion of Europe, which immensely increased its material wealth. At that time, a new relation to faith and reason was established that included the secularisation of religious property, the dissolution of monasteries, a new concept of good living, and a vision in which success was interpreted as God's dividend to the righteous. The good life was essentially based on the power to explore the world and its natural resources, and enjoy the resulting lifestyles.

Prosperity became increasingly associated with the ownership of private property and ever more diversified and luxurious goods and services. The material component of prosperity was further strengthened by the Anglo-Saxon idea of progress based on utilitarianism and on the moral and economic freedom from the state defended by John Stuart Mill in his book *On Liberty*, published in 1859.

There are signs of disaffection with the notion of prosperity associated with the consumerist lifestyles that support and are supported by economic growth. A growing number of people resent the stress, pollution, congestion, noise, and diminishing level of personal contact and friendship that result from the competitive spiral of the work and spend cycle. However, getting out of the cycle usually involves difficult options with regard to personal relationships, employment, and social status.

Since the beginning of the Great Acceleration after the end of World War II, people in the more industrialized countries have tended to own and consume more, but also to have on average considerably less leisure time. Among the most important factors that influence well-being are the partner/spouse and family relationships, health, personal and political freedom, friendship and community integration, employment, a stable financial situation, home, and home environment (SDC, 2009). Various studies have indicated that, beyond a certain level of GDP per capita of about 15 000 US \$ in Purchasing Power Parity of 1995 US \$, life satisfaction reaches a maximum and barely responds to further increases in the GDP per capita (WI, 2008; Donovan, 2002).

Some countries with high levels of life satisfaction, such as the Netherlands and Sweden, have a lower GDP per capita than countries such as the USA with a lower level of life satisfaction. However, in countries with very low incomes, life satisfaction increases steeply with GDP per capita. The important message here is that it is essential to pursue economic convergence if we want to have a more equitable and secure world. Developing countries should continue to have economic growth, but the more industrialized countries should reach a stable, non-growth economy. The way to reach this goal is to refocus on goods that are strongly correlated with life satisfaction and not on those that require ever-increasing consumption. In spite of these findings the conventional notion of prosperity based on material satisfaction and opulence is deeply ingrained on societies all over the world, in both developed and developing countries.

Prosperity has social, psychological, and material dimensions. All are indispensable, but different weights can be attached to each of them. A weaker emphasis on the material component would pave the way for a less uncertain future with diminished risks of resource scarcity, environmental degradation, and potential armed conflicts induced by the competition for natural resources. Amartya Sen (Sen, 1984) has clearly defined two alternative notions of prosperity with reduced dependence on material consumption. The first is characterized by the term utility, and centred on the satisfaction that commodities provide through their intrinsic quality and not through their quantity. Furthermore, this notion of prosperity is based on the awareness that the overriding pursuit of immediate gratification reduces long term security. The second is defined through the concept of flourishing capabilities. Here there is more emphasis on the social and psychological components of prosperity.

The capacity to flourish corresponds to the most important factors contributing to well-being.

A prosperous world would imply a more equitable world where people everywhere would have the capability to flourish in certain fundamental ways. The important point is to recognize that these capabilities must necessarily be bounded by the functioning of the Earth system, the finite nature of natural resources, and the sustainability of ecosystem services. The new notion of prosperity should ensure that humans can flourish in more equitable societies, and achieve higher levels of social cohesion and well-being, while reducing their material consumption. Such prosperity presupposes the effective acceptance and practice of intra- and intergenerational solidarity. It may seem an impossible task, but the alternatives are frightening: they are likely to lead to greater inequalities and to a more unstable and belligerent world with higher risk of repeated financial, economic, and environmental crises.

Epilogue I

The world is rapidly shifting to a multipolar mode. Power is moving from the more industrialized countries to the emerging economies and, to some extent, to the developing world. A significant number of countries in the first group are likely to go through a relatively long period of economic stagnation and high unemployment, driven largely by demographic disparities with the rest of the world, decreasing productivity of capital, and large debts and deficits created by profligate policies practised over many years. The continuous rise in government debt is like a Ponzi scheme, requiring an ever-increasing population to assume the burden, while the population is not in fact growing in most developed countries. The stagnation period will cause a lot of pain, and probably a general feeling of Western decline, at least in some Western countries, contributing to distrust with regard to politicians and the elites. A worst case scenario would be a new Western financial and economic crisis resulting to a large extent from the continuing incapacity of the political systems in the USA and the EU to reduce the deficits and debts of their countries.

Ideally, the downturn could be used positively as an opportunity to move toward the creation of a framework for a sustainable non-growth economy in the developed countries. However, there is no evidence whatsoever that there is currently any possibility of following this difficult pathway. On the contrary, from a psychological point of view the preferred response is to fight the downturn by promoting growth at any cost. Furthermore, if a period of growth returns, the dynamism and self-confidence it generates will also act against any serious attempt to achieve a non-growth economy. The most probable outcome in the medium to long term is that we will eventually be forced against our will into some sort of steady-state economy. Meanwhile, according to Joseph Stiglitz (Stiglitz, 2011a): "We have created an ersatz capitalism with unclear rules — but with a predictable outcome: future crises."

The unfolding power shift is likely to create considerable risks. At the beginning of 2011, the situation was characterized by what some called an international currency war, where various countries, particularly the emerging economies, drive down the value of their currencies against the US dollar in order to boost their exports. The USA is in dispute with China, accusing its government of dampening the value of the renmimbi and thereby making US exports more expensive. Various other countries, such as Brazil, Japan, and South Korea have intervened either directly or through policy measures to control the rise of their currencies.

The G20 meeting in October 2010 decided to give the countries with fast-growing economies a greater say at the International Monetary Fund and agreed to let the markets exert more influence in setting foreign exchange rates. Nevertheless, the international currency war may eventually lead to trade tensions that would decrease global trade and economic activity worldwide. Recently, the IMF has addressed the problem of defining circumstances under which capital controls could usefully form part of a policy response to the large current wave of capital inflows into emerging economies, since these tend to increase the value of a currency (IMF, 2011).

In their third summit, held on the island of Hainan on 14 April 2011, the BRIC countries plus South Africa called for a greater economic and political role in international affairs. Their share of the world economy increased from 17.7% in 2001 to 24.2% in 2009. According to the first annual Social and Economic Development report by the BRIC countries, published by the Chinese Academy of Social Sciences, their average annual economic growth in the first decade of the 21st century was over 8%, while that of the industrialized countries reached only 2.6% (WSWS, 2011). In their April 2011 summit, they called for "a restructuring of the World War II-era global financial system and an eventual end to the long reign of the US dollar as the world's reserve currency". Russia, China, and Brazil had previously agreed to use their own currencies in bilateral trade, instead of the US dollar.

There are indications that the current US policy of 'quantitative easing', which in effect corresponds to the printing of dollars to ease US economic troubles is causing the rise in food and energy prices in China and threatening some of Brazil's export sectors. Tension between the USA and China is likely to increase as China grows faster and acquires more power, in spite of the deep financial and economic ties between the two countries. It remains to be seen to what extent this tension spreads into the geostrategic and military domains.

Power is shifting from states to non-state actors, and this may promote or oppose cooperation for development. The actors in the first group include national and international non-governmental organizations, religious organizations, business corporations, advocacy groups, and civil society in general. The second includes extremist, criminal, and terrorist organizations, and networks empowered by the access to sophisticated technologies, which represent an increasing threat to world security. Pakistan is a sensitive example and likely to become a terrible conundrum for the West. It was an essential platform for the Afghanistan war but its relationship with the USA is deteriorating. The country has nuclear weapons and is a fertile ground for anti-American movements and terrorist groups, in part because it has been embarrassed and humiliated by continuous covert US operations and drone attacks in its territory, including the killing of Osama Bin Laden. Furthermore, Pakistan is increasingly reliant on its alliance with China. Much of the future global geostrategic power game is likely to involve the alignments and tensions between the USA, India, China, and Pakistan.

Peace and security are also threatened by a significant number of weak or failing states in the less developed countries. The number and intensity of conflicts generated by these countries may overwhelm the capacity for international conflict management and containment. It is crucially important to recognize that peace and security are closely related to development, and that they cannot be achieved in a world with an increasingly inequitable distribution of wealth. The poorer, less developed countries have higher risks of becoming failed states, and from there turning into breeding grounds for criminal and terrorist organizations.

In spite of these facts, shocking inequalities in development persist throughout the world. For part of the population in the developed countries, and a very small part of the population in the developing countries, material prosperity and well-being have never been better in the history of mankind. In a recent article the economist Joseph Stiglitz (Stiglitz, 2011) speaking about the growing inequality in the USA wrote:

> The upper 1% of Americans are now taking nearly a quarter of the nation's income every year. The top 1% have the best houses, the best education, the best doctors, and the best lifestyles, but there is one thing that money doesn't seem to have bought: an understanding that their fate is bound up with how the other 99% live.

The same reasoning also applies to the whole world, although probably on different time scales. In stark contrast to the fate of the world's upper economic bracket, poverty, hunger, disease, lack of safe drinking water, lack of basic sanitation, and lack of access to electricity continue to affect hundreds of millions of people. It is only with difficulty that most people in the developed world genuinely recognize these people as fellow human beings.

Social and economic inequalities are growing as much between as within countries. The current capitalist economic system has been highly successful in transforming the unstinting greed of a few into a mechanism for economic growth. However, the system tends to create extremes and to aggravate inequalities rather than ease them. From an ethical point of view, it is surprising to observe our individual, collective, and institutional inaction. There are many good initiatives around the world but their effective impact is still small. Social solidarity at the global level has a relatively low priority.

Leaders of rich and poor countries pledged to build a better world by 2015 through their agreement to fulfil the United Nations Millennium Development Goals. They agreed to halve extreme poverty and hunger from 1990 levels, to reduce by two-thirds the child mortality rate, by three-quarters the maternal mortality rate, and to achieve universal primary education. In September of 2010, the world leaders that participated in a Millennium Development Goals Review Summit recognized that progress so far makes it almost impossible to meet those targets by the deadline. To achieve the goals, it is essential that the wealthy nations contribute effectively, in particular through 0.7% of GDP in official development aid, a pledge that was agreed in 1970 but never fulfilled, except by a small handful of countries.

Another distinctive characteristic of the present time is the rising competition to secure natural resources. The major economic powers are increasing geostrategic pressure to ensure the supply of scarce natural resources. A clear sign are the restrictions that China has placed on its exports of some rare earth elements, vital for the development of the green economy. They are scarce in the world, and about 95% of the global production comes from China, but the Chinese have repeatedly reduced export quotas over the past five years, so that they are now well below world demand. In 2010, neodymium, a rare earth essential for making lightweight magnets for large wind turbines, hybrid cars, and electronic devices such as iPhones, cost about US $ 40 000 a ton in China, and about twice that price outside China due to export restrictions. Lanthanum is another rare earth, needed to make catalytic converters that reduce the pollution of gasoline-powered cars. In 2010 it cost about US $ 5 000 a ton in China, and ten times more outside the country.

We are at a turning point as regards the essential resources of food, water, and energy. Demand already exceeds what can be sustained at current levels of consumption. In the medium and long term, competition between states will be further increased by population growth and climate change impacts. The challenges of food, water, and energy are clearly interrelated, but we lack an overall integrated framework to manage them, particularly at the global level. The competition between the major states to secure reliable supplies may lead to a breakdown of cooperation. Scarcities are likely to hit the least developed countries the hardest, increasing the risk of internal or interstate conflicts that may become regional.

Food prices are becoming increasingly volatile, threatening food security. The severe drought in Russia in the summer of 2010 and the country's ban on wheat exports pushed prices up to two-year highs, reviving memories of the 2007–2008 global food crisis. The rising cost of agricultural food commodities at the beginning of 2011 was a contributing factor in the uprisings in North Africa and the Middle East. In some countries such as Morocco, Algeria, Tunisia, Egypt, Pakistan, and Indonesia, more than 36% of total household spending goes on food, making people in these countries very vulnerable to increasing food prices.

Volatile food prices are leading import-dependent countries to seek opportunities to secure supplies overseas, in particular through land acquisitions in developing countries. According to a recent World Bank report (WB, 2010), investors tend to buy arable land in countries with weak land governance and fail to link investment to the country's broader development strategy. The data on land transfers is currently sketchy, but it shows that they are quite significant. Between 2004 and 2009 the land area transferred in the Sudan was 3.9 million hectares, and in Ethiopia 1.2 million (WB, 2010).

Concerning water, a recent study (Vorosmarty, 2010) concludes that 80% of the world's population lives in areas where the fresh water supply is not secure. Rich countries have been able to safeguard drinking water supplies by means of huge investments in dams, canals, aqueducts, and pipelines, but poor countries are much more vulnerable because they cannot afford such infrastructure. Surface water represents only about one per cent of the available freshwater on Earth, while groundwater accounts for about 30%. A reduction in the availability of groundwater would have very serious consequences for a growing human population. According to a recent study (Bierkens, 2010), global groundwater extraction is about 1 000 to

1 100 km^3 per year, far exceeding the recharge rate. The increasing disparity means that the world is relying ever more on non-renewable groundwater.

As previously indicated, the energy sector probably involves the most serious risks in the near future. Energy prices, particularly those of oil, are likely to become a major obstacle on the way to robust global economic growth of the kind experienced in the period from 2001 to 2007.

The current situation is also critical as regards the environment, although the most damaging consequences are likely to occur in the medium and long term. There are increasing signs that climate change is getting out of control. In spite of the efforts made by President Barack Obama, there is very little chance that an effective climate and energy bill will be approved in the US Congress before the Durban UNFCCC Conference at the end of 2011, because of Republican Party opposition. According to a IEA report issued in May 2011, global CO_2 emissions reached a record 30.6 Gt in 2010, which represents an increase of 5% over the previous record year of 2008. If annual global emissions of CO_2 attain 32 Gt much before 2020, the prospects of limiting the global average temperature increase to 2°C become utopistic.

Meanwhile, the risks involved in uncontrolled climate change are becoming ever clearer. A study based on the analysis of NASA satellite data (Zhao, 2010) shows that the terrestrial net primary production has decreased in the decade from 2000 to 2009 due to large-scale droughts, particularly in the southern hemisphere. A continuous decline would pose very serious threats to food security. This is likely to happen in the future. Climate scenarios indicate that drought may endanger much of the world in the coming decades if we fail to significantly reduce the global emissions of greenhouse gases.

A recent review paper (Dai, 2010) from the US National Center for Atmospheric Research used an ensemble of 22 climate models to project a comprehensive index of drought conditions throughout the world. The conclusions indicate that most of the western hemisphere, along with large parts of Eurasia, Africa, and Australia, may suffer an increasing threat of extreme drought during this century. In the oceans, the amount of phytoplankton in upper layers has declined markedly over the last century (Boyce, 2010). The decline results from the stratification of those layers induced by ocean warming, and this in turn is caused by climate change. This is a significant ecological issue, because phytoplankton sits at the base of marine food chains.

The ecological footprint of humanity has doubled since 1966, and in 2007 the footprint exceeded the Earth's biocapacity by 50% (WWF, 2010), meaning that the Earth takes 1.5 years to produce the resources that humanity consumes in one year. The Living Planet Index that assesses the changing state of biodiversity fell by almost 30% between 1970 and 2007. There is an increasingly divergent trend between the tropical regions, where the index declined by about 60%, and the temperate regions where it increased by 30%. This discrepancy clearly shows the grievous consequences for biodiversity, and for the environment generally, of a profoundly unbalanced world marked by large inequalities between rich and poor countries. Loss of biodiversity is a very dangerous medium to long term threat. It represents a serious threat to entire ecosystems and economies, and ultimately to humans themselves, if it goes on unchecked. There are positive signs that investors and corporations are beginning to understand these risks, establishing partnerships with the public sector and NGOs to halt the loss of biodiversity.

The results obtained at the tenth meeting of the Conference of the Parties to the Convention on Biological Diversity held at Nagoya in October 2010 were also encouraging. The negotiations were difficult, but there was goodwill and readiness to compromise. Agreement was finally reached on the text of the Nagoya Protocol, which includes the decision to expand nature reserves to 17% of the world's land and to 10% of the world's waters. The EU and Brazil were very active in securing the agreement, while China and India showed willingness to compromise. However, the USA is not a signatory to the Convention. The reason for the success in Nagoya in 2010 and the relative failure in Copenhagen in 2009 and Cancun in 2010 is that CO_2 emissions have a much stronger coupling to economic growth through the energy sector than biodiversity loss. Competing countries require stringent tests on the implementation of CO_2 emission reductions to check compliance with mitigation commitments, because they are directly related to growth, but are less demanding as regards checking biodiversity targets. In fact the last biodiversity goals established by the United Nations were far from being implemented by the countries that made the commitments. Let us hope that the implementation of the Nagoya Protocol is strictly controlled.

Contrary to the naive confidence of the Promethean discourse, science and technology do not have the capacity to solve all the supply problems relating to food, water, energy, and other natural resources, and nor do they have the key to stemming the environmental degradation that would result from trying to sustain the current paradigm of economic growth indefinitely around the whole world. Science and technology are unable to bring about magical breakthroughs, because their achievements are ultimately bounded by the laws of nature. Let us consider a simple example. It is impossible to stop the greenhouse effect in our atmosphere and to stop its intensification as a result of large atmospheric emissions of greenhouse gases produced by some human activities. Science and technology can only help to reduce the emissions from such activities, or help to substitute them by others with lower emissions, or devise geoengineering solutions to counteract global warming.

Homo sapiens is the result of a remarkable evolutionary process. We do not know of other comparable processes in the Universe leading to intelligent beings. Since the extinction of *Homo neanderthalis* about 30 000 years ago, the cultural evolution of *Homo sapiens* accelerated in an extraordinary way. We colonized all the continents and evolved from primitive hunter–gatherer societies, through the invention of agriculture and the industrial revolution, to the current highly complex and predominantly urban civilization, heavily dependent today on science and technology.

We have explored the Earth from the depths of the oceans to the peaks of the highest mountains, from the tropical forests to the most inhospitable deserts. We have gone beyond the Earth and explored outer space. We have deciphered the fundamental laws of nature and we are able to observe, interpret, and understand phenomena happening at the subatomic scale, on Earth, in the Solar System, in our galaxy, in intergalactic space, and up to the most distant confines of space. We have reconstructed the history of the Universe from its origin and we are able to foresee how it will end. We know in great detail the history of life on Earth and how living organisms function, interact, and evolve.

Beyond these truly remarkable achievements, we dominate the biosphere and have profoundly changed the surface of our planet. It has been shown that our impact on the atmosphere, the cryosphere, the oceans, the lakes, the rivers, the aquifers, the soils, the ecosystems, and the biodiversity is growing, and is often devastating. The rate of exploitation of natural resources and interference with Earth subsystems, particularly the atmosphere, the oceans, and the biosphere, is not sustainable in the medium to long term future. If we do not take these warnings seriously and find ways to a sustainable development, critical economic and environmental situations will become more and more likely throughout the 21st century. There is a growing awareness of this challenge, especially in the developed countries, but progress to address it has been slow. Without adequate responses, the risk will grow unrelentingly and time will provide the inevitable adjustments, as Francis Bacon said at the beginning of the 17th century (Bacon, 1605): "He who will not apply new remedies must expect new evils, for time is the greatest innovator."

The number of issues on the international agenda regarding financial and economic systems, development, environment, peace, and security is increasing, and their acuteness, urgency, and complexity are exceeding the capacity of national governments and international organizations to cope with them. Moreover, rapid globalization has implied that localized crises and threats to security and stability are no longer locally containable, but constitute a risk to the global international system. Our common future depends crucially on our ability to collectively address the wide variety of pressing and interrelated global challenges, from financial and economic crises to nuclear arms proliferation, and to environmental degradation. To address these new global challenges successfully, a more innovative and robust system of global governance will be needed, able to take legal action and enforce compliance. This may seem a very demanding goal that would take many decades to achieve. However, if we do not pursue it, the risk of repeated critical situations and widespread conflicts is likely to increase dangerously.

World crises of the next 10 to 20 years are likely to be dealt with by the present institutions of global governance. It is also likely that the developed countries, especially the USA, will continue to play a central role in securing world peace, and also the stability of international financial and economic systems. Nevertheless, the world balance of economic power will continue to change relentlessly.

Most European countries face a systemic relative decline of their economy and standards of living, particularly the less developed ones. There are signs of a new protest movement that rejects the traditional political and parliamentary system and opposes it through the organization of demonstrations, protest marches, and sit-ins. In Germany, the protesters have been called 'Wutburger' or 'enraged citizens'. Their main message is that the current political parties are out of touch with the democratic needs of the 21st century, which include complex issues such as national debt, corruption, immigration, and energy and environmental challenges.

Dissatisfaction with political parties is visible in southern Europe, particularly in Greece, Portugal, and Spain where sit-ins in the squares of the main cities began in May 2011. Protesters complain against unemployment, economic hardships imposed by the government, and outdated and corrupt judiciary and political systems. Up to now these have been non-ideological movements that seek more direct forms of democracy and political representation. They are probably a sign of the medium and long term inadequacy of the current Western political and parliamentary system to deal with the major and pressing questions of this century. The relative economic decline of the USA is also inevitable and currently a very hot discussion topic (Nye, 2011; Time, 2011a).

To take advantage of this situation by developing a stable non-growth economy in the more industrialized world would be a major achievement in our cultural evolution. It would be a decisive contribution for sustainable development and economic convergence with the developing countries, and also an opportunity for cultural flourishing. China is likely to continue increasing its economic power and political influence around the world, and may eventually lead an economically self-sufficient group of Asian countries, establishing a new regional order. Still, there remains quite a lot of uncertainty because social and political movements seeking democratic representation may derail the current ascending economic growth. It all depends on whether the majority of people in China feel that their expectations of better living conditions are being fulfilled by the present regime.

Returning to global governance the most probable scenario is that the reformation of the institutions will be slow and gradual. In a less likely scenario, future crises in the next 10 to 20 years may provide the opportunity and the incentive to promote greater cooperation among states, and to effectively strengthen global governance. This response is clearly the most favourable to create an enduring pathway for sustainable development. In this situation, countries agree to share their political power with a global governance system capable of addressing the challenges of food, water, energy, other natural resources, environmental degradation, peace, and security in an integrated way. In a third, least probable scenario, crises may decrease the cooperation between countries and global governance may become increasingly inoperative and irrelevant. Competition among countries trying to maintain their hegemony and power and their high levels of economic growth by securing access to resources and markets will lead to growing tensions. Conflicts then become much more probable.

Beyond 2030, the uncertainties in the scenarios become increasingly great and it is practically impossible to assess their relative likelihood. In the medium to long term, it is difficult to see how the current largely unregulated financial system can provide the foundation for sustainable global development. Eventually, the unrestricted freedom to seek profit with ever more ingenious and sophisticated financial products is likely to become incompatible with the accessibility to decreasing natural resources and ecosystem services, and with the conservation of nature and the environment. Here humanity is very probably at an expected crossroads.

As a biological species we have distinctive features. The key to our remarkable success story in the biosphere is our ability to develop an increasingly complex social life that has decoupled us from nature. The driving forces of our cultural evolution were the ever more complex social bonds for protection, cohesion, and cooperation within the group, infighting to reach the leadership roles in the group, and strategies for aggression and warfare between groups. This evolution allowed us to adapt our lifestyles to the most diverse environments from the tropics to the Arctic, and from coastal zones to the highest mountains, and to transform and use them to our benefit. Now we are reaching an extreme situation in this paradigm. We may believe that we can go on extrapolating it, even beyond the Earth. A more likely outcome is that we may have to pay for our success by learning to live in a sustainable way on Earth.

Epilogue II

In this long journey back through the history of civilizations, all the way up to the contemporary problems of development, natural resources, and the environment, something important has been missing. If this omission corresponded to a real deprivation it would create a feeling of imbalance and angst and an exaggerated sense of disquiet and even anguish. There is something essential in human nature that has not been brought to the fore.

We do not like to be faced with future scenarios filled with uncertainty and risk, and which some consider overstated and catastrophic. For the majority of people, the problems that have to be solved in the next few days, weeks, and months are overwhelming and constitute a heavy enough burden without the negative visions of a more distant future that will mainly affect only subsequent generations. We can hide behind the selfish assumption that our contribution to solving the problems of sustainability is minimal and hence practically unnecessary. But this is clearly erroneous.

Each of us is truly essential in the quest for solutions. We know that we should believe, appeal to, and promote intragenerational and intergenerational solidarity. We know that hunger, poverty, and misery afflict hundreds of millions of people and that, without some form of sustainable development, critical situations are likely to become more frequent and even to threaten the stability of our civilization. We know that there are profound inequalities in the world. We may say that our potential contribution to facing and solving these challenges is practically irrelevant. Again this is erroneous.

Some may argue that the inequalities have always been present, right through the history of civilizations, and that it would be impossible ever to eradicate them completely from human life. Whatever our vision about the present and the future, we are all more or less conscious that we live in a risk society and that the risk is likely to increase in the future. But again some may argue that there is no significant difference with past degrees of risk, and therefore that there is no reason for special concern.

How is it then possible to assume the human condition and live with its inexhaustible supply of deficiencies, miseries, frustrations, anguishes, failures, accidents, catastrophes, conflicts, wars, uncertainties, and risks? There are various experiences that can help us, in particular religion and the many forms of spirituality. However, an important one has not yet been mentioned. The omission was *art*, with its history and its capacity to stimulate and convey emotions and ideas that can make us transcend our condition. Art cannot be separated either from history or from the future pathway of the current globalizing civilization. Art in its various forms — plastic arts, music, opera, theatre, cinema, dance, poetry, literature, and architecture — is a fundamental part of our culture, like religion and spirituality, science and technology. Without art, the world would be incomprehensible, empty, much more hostile, and in fact unimaginable. Artistic creation is one of the defining characteristics of our species.

Since about 40 000 years ago, *Homo sapiens* has created a remarkable diversity of representations and objects with symbolic, ornamental, ritualistic, and aesthetic value that constituted pioneering artistic expressions. Each of the great civilizations — Egyptian, Persian, Indian, Chinese, Greek, Roman, Mayan, Incan, and European, among many others — developed forms of art with unique and characteristic styles. To appreciate the monumental and artistic remains of those civilizations and the magnificent expressions of Islamic, oriental, mediaeval, Renaissance, and Modern art, is an immense pleasure that can help us to understand the human odyssey and to mitigate the sorrows and miseries of our individual and collective lives.

Art is a way to surpass and exchange the hard reality with another one, more firmly built upon human values, and geared to promote new emotions that range from faith to aesthetic pleasure, the enjoyment of harmony, eroticism, argumentation, social criticism, and the deconstruction of the supposed real world, of social conventions, preconceptions, certainties and beliefs, to repulsion and nausea. It is the irrepressible impulse to create with the feeling of absolute freedom, and to reach for and express the deepest and most secret essence of our changing human identity.

References

ACIA (2004): Impacts of a Warming Arctic. Arctic Climate Impact Assessment, Cambridge University Press

ACUNU (2000): Global Challenges for Humanity: United Nations Millennium Summit and Forum — Special Edition, Washington DC

Adeel, Z., et al. (2005): Ecosystems and Human Well-Being: Desertification Synthesis, World Resources Institute, Washington, DC

Alfe, D., M.J. Gillan, L. Vocadlo, J. Brodholt, and G.D. Price (2002): The ab initio simulation of the Earth's core. Phil. Trans. R. Soc. Lond. A **360**, 1227-1244

Alley, R.B., et al. (2005): Ice-sheet and sea-level changes. Science **310**, 456–460

Allison D.B. et al. (1999): Annual deaths attributable to obesity in the United States. Journal of the American Medical Association **282** (16), 1530–1538

Allman, J.M. (1999): Evolving Brains, Scientific American Library Series, No. 68

Ambrose, S.H. (2001): Paleolithic technology and human evolution. Science **291**, 1748-1753

Ambrose, S.H. (1998): Late Pleistocene human population bottlenecks, volcanic winter and differentiation of modern humans. Journal of Human Evolution **34**, 623–651

Archer, D. et al. (2009): Atmospheric lifetime of fossil fuel carbon dioxide, Annual Review of Earth Planetary Sciences **37**, 117–134

Armstrong, K. (2001): Buddha, Penguin Books

Arrhenius, S. (1896): On the influence of carbonic acid in the air upon the temperature of the ground, The London, Edinburgh, and Dublin Philosophical Magazine and Journal of Science, 5th ser. (April), pp. 237–276

Arrhenius, S. (1908): Worlds in the Making: The Evolution of the Universe, New York: Harper & Bros

Ashman K.M., and P.S. Baringer (Eds) (2001): After the Science Wars, Routledge

Atran, S. (2006): The moral logic and growth of suicide terrorism. The Washington Quarterly **29:2**, 127–147

Aubreville, A. (1949): Climats, Forêts et Desertification de l'Afrique Tropicale, Société d'Editions Géographiques, Maritimes et Coloniales, Paris

Auty, R.M. (1993): Sustaining Development in Mineral Economies, Routledge

Bacher, M. (2005): Are humans still evolving? Science **309**, 234–237

Bacon, F. (1605): The Advancement of Learning, Book 1, Chapter 5, Section 3

Bahn, P.G., and J. Flenley (1992): Easter Island, Earth Island, Thames and Hudson, New York

Bairoch, P. (1997): Victoires et Déboires. Histoire Economique et Sociale du Monde du XVI Siècle à nos Jours, Gallimard, Paris

Balmford, A. (2002): Economic reasons for conserving wild nature. Science **297**, 950–953

Balter, M. (2004): Dressed for success: Neanderthal culture wins respect. Science **306**, 40–41

Bandourian, R., J.B. McDonald, and R.S. Turley (2002): A Comparison of Parametric Models of Income Distribution Across Countries and Over Time, Luxemburg Income Study Working Paper No. 305, Syracuse University, Maxwell School, Syracuse, USA

351

Barbier, E.B. (2009): Rethinking the Economic Recovery: A Global Green New Deal, UNEP

Barlett, R. (1993): The Making of Europe. Conquest, Colonization and Cultural Change 900–1350, Princeton University Press

Barrow, J.D., and F.J. Tipler (1986): The Anthropic Cosmological Principle, Oxford University Press

BBC (2009): BBC News, 15 January 2009

BBC (2009a): Wide dissatisfaction with capitalism — Twenty years after the fall of the Berlin wall, www.worldpublicopinion.org

Beck, U. (1992): The Risk Society. Towards a New Modernity, Sage, London

Becker, L., et al. (2004): Bedout: A possible end-Permian impact crater offshore of northwestern Australia. Science **304**, 1469–1476

Beckerman, W. (1974): In Defense of Economic Growth, Cape, London

Beckerman, W. (1995): Small is Stupid: Blowing the Whistle on the Greens, Duckworth, London

Bell, R. (1926): The Origin of Islam in Its Christian Environment, Macmillan, London

Benestad, R.E. (2002): Solar Activity and Earth's Climate, Springer Verlag, Berlin

Berger, A., and M.F. Loutre (2002): Science **297**, 1287

Berry, N., M. Corbin, C. Hellman, D. Smith, R. Stohl, and T. Valasek (2003): Military Almanac 2001–2002, CDI, Washington, DC

Besansky, N.J., et al. (2010): Distinct clones of *Yersinia pestis* caused the black death. PLoS Pathogens, 6(10):e1001134DOI

BI (2000): BirdLife International, Threatened Birds of the World, BirdLife International and Lynx Editions, Barcelona and Cambridge

Bierkens, M.F.P., et al. (2010): A worldwide view of groundwater depletion. Geophysical Research Letters, DOI 10.1029/GL044571

Bjornsson, J. (1998): The Potential Role of Geothermal Energy and Hydro Power in the World Energy Scenario in Year 2020. In: Proceedings of the 17th WEC Congress, Houston, Texas, Cambridge University Press

Bloom, J.M. (2001): Preprint, Yale University Press

Bookchin, M. (1982): The Ecology of Freedom: The Emergence and Dissolution of Hierarchy, Cheshire, Palo Alto

Boserup, E. (1981): Population and Technological Change: A Study of Long Term Trends, University of Chicago Press

Boyce, D.G., M.R. Lewis, and B. Worm (2010): Global phytoplankton decline over the past century. Nature **466**, 591–596

Bowles, S. (2009): Did warfare among ancestral hunter–gatherers affect the evolution of human social behaviour? Science **324**, 1293-1298

BP (2010): Statistical Review of World Energy

Brain, C.K., and A. Sillent (1988): Evidence from the Swatkrans cave for the earliest use of fire. Nature **336**, 464–466

Braungart, M. and W. McDonough (2002): Cradle to Cradle. Remaking the Way We Make Things, North Point Press

Brinkhoff, Th. (2006): The Principal Agglomerations of the World, www.citypopulation.de

Brockman, J. (1995): A Universe in Your Backyard, A. Guth. In: The Third Culture: Beyond the Scientific Revolution, Simon & Schuster

Brown, P., et al. (2004): A new small-bodied hominin from the late pleistocene of Flores, Indonesia. Nature 1055–1061

Brown, L.R., C. Christofer, and S. Postel (1992): Saving the Planet: How to Shape an Environmentaly Sustainable Global Economy, Earthscan, London

Brown, L.R. (2003): Plan B: Rescuing a Planet Under Stress and a Civilization in Trouble, Earth Policy Institute

Brown, L.R. (2006): Plan B 2.0, W.W. Norton & Company, New York

Brown, L.R. (2011): World on Edge. How to Prevent Environmental and Economic Collapse, W.W. Norton & Company, New York

Burger, W.C. (2003): Perfect Planet, Clever Species, Prometheus Books, New York

Burke, L., K. Reytar, M. Spalding, and A. Perry (2011): Reefs at Risk Revisited, World Resources Review

Calvin, W.H. (2002): A Brain for All Seasons: Human Evolution and Abrupt Climate Change, University of Chicago;

Campbell, M.J.C., A. Ezeh, and N. Prata (2007): Science 315, 1501–1502

Carter, B. (1974): Large number coincidences and the anthropic principle in cosmology, IAO Symposium 63, Confrontation of Cosmological Theories with Observational Data, Reidel, pp. 291–298

Cavalli-Sforza, L., and F. Cavalli-Sforza (1995): The Great Human Diasporas, Addison-Wesley

Changeux, J.-P. (1983): L'homme neuronal, Hachette

Changeux, J.-P. (2007): L'Homme artificiel, Colloque Annuel, Collège de France, Odile Jacob

Chauvin, G., et al. (2004): A giant planet candidate near a young dwarf. Astronomy and Astrophysics 425, L29–L32

Clapp, B. (1994): An Environmental History of Britain, Longman, London

Clausius, R. (1865): Ueber verschiedene für die Anwendung bequeme Formen der Hauptgleichungen der mechanischen Wärmetheorie. Annalen der Physik und Chemie 125, 353

Cobb, C., M. Glickman, and C. Cheslog (2001): The Genuine Progress Indicator: 2000 Update, Redefining Progress, San Francisco

Cobb, C., G.S. Goodman, and M. Wakernagel (1999): Why Bigger Isn't Better: The Genuine Progress Indicator, Redefining Progress, San Francisco

Coffey, P. (2008): Cathedrals of Science: The Personalities and Rivalries that Made Modern Chemistry, Oxford University Press, New York.

Cohen, B.A., T.D. Swindle, and D.A. Kring (2000): Support for the lunar cataclysm hypothesis from lunar meteorite impact melt ages. Science 1754–1756

Cohen, J.E. (1995): How Many People Can the World Support? W.W. Norton, New York

Cohen, J.E. (2003): Human population: The next half century. Science 302, 1172–1175

Collis, J. (1984): The European Iron Age, Batsford, London

Conard, N.J. (2009): A female figurine from the basal Aurignacian of Hohle Fels Cave in southwestern Germany. Nature 459, 248–252

Constanza, R., L.J. Graumlich, and W. Steffen (Eds) (2007): Sustainability or Collapse? MIT Press Cooper, W.H., 2009, Russia's economic performance and policies and their implications for the United States, Congressional Research Service, USA

Coppens, Y. (2004): Human Origins: The Story of Our Species, Hachette Illustrated

Cox, T.J. and A. Loeb (2007): The collision between the Milky Way and Andromeda. Monthly Notices of the Royal Astronomical Society 386, 461

Crutzen, P.J., A.R. Mosier, K.A. Smith, and W. Winiwarther (2006): N_2O release from agro-biofuel production negates climate effect of fossil fuel derived 'CO_2' savings. In publication

Crutzen, P.J., and J.W. Birkins (1982): The atmosphere after a nuclear war. Twilight at noon. Ambio, Vol. II, No 2–3, 114; Turco, R.P., O.B. Toon, T.P. Ackerman, J.B. Pollack, and C. Sagan (1983): Nuclear winter: Global consequences of multiple nuclear explosions. Science 222, 1283-1292

CSIS (2010): Center for Strategic and International Studies, Rare earth elements: A wrench in the supply chain

D'Eaubonne, F. (1974): Le Feminisme ou la Mort, Pierre Horay, Paris

Dai, A. (2010): Drought under Global Warming: A Review, Wiley Interdisciplinary Reviews: Climate Change

Dart, R. (1925): Australopithecus africanus: The man-ape of South Africa. Science 115, 195–199

Darwin, C. (1862): On the various contrivances by which British and foreign orchids are fertilized by insects, London, John Murray

Davies, P. (2007): Energy in Perspective, BP Statistical Review of World Energy

Dawson, J. (2007): Future of US nuclear weapons. A tangle of visions, science, and money. Physics Today 60 (2), 24–26

De Gruijl, F.R. (1995): Impacts of a projected depletion of the ozone layer. Consequences 1 (2), 12–21

Davis, S.J. and K. Caldeira (2010): Consumption-based accounting of CO_2 emissions, Procee-dings of the National Academy of Sciences, www.pnas.org/content/107/12/5687

Deffeyes, K.S. (2001): Hubert's Peak: The Impending World Oil Shortage, Princeton University Press

Der Spiegel (2010): 9 January issue

Der Spiegel (2010a): 16 September issue

Diamond, J. (2005): Collapse. How Societies Choose to Fail or Succeed, Viking

Donovan, N. and Halpern, D. (2002): Life satisfaction: The state of knowledge and implications for government, London, Cabinet Office

Doran, P.T. and M.K. Zimmerman (2009): Examining the scientific consensus on climate change. EOS **90** (3), 22–23

Drexler, E. (1986): Engines of Destruction, Anchor Books, New York

Dryzek, J.S. (1987): Rational Ecology: Environment and Political Economy, Basil Blackwell

Dryzek, J.S. (2005): The Politics of the Earth, Oxford University Press

Dumont, R. (1973): L'Utopie ou la mort, Editions Seuil

Duve, C. (2002): Life Evolving. Molecules, Mind and Meaning, Oxford University Press

Easterly, W. (2006): The White Man's Burden, The Penguin Press, New York

Edelman, G.M. (1987): Neural Darwinism, Basic Books; L'Homme Neuronal, J.P. Changeux, Fayard (1983)

Ehrlich, P.R., and A.H. Ehrlich (1991): Healing the Planet, Addison-Wesley

Ehrlich, P., and A. Ehrlich (2004): One with Nineveh: Politics, Consumption, and the Human Fu-ture, Island Press

EIA (2011): USA Energy Information Administration, www.eia.gov

El Pas (2009): 28 November

ENS (2011): European Nuclear Society, www.euronuclear.org

EP (2005): Technological Assessment of Converging Technologies, European Parliament, IP/A/STOA/ST/2006-6

ESI (2005): Environmental Sustainability Index, http://sedac.ciensin.columbia.edu/es/esi/

Ettinger, R.C.W. (1974): Man into Superman, Avon

Fagan, B. (2004): The Long Summer, Basic Books, New York

FAO (2001): The Global Forest Resources Assessment 2000, Food and Agriculture Organization

FAO (2004): The State of Food Insecurity in the World, Food and Agriculture Organization

FAO (2005): The Global Forest Resources Assessment, Food and Agricultural Organization

FAO (2006): Livestock's Long Shadow, Food and Agricultural Organization

FAO (2008): The State of Food Insecurity in the World, Food and Agricultural Organization

FAO (2009): State of the World Fisheries and Aquaculture 2008, Food and Agricultural Organi-zation

FAO (2010): The State of Food and Agriculture 2009. Livestock in the balance, Food and Agri-culture Organization

Farman, J.C., B.G. Gardiner, and J.D. Shanklin (1985): Large losses of total ozone in Antarctica reveal seasonal ClO_x/NO_x interaction. Nature **315**, 207-210

Ferguson, N. (2008): The Ascent of Money. A Financial History of the World, Penguin Books

Fernández-Jalvo, Y., et al. (1996): Evidence of early cannibalism. Science **271**, 277–278

Financial Times (2007): 13 March issue

FH (2010): Freedom House, Freedom in the World 2010

Finkelstein, I., and N.A. Silberman (2002): The Bible Unearthed: Archaeology's New Vision of Ancient Israel and the Origin of Its Sacred Texts, Free Press

Fisher, D.R. and W.R. Freudenburg (2001): Ecological modernization and its critics: Assessing the past and looking toward the future. Society and Natural Resourses **14**, 701–709

FM 2030 (1989): (Fereydun M. Esfandiary) Are you a transhuman? Monitoring and Stimulating your Personal Rate of Growth in a Rapidly Changing World, Viking Adult; Ettinger, Robert, 1974, Man and Superman, Avon

Fournier, V. (2008): Escaping from the economy: The politics of degrowth. International Journal of Sociology and Social Policy **38**, 528–545

Foucault, M. (1980): Power/Knowledge, Selected Interviews and Other Writings, 1972–1977, Harvester, Brighton

Franck, S., C. Bounama, and W. von Bloh (2006): Causes and timing of future biosphere extinctions. Biogeosciences 3, 85–92

Frank, A.G. (1998): ReOrient: Global Economy in the Asian Age, University of California Press

French, H. (2000): Vanishing Borders. Protecting the Planet in the Age of Globalization, W.W. Norton and Company, New York

Freud, S. (1929): Civilization and Its Discontents, W.W. Norton & Co

Freud, S. (1939): Moses and Monotheism, Vintage Books

Friedman, B.M. (2005): The Moral Consequences of Economic Growth, Alfred A. Knopf

Friedman, T.L. (2008): Hot, Flat, and Crowded, Allen Lane

FT (2009): Financial Times, 3 February issue

FT (2010): Financial Times, 23 June issue

Fukuyama, F. (2004): State Building, Profile Books, London

Gardner, H. (2007): An Embarrassment of Riches, Foreign Policy, May/June

Georgescu-Roegen, N. (1971): The Entropy Law and the Economic Process, Harvard University Press

Gibson, D.G., et al. (2010): Creation of a bacterial cell controlled by a chemically synthesized genome. Science 329, 52–54

Gilbert, W. (1986): Origin of life: The RNA world. Nature 319, 618

Gini, C. (1912): Variabilita e Mutabilita, Studio Economicogiuridici, Anno III, Parte 2a, Universita di Cagliari

Gleick, P.H. (2004): The World's Water 2004–2005, Island Press, Washington DC

Goebel, T. (1999): Pleistocene human colonization of Siberia and peopling of the Americas. An evolutionary approach. Evolutionary Anthropology 8 (6), 208–227

Gorindasamy, B., and K. Caldeira (2000): Geoengineering Earth's radiation balance to mitigate CO_2-induced climate change. Geophysical Research Letters 27, 2141–2144

Gould, S.J. (1989): Wonderful Life: The Burgess Shale and the Nature of History, W.W. Norton & Co

Green, R.E., et al. (2010): A draft sequence of the Neanderthal genome. Science 328, 710–722

Gregory, J.M., P. Huybrechts, and S.C.B. Raper (2004): Threatened loss of the Greenland ice-sheet. Nature 428, 616

Gross, P., and N. Levitt (1994): Higher Superstition: The Academic Left and Its Quarrels with Science, Johns Hopkins University Press, Baltimore

Gruijl, F.R. (1995): Impacts of a projected depletion of the ozone layer. Consequences 1 (2), 12–21

Guha, R. (1997): The environmentalism of the poor. In: R. Guhua and J. Martinez-Alier (Eds), Varieties of Environmentalism, Earthscan, pp. 3–21

Guinnessy, P. (2003): Pentagon revamps nuclear doctrine. Physics Today 56, 27–28

Hajer, M. (1995): The Politics of Environmental Discourse: Ecological Modernization and the Policy Process, Oxford University Press

Halper, S. (2010): The Beijing Consensus: How China's Authoritarian Model Will Dominate the Twenty-First Century, Basic Books, New York

Hansen, J., A. Lacis, R. Ruedy, and M. Sato (1992): Potential climate impact of Mount Pinatubo eruption. Geophysical Research letters 19, 215

Hardin, G. (1977): The tragedy of the commons. Science 162, 1243–1248

Hardin, G. (1993): Living Within Limits: Ecology, Economics, and Population Taboos, Oxford University Press

Harlan, J.R. (1995): The Living Fields: Our Agricultural Heritage, Cambridge University Press

Harrell, J.A., and V.M. Brown (1992): The world's oldest surviving geological map — the 1150 BC Turin papyrus from Egypt. Journal of Geology 100, 3–18

Harremoes, P., et al. (Ed) (2002): The Precautionary Principle in the 20th Century, Earthscan

Harris, R. (1986): The Origin of Writing, Duckworth, London

Hasselmann, K., M. Latif, G. Hooss, C. Azar, O. Edenhofer, C.C. Jaeger, O.M. Johannessen, C. Kemfert, M. Welp, and A. Wokaun (2003): The challenge of long-term climate change. Science **302**, 1293–1295

Hay, P. (2002): Main Currents in Western Environmental Thought, University of New South Wales, Sidney

Haynes, J. (1999): Power, politics and environmental movements in the third world. Environmental Politics **8** (1), 222–242

Heilbroner, R. (1991): An Inquiry into the Human Prospect: Looked at Again for the 1990s, Norton, New York

Hesse, M. (1992): Need a constructed reality to be non-objective? Reflections on science and society. In: The End of Science? Attack and Defence, R.Q. Elvee (Ed), University Press of America, Lanham

Hibbard K.A., et al. (2007): Group Report: Decadal-scale interactions of humans and the environment. In: Sustainability or Collapse, R. Constanza, L.J. Gramlich and W. Steffen (Eds), MIT Press

HIIK (2005): Heidelberg Institut für Internationale Konfliktforschung e.V. am Institut für Politische Wissenschaft der Universität Heidelberg, www.hiik.de

Hill, D. (1992): Islamic Technology: An Illustrated History, Routledge

Hiwatari R., et al. (2005): Demonstration tokamak fusion power plant for early realization of net electrical power generation. Nuclear Fusion **45**, 96–109

Hofferth, M.I., et al. (2002): Advanced technology paths to global climate stability: Energy for a greenhouse planet. Science **298**, 981–987

Holton, G. (1993): Science and Anti-Science, Harvard University Press

Hong, S., et al. (1996): History of ancient copper smelting pollution during Roman and medieval times recorded in Greenland ice. Science **272**, 246–248

Hong, S., et al. (1994): Greenland evidence of hemispheric lead pollution two millennia ago by Greek and Roman civilizations. Science **265**, 1841-1843

Howarth, R.W., R. Santoro and A. Ingraffea (2011): Methane and the greenhouse-gas footprint of natural gas from shale formations, Climate Change **106**, 679–690

Huber, J. (1982): Die Verlorene Unschuld der Okologie. Neue Technologien und Superindustriellen Entwicklung, Fisher Verlag, Frankfurt am Main

Huber, J. (1985): Regenbogengesellschaft. Okologie und Socialpolitik, Fisher Verlag, Frankfurt am Main

Huebner, J. (2005): A possible declining trend for worldwide innovation. Technological Forecasting and Social Change, 72, 980–986

Hulme, M. (2001): Climate perspective on Sahelian dessiccation: 1973–1998. Global Environmental Change **11**, 19–29

Hunt, T.L., and C.P. Lipo (2006): Late colonization of Easter Island. Science **311**, 1603–1606

Huntington, S.P. (1991): The Third Wave: Democratization in the Late Twentieth Century, Oklahoma University Press

Huntington, S.P. (1993): The Clash of Civilizations, Foreign Affairs **72** (3), 22–49

Huntington, S.P. (1996): The Clash of Civilizations and the Remaking of the World Order, Simon & Schuster, New York

Huxley, A. (1932): Brave New World, Chatto and Windus, London

Huxley, J. (1957): New Bottler for New Wine, Chatto and Windus, London

Huyssteen, J.W. (1998): Duet or Duel? Theology and Science in a Postmodern World, Trinity Press International

IAEA (2006): Nuclear Power Reactors in the World, International Atomic Energy Agency

IAEA (2007): Combating Illicit Trafficking in Nuclear and Other Radioactive Material, International Atomic Energy Agency

IEA (2004a): World Energy Outlook, International Energy Agency

IEA (2004b): Renewables in Global Energy Supply, International Energy Agency

IEA (2004c): Biofuels for Transport. An International Perspective, International Energy Agency

IEA (2006): Key World Energy Statistics, International Energy Agency

IEA (2008): World Energy Outlook, International Energy Agency
IEA (2008a): Energy Technology Perspectives, International Energy Agency
IEA (2009): World Energy Outlook, International Energy Agency
IEA (2009a): Key World Energy Statistics, International Energy Agency
IEA (2009b): World Energy Outlook 2009: Post-2012 Climate Policy Framework, International Energy Agency
IEA (2010): Technological Roadmap: Solar Photovoltaic Energy, International Energy Agency
IEA (2010a): Technological Roadmap: Concentrating Solar Power, International Energy Agency
IEPC (2010): Interactive Extra-Solar Planets Catalog, http://exoplanet.eu/
IGAS (2006): Institute for the Analysis of Global Security, www.iags.org
IHGSC (2001): International Human Genome Sequencing Consortium, Initial sequencing and analysis of the human genome project. Nature **409**, 860–921
IHS (2007): Information Handling Services, Report of 18 April
IMF (2010): World Outlook Database, 2010, International Monetary Fund
IMF (2010a): Sub-Saharan Africa. Back to High Growth? International Monetary Fund
IMF (2011): Managing Capital Inflows: What Tools to Use? International Monetary Fund
International Herald Tribune (2006): 19 September issue
International Herald Tribune (2007): 30 August issue
International Herald Tribune (2009): 23 January issue
IPCC (2000): Intergovernmental Panel on Climate Change, Emissions Scenarios. A Special Report of Working Group III of the IPCC, Cambridge University Press
IPCC (2001): Intergovernmental Panel on Climate Change, Contributions of Working Groups I, II and III to the IPCC Third Assessment Report, Cambridge University Press
IPCC (2007): Intergovernmental Panel on Climate Change, Contributions of Working Groups I, II and III to the IPCC Fourth Assessment Report, Cambridge University Press
IUCN (1980): International Union for Conservation of Nature and Natural Resources, World Conservation Strategy: Living Resources Conservation for Sustainable Development, Gland, Switzerland
Jackson, J.B.C. (2001): Historical overfishing and the recent collapse of coastal ecosystems. Science **293**, 629
Jackson, T. (2009): Prosperity without growth. A transition to a sustainable economy, UK Sustainable Development Commission
Jacque, L.L. (2010): International Herald Tribune, 11–12 December issue
Jacques, M. (2009): When China Rules the World, The Penguin Press
Jahn, D. (1998): Environmental performance and policy regimes: Explaining variations in 18 OECD countries. Policy Sciences **31**, 107–131
Janicke, M. (1985): Preventive Environmental Policy as Ecological Modernisation and Structural Policy, Wissenschaftszent, Berlin
Jaspers, K. (1949): Vom Ursprung und Zeit der Geschichte, Piper Verlag, Munchen
Jenkins, A. and D. Holland (2007): Melting of floating ice and sea level rise. Geophysical Research Letters **34**, L16609, doi:10.1029/2007 GL030784
Jenkins, M. (2003): Prospects for biodiversity. Science **302**, 1175–1177
Jerison, H.J. (1991): Brain Size and the Evolution of Mind, 59th James Arthur Lecture, American Museum of Natural History
Johansen, B.E. (2002): The Global Warming Desk Reference, Greenwood Press
Johanson, D., and M. Edey (1981): Lucy: The Beginnings of Humankind, Simon and Schuster
Kalas, P., et al. (2008): Optical images of an extrasolar planet 25 light-years from Earth. Science **322**, 2348–1352
Karachentesev, I.D., and O.G. Kashibadze (2006): Masses of the local group and of the M81 group estimated from distortions in the local density field. Astrophysics **49**, 3–18
Kasting, J.F. (1993): Earth's early atmosphere. Science **259**, 920
Katsh, A.I. (1954): Judaism in Islam: Biblical and Talmudic Backgrounds of the Koran and its Commentaries Suras II and III, Bloch Publishing, New York
Kenrick, P., and P.R. Crane (1997): Origin and early evolution of plants on land. Nature **389**, 33–34

Kenworthy, J., and F. Laube (2002): Urban transport patterns in a global sample of cities and their linkages to transport infrastructures, land use, economics and environment. World Transport Policy and Practice **8** (3), 5–20

Kerr, R.A. (1998): Sea-floor dust shows drought felled Akkadian Empire. Science **279**, 325–326

Kerr, R.A. (2006): A worrying trend of less ice, higher seas. Science **311**, 1698–1701

Kevles, D.J. (1995): The Physicists. The History of a Scientific Community in Modern America, Harvard University Press

Klare, M.T. (2009): The Era of Extreme Energy, www.countercurrents.org

Korycansky, D.G., G. Laughlin, and F.C. Adams (2001): Astronomical engineering: A strategy for modifying planetary orbits. Astrophysics and Space Sciences **275**, 349–366

Kramer, D. (2009): Nuclear weapons at a crossroads as Obama takes office. Physics Today **62**, 19

Krasner, D., D. Stephen, and C. Pascual (2005): Addressing state failure. Foreign Affairs **84** (4), 153–163

Kuhn, T. (1962): The Structure of Scientific Revolutions, 2nd edn, University of Chicago Press (1970)

Kurzweil, R. (2005): The Singularity Is Near: When Humans Transcend Biology, Viking Books

Kuznets, S. (1955): Economic growth and income inequality. American Economic Review **49**, 1–28

Lamb, H. (1996): The rise of global governance, Environmental Conservation Organization

Lamont, L. (1965): Day of Trinity, Atheneum, New York

Larsen, C.S. (1996): Biological changes in human populations with agriculture. Annual Review of Antropology **24**, 185–213

Latour, B., and S. Woolgar (1979): Laboratory Life: The Social Construction of Scientific Facts, Sage, London

Latour, B. (1987): Science in Action: How to Follow Scientists and Engineers Through Society, Harvard University Press

Lawton, J.M. and R.M. May (2005): Extinction Rates, Oxford University Press

Le Monde (2006): L'Atlas du Monde Diplomatique

Le Monde (2006a): Manière de Voir, Confucius, Mao, Le Marché … Jusqu'où Ira la Chine? Le Monde Diplomatique, No 85, February/March

Le Monde (2007): 25 August issue

Le Monde (2009): 21 July issue

Le Monde (2010): 30 March issue

Le Monde (2010a): 23 July issue

Le Quéré, C. et al. (2009): Trends in the sources and sinks of carbon dioxide. Nature Geoscience **2**, 831–836

Leakey, R.E., et al. (1992): Origins Reconsidered, Doublesday, New York

Leakey, M.D., and R.L. Hay (1979): Pliocene footprints in the Laetolil beds at Laetoli, Northern Tanzania. Nature **278**, 317

Leakey, R., and R. Lewin (1994): The Origin of Mankind, Basic Books

Lenczowski, G. (1989): American Presidents and the Middle East, Duke University Press

Lev-Yadun, S., A. Gopher, S. Abbo (2000): The cradle of agriculture. Science **288**, 1602–1603

Lewin, R., and R.A. Foley (2004): Principles of Human Evolution, Blackwell, Oxford

Lineweaver, C.H., Y. Fenner, and B.K. Gibson (2004): The galactic habitable zone and the age distribution of complex life in the Milky Way. Science **303**, 59–62

Lissauer, J.J. (1993): Planet formation. Ann. Rev. Astr. Astrophysics **31**, 129

Lomborg, B. (2001): The Skeptical Environmentalist. Measuring the Real State of the World, Cambridge University Press

Lomborg, B. (2010): Smart Solutions to Climate Change. Comparing Costs and Benefits, Cambridge University Press.

Lovelock, J.E., and M. Whitfield (1982): The lifespan of the biosphere. Nature **296**, 561-563

Lynn, R. (2001): Eugenics: A Reassessment (Human Evolution, Behaviour and Intelligence), Praeger Publishers

Lyotard, J.F. (1979): La Condition Postmoderne: Rapport sur le Savoir, Les Editions de Minuit

Maddison, A. (2001): The World Economy: A Millennial Perspective, OECD, Paris

Maddison, A. (2010): Percentage of World GDP, Visualizing Economics

Mahadevan, T.M.P. (1956): Outlines of Hinduism, Chetana, Bombay

Maisels, C.K. (2005): The Emergence of Civilization, Taylor and Francis

Malthus, T.R. (1798): An Essay on the Principle of Population, as it Affects the Future Improvement of Society with Remarks on Speculations of Mr. Godwin, M. Condorcet, and Other Writers, printed for J. Johnson in St. Paul's Chuchyard

Manes, C. (1990): Green Rage: Radical Environmentalism and the Unmaking of Civilization, Little Brown and Company

Mann, M.E. and K.A. Emanuel (2006): Atlantic hurricane trends linked to climate change. EOS 87, 233–244

Mann, T. (1983): Diaries 1918–1939, A. Deutsch, London

Markandaia, A., P. Haron, L. Bellu, and V. Cistulli (2002): Environmental Economics for Sustainable growth: A Handbook for Practitioners, Edgar Elgar

Marois, C., et al. (2008): Direct imaging of multiple planets orbiting the star HR 8799. Science 322, 1348–1352

Marshall, E.A. (1915): Edison's Plan for Preparedeness, New York Times Magazine

Martin, P.S., and R.G. Klein (Eds) (1984): Quaternary Extinctions, University of Arizona Press

Mather, A.S. (1990): Global forest resources, Belhaven Press

Mayor M., and D. Queloz (1995): A Jupiter-mass companion to a solar-type star. Nature 378, 355–359

Mayr, E. (2001): What Evolution Is, Basic Books

McBrearty, S., and N.G. Jablonski (2005): Nature 437, 105–108

McCluney, R. (2005): Renewable energy limits. In: The Final Energy Crisis, A. McKillop and S. Newman (Eds), Pluto Press

McCullum, H. (2006): Fuelling Fortress America. A Report on the Athabasca Tar Sands and US Demands for Canada's Energy, Canadian Centre for Policy Alternatives and Parkland Institute

McGuffie, K.A. Henderson-Sellers, and H. Zhang (1998): Modelling climate impacts of future rainforest destruction. In: Maloney, B.K. (Ed), Human Activities and the Tropical Rainforest, Kluwer, pp. 169-193

McKinsey (2010): Impact of the Financial Crisis on Carbon Economics, Version 2.1 of the Global Greenhouse Gas Abatement Cost Curve, McKinsey & Company

McKinsey (2011): Urban World. Mapping the Economic Power of Cities, McKinsey & Company

McNeill, J. (2000): Something New Under the Sun: An Environmental History of the Twentieth-Century World, Norton

MDG (2000): Millennium Development Goals, United Nations, www.un.org/millenniumgoals/

Meadows, D.L., and J. Randers (1992): Beyond the Limits: Confronting Global Collapse, Envisioning a Sustainable Future, Chelsea Green

Meadows, D.L., J. Randers, and D.L. Meadows (2004): Limits to Growth: The 30-Year Update, Chelsea Green

Meadows, L.H., D.L. Meadows, J. Randers, and W.W. Behrens III (1972): The Limits to Growth, Universe Books, New York

Meitner, L., et al. (1939): Desintegration of uranium by neutrons: A new type of nuclear reaction. Nature 143, 239–240

MIGREUROP (2007): www.migreurop.org

Milankovitch, M. (1930): Matematische Klimalehre und Astronomische Theorie der Klimaschwankungen. In: Hanbuck der Klimatologie, W. Koppen and R. Geiger (Eds), Vol.1, Part A 1–176, Berlin, Borntraeger

Miller G.F., et al. (1999): Pleistocene extinctions of Genyornis newtoni: Human impact on Australian megafauna. Science, 283, 205–208

Mojzsis, S.J. (1996): Evidence for Life on Earth before 3800 million years ago. Nature 384, 55

Monk, R. (1990): Ludwig Wittgenstein: The Duty of a Genius, Free Press

Molina, M.J., and F.S. Rowland (1974): Stratospheric sink for chlorofluoromethanes: Chlorine atom catalysed destruction of ozone. Nature 249, 810–812

Monod, J. (1971): Chance and Necessity: An Essay on the Natural Philosophy of Modern Biology, Alfred A. Knopf

Moore, G.E. (1965): Cramming more components onto integrated circuits. Electronics Magazine **38**, No. 8, 19 April

Morelli, G., et al. (2010): *Yersinia pestis* genome sequencing identifies patterns of global phylogenic diversity. Nature Genetics **42** (12), 1140–1143

Myers, R.A., and B. Worm (2003): Rapid worldwide depletion of predatory fish communities. Science **423**, 280

Naess, A. (1989): Ecology, Community and Lifestyle, Cambridge University Press

NASA (2007): www.neo.jpl.nasa.gov/risk/

Nature (2010): Science scorned (editorial). Nature **467**, 133

New York Times (1990): 19 April issue

Newsweek (2011): 2 May issue

Nielsen, R. (2006): The Little Green Handbook, Picador, New York

Nigosian, S.A. (1990): World Faiths, St. Martin's Press, New York

NOAA (2011): Earth System Research Laboratory, Global Monitoring Division, National Oceanic & Atmospheric Administration

Noonan J.P., et al. (2006): Science **314**, 1113–1118

Nordhaus, W.D. (2002): The Economic Consequences of a War with Iraq: Costs, Consequences and Alternatives, Committee on International Security Studies, American Academy of Arts and Sciences, Cambridge, MA, pp. 51–86

Norris, R.S., and W.M. Arkin (2000): Global nuclear stockpiles 1945–2000. The Bulletin of Atomic Scientists **56** (2), 75

Norval, M. et al. (2007): The effects on human health from stratospheric ozone depletion and its interactions with climate change. Photochemical and Photobiological Sciences **6** (3), 232–251

Noss, R.F. (1992): Issues of scale in conservation biology. In: P.L. Fiedler and S.K. Jain (Eds), Conservation Biology, Chapman Hall, New York; Wilson (1988)

NSF (2003): Converging Technologies for Improving Human Performance: Nanotechnology, Biotechnology, Information Technology and Cognitive Science, National Science Foundation, Kluwer Academic Publishers

NSSUS (2002): National Security Strategy of the United States, Washington

Nye, Jr., J.S. (2011): The future of power, Public Affairs

NYT (2004): New York Times, 3 November 2004

NYT (2011): New York Times, 22 January issue

OCHA (2007): OCHOA-oPt Protection of Civilians, United Nations Office for the Coordination of Humanitarian Affairs

OCHOA (2009): Monitoring disaster displacement in the context of climate change, United Nations Office for the Coordination of the Humanitarian Affairs

Oerlemans, J. (2005): Extracting a climate signal from 169 glacier records. Science **306**, 1686

Ophuls, W. (1977): Ecology and the Politics of Scarcity, W.H. Freeman

Oreskes, N. (2004): The scientific consensus on climate change. Science **306**, 1686

Owen, R. (1858): Supplementary Appendix to the Life of Robert Owen, Published by Effingham Wilson, London

Page, M. (2002): The first global village, Casa Das Letras

Pádua, J.A. (2004): Nature conservation and nature building in the thought of a Brazilian founding father: José Bonifácio (1763–1838), Working Paper CB5-53-04, Center for Brazilian Studies, University of Oxford

Palmer, T. (2005): Global warming in a nonlinear climate — Can we be sure. Europhysics News, European Physical Society, March/April, 42-46

Pape, R.A. (2005): Dying to Win: The Strategic Logic of Suicide Terrorism, New York, Random House

PBS (2004): Is Wal-Mart Good for America?, Public Broadcasting Service, Frontline, 16 November

PCI (2010): Post Carbon Institute, Energy Bulletin

Pearsall, D. (1992): The origin of plant cultivation in South America. In: The Origins of Agriculture, C.C. Wesley and P. Watson (Eds), Smithsonian Institute Press

Perman, R., Y. Ma, J. McGilvray, and M. Common (1999): Natural Resource and Environmental Economics, Longman

Plumwood, V. (1995): Has democracy failed ecology? Environmental Politics **4**, 134–168

Polkinghorne, J. (1998): Belief in God in an Age of Science, Yale University Press

Pope, R.A. (2005): Dying to Win: The Strategic Logic of Suicide Terrorism, Random House, New York

Porter, R. (1998): The Greatest Benefit to Mankind. A Medical History of Humanity, W.W. Norton, New York

Potter, C.W. (2001): A history of influenza. Journal of Applied Microbiology **91** (4), 572–579

Powell, A., S. Shennan, and M.G. Thomas (2009): Late Pleistocene demography and the appearance of modern human behaviour. Science **324**, 1298-1301

Rahmstorf, D. (2007): A semi-empirical approach to projecting future sea-level rise. Science **315**, 368–370

Ramo, J.C. (2004): The Beijing Consensus, The Foreign Policy Centre, London

Rashed, R. (1990): A pioneer in anaclastics: Ibn Sahl on burning mirrors and lenses. Isis **81**, 464–491

Raup, D., and J.J. Sepkosky Jr. (1982): Mass extinctions in the marine fossil record. Science **215**, 1501–1503 Renner, M.,2001,War trends mixed, in Vital signs 2001 :The trends that are shaping our future, L.Starke (editor),WWI, W.W.Norton, New York

Reeves, H., et F. Lenoir (2003): Mal de Terre, Editions du Seuil

REN21 (2006): Renewables Global Status Report, 2006 Update, www.ren21.net

Ridel, B. (2007): Al-Qaeda strikes back. Foreign Affairs **86** (3), 24–40

Rideout III, W.M., K. Eggan, and R. Jaenisch (2001): Nuclear cloning and epigenetic reprogramming of the genome. Science **93**, 1093–1098

Ritvo, H. (2003): Fighting for Thirlmere — The roots of environmentalism. Science **300**, 1510

Rockstrom, J. et al. (2009): A safe operating space for humanity. Nature **461**, 472–475

Rolt, L.T.C., and J.S. Allen (1997): The Steam Engine of Thomas Newcomen, Moorland Publishing Company

Rosen, C.M. (1995): Businessmen against pollution in nineteenth-century Chicago. Business History Review **69**, 351–397, cited by John McNeill (McNeill, 2000)

Rothschild, W., and K. Jordan (1903): A revision of the lepidopterous family *Sphingidae*, London and Aylesbury, Hazell, Watson and Viney, pp. 30-32

Roubini, N. (2010): Crisis Economics. A Crash Course in the Future of Finance, Allen Lane

RS (2009): Geoengineering the climate: Science, governance and uncertainty. The Royal Society, UK

RS (2011): Knowledge, networks and nations: Global scientific collaboration in the 21st century, The Royal Society, UK

Runge, C.F., and B. Senauer (2007): How biofuels could starve the poor. Foreign Affairs, May/June 2007, 41–53

Ruse, M. (2008): Evolution: A challenge standing on shaky clay. Science **322**, 47–48

Sackmann, I.J., A.I. Boothroyd, and K.E. Kraemer (1993): Our Sun III. Present and future. Astrophysical Journal **418**, 457-458

Sachs, W. (1992): The Development Dictionary, ed. by W. Sachs, Zen Books

Samuels, D. (2005): The end of the plutonium age. Discover, November 2005, 42

Santos, F.D. (2009): Energy and climate change: Innovation and public policy, in Brimmer, E., ed., Power Politics: Energy Security, Human Rights and Transatlantic Relations, Washington D.C., Center for Transatlantic Relations, pp. 69–80

Sayer, J., and B. Campbell (2003): The Science of Sustainable Development: Local Livelihoods and the Global Environment, Cambridge University Press

Sayili, A. (1960): The observatory in Islam and its place in the general history of the observatory. Turkish Historical Society Series VII, No 38, Turk Tarih Kurumu Basimevi, Ankara

Schiestl, F.P., M. Ayasse, H.F. Paulus, C. Lofstedt, B.S. Hansson, F. Ibarra, and W. Francke (1999): Orchid pollination by sexual swindle. Nature **399**, 421

Schopf, J.W. (1999): The Cradle of Life, Princeton University Press

Schroder, K.P. and R.C. Smith (2008): Distant future of the Sun and Earth revisited. Monthly Notices of the Royal Astronomical Society **386**, 155–163

SDC (2009): Sustainable Development Commission, UK, Prosperity without growth: The transition to a sustainable economy

Seife, C. (2000): Zero: The Biography of a Dangerous Idea, Penguin

Sen, A. (1984): The living standard, Oxford Economic Papers **36**, 74–90

SEP (2008): Stanford Encyclopedia of Philosophy

SEPA (2003): State Environmental Protection Administration of China, China's Environmental
Situation

Shiklomanov, I. (1993): World fresh water resources. In: Water in Crisis: A Guide to the World's
Fresh Water Resources, P.H. Gleick (Ed), Oxford University Press

Shelley, M. (1818): Frankenstein: Or, the Modern Prometheus, Harding, Mavor & Jones, London

Shelley, M. (1826): The Last Man, Henry Colburn, London

Shiva, V. (2002): Vedic Ecology. Practical Wisdom for Surviving the 21st Century, Mandala Publishing Group, Novato, California

Siegenthaler et al. (2005): Stable carbon cycle–climate relationship during the late Pleistocene.
Science **310**, 1313–1317

Simon, J., and H. Kahn (Eds) (1984): The Resourceful Earth: A Response to Global 2000, Basil
Blackwell, New York

Simon, J. (1981): The Ultimate Resource, Princeton University Press

Simon, J. (1996): The Ultimate Resource II, Princeton University Press

Singer, S.F. (1998): Testimony before the House Small Business Committee on 29 July 1998

SIPRI (2003): SIPRI (Stockholm International Peace Research Institute) Yearbook, Armaments,
Disarmaments and International Security, Oxford University Press

SIPRI (2006): The SIPRI (Stockholm International peace Research Institute) Military Expenditure Database, www.sipri.org

SIPRI (2009): The SIPRI (Stockholm International peace Research Institute) Military Expenditure Database, www.sipri.org

SIPRI (2010): The SIPRI (Stockholm International peace Research Institute) Military Expenditure Database, www.sipri.org

Smith, A. (1776): An Inquiry into the Nature and Causes of the Wealth of Nations

Smith, A.K. (1965): A Peril and a Hope: The Scientists' Movement in America 1945–1947, Chicago

Smith, B.D. (1997): The initial domestication of *Cucurbita pepo* in the Americas ten thousand
years ago. Science **276**, 932–934

Sokal, A. (1996): Transgressing the boundaries: Toward a transformative hermeneutics of quantum gravity. Social Text **46–47**, 217–252

Sorenson, B. (1995): History of, and recent progress in, wind-energy utilization. Annual Review
of Energy and the Environment **20**, 387–424

Spengler, O. (1918): The Decline of the West, Oxford University Press (1991)

Spergel, D.N. et al., (2003): First-year Wilkinson Microwave Anisotrophy Probe (WMAP) observations: Determination of cosmological parameters. Astrophysical Journal Supplement Series
148, 175-194

Stanley, S. (1998): Children of the Ice Age: How a Global Catastrophe Allowed Humans to
Evolve, W.H.Freeman

Steig, E.J., et al. (2009): Warming of the Antarctic ice-sheet surface since the 1957 International
Geophysical Year. Nature **457** (7228) 459–462

Stern, N. (2006): Stern Review on the Economics of Climate Change, HM Treasury, London

Stiglitz, J.E. (2002): Globalization and Its Discontents, Penguin Books

Stiglitz, J. (2011): Vanity Fair, May issue

Stiglitz, J. (2011a): Freefall. Freemarkets and the Sinking of the Global Economy, Allen Lane

Stock, G. (2003): Redesigning Humans. Choosing our Genes, Changing our Future, Mariner
Books

Stone, E.J. et al. (2010): The effect of more realistic forcings and boundary conditions on the modelled geometry and sensitivity of the Greenland ice-sheet. The Cryosphere Discussions **4**, 233–285

Sussman, R.W. (1991): Primate origins and the evolution of angiosperms. American Journal of Primatology **23**, 209-223

Svensmark, H. (2007): Cosmoclimatology: A new theory emerges. Astronomy and Geophysics **48**, 18–24

Szaro, R., and B. Shapiro (1990): Conserving our Heritage. America's Biodiversity, The Nature Conservancy, Arlington, VA

Tallmadge, J. (1997): Meeting the Tree of Life: A Teacher's Path, University of Utah Press

Testot, L. (2008): Les religions dans le monde: Que disent les chiffres? Sciences Humaines No 168, November 2008

The Economist (2006): 16 September issue

The Economist (2007): 8 September issue

The Economist (2010): 9 December issue

The Millennium Project (2010): 2010 State of the Future

The National (2011): 13 January issue

The Times (2005): 17 November issue

The Times of India (2011): 22 February issue

Thomas, D.W. (1965): Documents from Old Testament Times, Harper and Row, New York

Thomas, P.J., C.F. Chyba, C.P. McKay (Eds) (1997): The origin of the atmosphere and of the oceans, A. Delsemme. In: Comets and the Origin and Evolution of Life, Springer, Berlin

Time (2007): 10 September issue

Time (2011): 28 February issue

Time (2011a): 14 March issue

Tinetti, G. (2007): Water vapour in the atmosphere of a transiting planet. Nature **448**, 169–171

Trauth, M.H., M.A. Maslin, and M.R. Strecker (2005): Late Cenozoic moisture history of East Africa. Science **309**, 2051–2053

Turco, R.P., O.B. Toon, T.P. Ackerman, J.B. Pollack, and C. Sagan (1983): Nuclear winter. Global consequences of multiple nuclear explosions. Science **222**, 1283–1292

Turco, R.P. (1997): Earth Under Siege: From Air Pollution to Global Change, Oxford University Press; De Gruijl, F.R. (1995): Impacts of a projected depletion of the ozone layer. Consequences **1** (2), 12-21

Turner, G.M. (2008): A comparison of *The Limits to Growth* with 30 years of reality. Global Environmental Change **18**, 394–411

Udry, S. (2007): The HARPS search for southern extra-solar planets. XI. Super-Earths (5 & $8M_{Earth}$) in a 3-planet system. Astronomy and Astrophysics, preprint

UNAIDS/WHO (2005): AIDS Epidemic Update, Joint United Nations Programme on HIV/AIDS

UNCCD (1996): United Nations Convention to Combat Desertification, www.unccd.int

UNDESA (2008): United Nations Department of Economics and Social Affairs/Population Division, World Population Prospects: The 2008 Revision

UNDP (1999): Human Development Report 1999, United Nations Development Programme

UNDP (2002): Human Development Report 2002, United Nations Development Programme

UNDP (2003): Human Development Report 2003, United Nations Development Programme

UNDP (2007): Human Development Report 2007, United Nations Development Programme

UNEP (1992): Convention on Biological Diversity, United Nations Environment Programme, Nairobi

UNEP (1995): Global Biodiversity Assessment, United Nations Environmental Programme, Cambridge University Press

UNEP (2000): Malmo Ministerial Declaration, Global Ministerial Environment Forum, United Nations Environment Program

UNEP (2008): Vital Water Graphics, United Nations Environmental Programme

UNEP (2009): Global Green New Deal, An Update for the G20 Pittsburgh Summit, United Nations Environmental Programme

UNEP (2009a): Climate in Peril, United Nations Environmental Programme

UNFPA (2010): United Nations Population Fund

UNFCCC (2007): Climate change: Impacts, vulnerabilities and adaptation in developing countries. United Nations Framework Convention on Climate Change

UNGA (2009): United Nations General Assembly Resolution A/RES/64/236

UNH (2010): State of the world's cities 2010/2011 — cities for all: Bridging the urban divide. United Nations Human Settlements Programme, UN-Habitat

UNH (2011): Urbanisation facts and figures. United Nations Human Settlements Programme, UN-Habitat

UNHCR (2006): United Nations High Commissioner for Refugees, www.unhcr.org

UNICEF (2008): State of the World's Children 2008, UNICEF, January 2008

UNMC (2002): United Nations Millennium Campaign, www.millenniumcampaign.org

UNMDG (2000): United Nations Millennium Development Goals, http://millenniumgoals/

UNPD (2004): World Population in 2300, United Nations Population Division, New York

UNPD (2007): World Population Prospects: The 2006 Revision, United Nations Population Division, New York

UNPD (2005): World Population Prospects: The 2004 Revision, United Nations Population Division, New York, 2005

USDOE (2010): U.S. Department of Energy, Critical Materials Scarcity

USDS (2006): United States Department of State, Office of the Coordinator for Reconstruction and Stabilization, www.state.gov/s/crs/

USLP (2011): Iraq war facts, results and statistics at April 26, 2011. US Liberal Politics

Valencia, R., H. Balslev, and G. Paz y Miño (1994): High tree alpha-diversity in Amazonian Ecuador. Biodiversity and Conservation 3, 21–28

Van de Walle, N. (2001): African Economies and the Politics of Permanent Crisis 1979–1999, Cambridge University Press

Vermeer, M. and S. Rahmstorf (2009): Global sea level linked to global temperature. Proceedings of the National Academy of Sciences 106 (51), 21527–21532

Virk, Z. (2003): Europe's debt to the Islamic world. Review of Religions 98 (07), 36–52

Vorosmarty, C.J., et al. (2010): Global threats to human water security and river biodiversity. Nature 467, 555–561

Wachtershauser, G. (1998): Origin of life in an iron–sulfur world. In: The Molecular Origins of Life, A. Brack (Ed), Cambridge University Press

Wackernagel, M., et al. (2002): Tracking the ecological overshoot of the human economy. Proceedings of the National Academy of Sciences 99 (14), 9266

Walker, A., and R. Leakey (1993): The Nariokotome Homo erectus Skeleton, Harvard University Press

Walsh, M. (1990): Global trends in motor vehicle use and emissions. Annual Review of Energy and the Environment 15, 217–243

Wang, X., and D.L. Mauzerall (2006): Evaluating impacts of air pollution in China on public health: Implications for future air pollution and energy policies. Atmospheric Environment 40, 1706–1721

Ward, P., and D. Brownlee (2002): The Life and Death of Planet Earth, Time Books

Ward, P. (2001): Future Evolutions, W.H. Freeman

Washington Post (2007): 9 March issue

Watson, L. (1995): Dark Nature: A Natural History of Evil, Hodder and Stoughton

WB (2006): China Quick Facts, The World Bank

WB (2008): Global Purchasing Power Parities and Real Expenditures. The World Bank 2005 International Comparison Program

WB (2010): Rising Global Interest in Farmland. Can it Yield Sustainable and Equitable Benefits? The World Bank

WB (2010a): The World Bank, PPP GDP 2009

WBCSD (2009): Water Facts and Trends. World Business Council for Sustainable Development

WBCSD (2010): Vision 2050: The New Agenda for Business. World Business Council for Sustainable Development

WBCSD (2009): Water Facts and Trends. World Business Council for Sustainable Development

WCC (1974): World Council of Churches, cited in: A global view. In: Partnerships in Practice, Elizabeth Dowdeswell (1994), Department of the Environment, London

WCED (1987): World Commission on Environment and Development. Our Common Future, Oxford university Press

Weber, M. (1918): Science as a vocation. In: H.H. Gerth, C. Mills, and C. Wright (Eds), From Max Weber: Essays in Sociology, Routledge and Kegan Paul, London

Weber, M. (1946): From Max Weber. Essays in Sociology, H.H. Gerth and C.W. Mills (Eds), Oxford University Press (1958)

Weinberg, S. (1993): Dreams of a Final Theory: The Scientist's Search for the Ultimate Laws of Nature, Random House

Weisz, P.B. (2004): Basic choices and constraints on long-term energy supplies. Physics Today, July 2004, 47–51

Wetherill, G.W. (1990): Formation of the Earth. Ann. Rev. Earth Planet Science 18, 205-56

White, Lynn, Jr., (1967): The historical roots of our ecological crisis. Science 155, 1203-1207

Whitehead, A.N. (1953): Science and the Modern World, Cambridge University Press, Cambridge

WHO (2002): Fact Sheet No 104, World Health Organization

WHO (2003): The World Health Report, World Health Organization

WHO (2007): Maternal mortality ratio falling to slow to meet goal, World Health Organization

WHO (2008): Safer water, better health, World Health Organization

WHO (2010): Obesity and Overweight, World Health Organization

WI (2008): Worldwatch Institute, State of the World 2008

Williams, M. (2003): Deforesting the Earth. From Prehistory to Global Crisis, The University of Chicago Press

Wilson, E.O., and F.M. Peters (Eds) (1988): Biodiversity, National Academy Press, Washington DC

Wilson, E.O. (1975): Sociobiology: The New Synthesis, Harvard University Press (2000)

Wittgenstein, L. (1990): Raymond Monk, Penguin Books, New York

Woese, C.R. (1987): Bacterial evolution. Microbiol. Rev. 51, 221

Woese, C.R. (1998): The universal ancestor. Proceedings of the National Academy of Sciences 95, 6854-6859

Wolfe, L.M. (1945): Son of Wilderness: The Life of John Muir, Alfred A. Knopf

Wolfe-Simon, F., et al. (2010): A bacterium that can grow by using arsenic instead of phosphorus. Science DOI:10.1126/science. 1197258

Wouters, B., D. Chambers, and E.J.O. Schrama (2008): GRACE observes small-scale mass loss in Greenland. Geophys. Res. Lett. 35, L20501, doi:10.1029/2008GL034816

WRI (1992): Global Biodiversity Strategy, World Resources Institute

Wright, H. (1966): Explorer of the Universe: A Biography of George Ellery Hale, New York

Wrigley, E.A. (1969): Population and History, McGraw-Hill

WSWS (2011): BRICS summit denounces 'use of force' against Libya. World Socialist Web Site, www.wsws.org

WWF (2010): The Living Planet Report 2010, World Wide Fund for Nature

Xiaochuan, Z. (2009): Reform the International Monetary System, 23 March 2009, People's Bank of China, www.pbc.gov.cn

Zhao, M. and S. Running (2010): Drought-induced reduction in global terrestrial net primary production from 2000 to 2009. Science 329, 940–943

Zhijun, Z. (1998): The middle Yangtze region in China is one place where rice was domesticated. Phytolith evidence from the Diaotonghuan cave, Northern Jiangxi. Antiquity 72, 885–897

Zuckerman, P.A. (1996): Beyond the Holocaust: Survival or Extinction? A Survival Manual for Humanity, Human Progress Network, Washington, USA, www.hpn.org

Zwally, H.J., et al. (2006): Mass changes in the Greenland and Antarctic ice sheets and shelves and contributions to sea-level rise: 1992–2002. Journal of Glaciology 51, 509–527

Index

meat consumption, 250
nuclear fission, 330
perception of Middle East conflicts, 242
public debt, 253
qualified labour, 253
role of IMF, 255
sustainable development, 263
Emerson, R.W., 299
emissions trading, 185
emmer wheat, 86
employment, 194, 199, 211, 230, 304, 336, 337
opportunities in Arab countries, 249
vulnerable, 231
encephalization, 71–76, 78
quotient, 70, 74
energy, 34, 121, 160, 175, 194, 196, 199, 206, 231, 263, 325, 342, 345, 346
and Promethean discourse, 305
and sustainable development, 329–335
carrier, 141
chemical, 121, 122
commercialized, 318
conservation, 295
consumption, 331, 334
converter, 122, 334
demand, 330, 331
efficiency, 272, 273, 323, 330, 334
electrical, 122
from oceanic surface currents, 144
from oceans, 334
future scenarios, 329–335
geothermal, 127, 133, 140, 144, 334
global consumption, 330, 333
global crisis, 272
global demand, 259, 334
gravitational potential, 121
intensity, 331
kinetic, 121
low carbon, 323
mechanical, 121
nuclear, 122
prices, 321, 330, 331, 340
radiant, 122
renewable, 127, 139–147, 260, 261, 268, 272, 273, 330, 334
research, 273, 334
risks, 343
scarcity, 280
solar, 127, 133, 140–143, 334
storage, 334
sustainable use, 299
technologies, 260
thermal, 121, 122

tidal, 127, 133, 140, 144
trade, 333
wave, 127, 133, 140, 144
wind, 127, 133, 140, 143, 334
energy return on investment (EROI), 132
Energy–Climate Era, 263
Engels, F., 191
engine
combustion, 279
diesel, 125
electric, 122
heat, 125
internal combustion, 122, 124, 125, 145
steam, 123–125, 279
Stirling, 125
England, 123
environmental movements, 300
industrial revolution, 166
Enlightenment, 7, 191, 228, 277, 299
antithesis to, 24
romantic rejection of, 311
Enola Gay, 18
entropy, 125, 126, 142
in ecological economics, 326
environment, 26, 206, 225, 343, 345, 349
and economic growth, 304
and economy, 325
and human development, 297, 298
awareness of, 300, 301, 307, 316
commitment and action, 324
conservation, 299
cumulative global risk, 227
global change, 227, 228, 272, 319
legislation, 212
patron saint, 313
protection, 310, 311, 318, 320, 323, 327
state of, 300
sustainability, 229
threat from GMOs, 312
undesirable practices, 212
environmental crisis, 254, 265, 268, 336, 338, 345
environmental degradation, 90, 167, 199–201, 225, 227, 240, 260–264, 300, 302, 321, 345, 346
and deep ecology, 307
and ecological modernization, 328
and economic growth, 298, 329
and green economy, 321
and patriarchal society, 309
and Promethean discourse, 306
and religion, 313
and technology, 311
and world economic output, 319

Arab population, 246, 247
British Mandate, 246
Christian minority, 245
civil war, 246
Islamic majority, 245
Jewish population, 245, 246
legislative elections, 247
Partition Resolution, 246
refugees, 247
rocket attacks on Israel, 247
United Nations Special Committee, 246
White Paper 1939, 246
Palestinian State, 246
Pali Canon, 107
palm oil, 147, 263
palm tree, 192
pandemic, 225, 275
Pangaea, 66, 76, 285
Pannotia, 285
paper making, 116
Papin, D., 123
Papua New Guinea, 180
papyrus, 116
parchment, 116
Paris, 167
Parsons, W.S., 18
particle accelerator, 294
Parvati, 106
Pascal, B., 8
Pashtuns, 236
  grand assembly, 236
pasture, 155
Patagonia, 285
Paul the Apostle, 99
Pauli exclusion principle, 290
Pax Americana, 210
Pax Britannica, 207, 210
PCB, 153
pea, 86
peace, 213, 218, 229, 238, 241, 245, 258, 314, 318, 341, 345, 346
peak oil, 129, 331, 333
peasant rebellions, 257
Pech Merle cave, 93
Peking Man, 73
Permian period, 66
Permian–Triassic extinction, 66
peroxisome, 60
persecution, 198, 220
Persia, 100, 119
Persian Empire, 99, 350
Peru, 221
pesticides, 26, 128, 145, 153, 157, 160, 226, 311

petroleum, 124, 127, 222
Phanerozoic, 66
Pharaoh, 95
pharmaceutical research, 204
Philippines, 144, 176
Philistines, 97
philosophy, 2, 9, 102, 106
  Cartesian, 4
  in China, 109–110
  in Japan, 111
  of science, 25
Phoenicians, 92
phosgene, 216
phosphorus, 54, 153, 157
  cycle, 196, 273, 311
photobioreactors, 334
photon, 12, 42, 142
photonic devices, 334
photosynthesis, 51, 59, 66, 122, 157, 165, 170, 214, 278, 287
  end of, 287
  optimization, 334
photovoltaic cell, 122, 133, 142, 334
phylogenetic tree, 58
physical science, 102
physics, 3, 186, 265
  laws of, 4, 32, 39, 41, 45, 125, 187, 267, 274, 281, 344
phytoplankton, 70, 153, 185, 343
pig, 87
Pinchot, G., 299
piracy, 241
placenta, 69
plague, 217
planet, 44
planetary engineering, 292
planetary motion, 12, 118
planetary nebular, 291
planetary science, 56
planetesimals, 47, 49
plankton, 170
planning, 77, 265
plants, 287
  breeding, 312
  C3 and C4, 287
  longevity, 287
plasma, 138
plasmodium parasites, 204
plastics, 153
plate tectonics, 70, 172, 285, 289
  computer models, 285
platinum, 260
Pleistocene period, 84, 85
Plumwood, V., 309